James W. Moore   S. Ramamoorthy

Organic Chemicals in Natural Waters

Applied Monitoring and Impact Assessment

With 81 Figures

Springer-Verlag
New York Berlin Heidelberg Tokyo

James W. Moore

Alberta Environmental Centre
Vegreville, Alberta
TOB 4L0 Canada

S. Ramamoorthy

Alberta Environmental Centre
Vegreville, Alberta
TOB 4L0 Canada

Library of Congress Cataloging in Publication Data
Moore, James W., 1947–
Organic chemicals in natural waters.
(Springer series on environmental management)
Bibliography: p.
Includes index.
1. Water chemistry. 2. Organic chemicals—Analysis. I. Ramamoorthy, S. II. Title.
III. Series.
GB9855.M653 1984 628.1′6 84-10507

Softcover reprint of the hardcover 1st edition 1984

Typeset by Ampersand Inc., Rutland, Vermont

9 8 7 6 5 4 3 2 1

ISBN-13: 978-1-4613-9540-9 e-ISBN-13: 978-1-4613-9538-6
DOI: 10.1007/978-1-4613-9538-6

Series Preface

This series is dedicated to serving the growing community of scholars and practitioners concerned with the principles and applications of environmental management. Each volume is a thorough treatment of a specific topic of importance for proper management practices. A fundamental objective of these books is to help the reader discern and implement man's stewardship of our environment and the world's renewable resources. For we must strive to understand the relationship between man and nature, act to bring harmony to it, and nurture an environment that is both stable and productive.

These objectives have often eluded us because the pursuit of other individual and societal goals has diverted us from a course of living in balance with the environment. At times, therefore, the environmental manager may have to exert restrictive control, which is usually best applied to man, not nature. Attempts to alter or harness nature have often failed or backfired, as exemplified by the results of imprudent use of herbicides, fertilizers, water, and other agents.

Each book in this series will shed light on the fundamental and applied aspects of environmental management. It is hoped that each will help solve a practical and serious environmental problem.

Robert S. DeSanto
East Lyme, Connecticut

Preface

This is the second of two volumes on monitoring and impact assessment of chemical pollutants in natural waters. Our intention is to provide a review of data, methods, and principles that are of potential use to individuals involved in environmental management and research. The first volume dealt with the most common heavy metals in natural waters, including arsenic, cadmium, chromium, copper, lead, mercury, nickel, and zinc. The second volume considers organic compounds outlined in the priority pollutant list (EPA) and Environmental Contaminants Act (Canada) and includes aliphatic compounds, aromatic compounds, chlorinated pesticides, petroleum hydrocarbons, phenols, polychlorinated biphenyls, and polychlorinated dibenzo-*p*-dioxins. Most of these chemicals are widespread in the environment and toxic to fish and humans; many are mutagenic, carcinogenic, and teratogenic. As in the first volume, a multidisciplinary approach is emphasized. There are extensive reviews of the chemistry, production, uses, discharges, behavior in natural waters, uptake, and toxicity of organics. This is followed by a description of criteria for prioritizing chemical hazards posed to users of aquatic resources. Several recommendations are made with the intention of improving current monitoring techniques.

We would like to acknowledge the assistance of staff from the Alberta Environmental Centre in the preparation of this volume. We relied heavily on Sita Ramamoorthy for the compilation and indexing of the literature. Sita Ramamoorthy and Jim Bradley also proofread the various drafts. Mrs. Diana Lee from the library handled all of our literature requests, Mrs. Arhlene Hrynyk arranged for typing of the drafts, and Mr. Terry Zenith was responsible for figure preparation. Finally, we would like to acknowledge Dr. R.S. Weaver (Executive Director, Alberta Environmental Centre) and Dr. L.E. Lillie (Head, Animal Sciences Wing) for their support during this project.

Contents

1
Introduction

Multidisciplinary Studies

Priorities in environmental research and management change. Ten years ago, we were largely concerned with eutrophication and warm water discharges into rivers. Although most scientists and managers were aware of the symptoms of poisoning by mercury, cadmium, and organo-chlorines, chemical disease was an area that the majority of environmentalists did not study. Consequently, there has been a surprising number of cases involving the exposure of humans and other organisms to chemicals. Some of the most significant examples include the Love Canal and Michigan incidents (Smith, 1980) and the closure of commercial fisheries in the lower Great Lakes. In retrospect, eutrophication and warm water do not seem that important, and it is easy to suggest that research money should have been spent in other areas.

The foregoing scenario would have been largely avoided if researchers and managers had taken a broader outlook on problem solving. All of us have to avoid the tendency to be narrow in our thinking. In most cases, we have to make a conscious attempt to expand our knowledge into different disciplines. The need for a broad outlook is generally more important in environmental studies than in other areas. The complexity and diversity of environmental problems require knowledge of chemistry, life sciences, and engineering. The effective implementation of recommendations from such studies also requires an empathy for social, political, economic, and legal factors. In short, perturbations on the sum of conditions that influence an organism or population (the environment) cannot be effectively studied using a restricted approach.

Multidisciplinary research involves the use of several disciplines to reach a common goal. The disciplines have to be diverse in nature, spanning the physical, chemical, life, and engineering sciences. Each component should contribute substantially to the common goal and not be restricted to a service role. Active and strong input from individuals with diverse training will produce a broad understanding of the long-term consequences of specific environmental problems. This will in turn lead to improvement in the means of prioritizing research needs.

We hope that the need for multidisciplinary studies is apparent to environmentalists. If not, the example of the hazards posed by 2,3,7,8-tetrachlorodibenzo-*p*-dioxin (TCDD) in water should perhaps be considered. Short-term exposures of coho salmon to TCDD showed that there was no measurable impact on food consumption, weight gain, or survival during a 60-day postexposure period (Miller *et al.*, 1979). This type of testing is far more sophisticated than the routine monitoring programs conducted in many government laboratories and should have been reasonably well suited for the protection of aquatic resources. However, more detailed studies have shown that growth and survival of the salmon decreased after a 114-day exposure to higher concentrations of TCDD. In addition, the body burden of TCDD increased with both the level and duration of exposure. Similarly, although there is histological evidence of liver degeneration in rainbow trout fed TCDD at a concentration of 2.3 mg g^{-1} food (Hawkes and Norris, 1977), histological methods are relatively insensitive in testing for TCDD toxicity (Wallace, 1979). On the other hand, the mutagenic and carcinogenic properties of TCDD to higher animals can be directly attributed to its induction of aryl hydrocarbon hydroxylase and accumulation of the intermediates (Kouri *et al.*, 1973). It is also important to point out that laboratory exposures of TCDD to animals may not reflect the chemical nature of the environment, thereby neglecting potential synergistic actions (Franklin, 1976; Gori, 1980). Obviously, the foregoing investigations would be largely unnecessary if TCDD is rapidly photodecomposed in the environment. We hope that environmentalists who have read this section will realize that unidiscipline investigations could have led to misevaluations of the hazards of TCDD.

Objectives

Our objective is to provide the manager and scientist with information on organic chemicals that can be used during monitoring, impact assessment, and decision-making processes. Initially, there is a treatment of the chemistry of organics in natural waters and tissues. This is followed by a more detailed discussion on the production, source, use, and discharge of organics into water, air, and land. Such information is intended to help the manager predict the potential for contamination in natural waters. The next section describes the effect of the environment on the chemistry, toxicity, and fate of organics, followed by a review of residues in water, sediments, precipitation, effluents, emissions, and tissues in lakes, rivers, and coastal marine waters. This will allow comparisons to be made between the extent of contamination in a manager's jurisdiction and other areas. The toxicity of organics to different groups of organisms, including aquatic plants, invertebrates, fish, and humans is then reviewed.

The chemical formulas of compounds cited in this book are listed in Appendix A. Appendices B and C are glossaries of physical and chemical terms and scientific names of fish, respectively. Appendix D consists of equations for the evaluation of physico-chemical fate processes, referred to in Chapter 2.

References

Franklin, M.R. 1976. Methylene dioxyphenyl insecticide synergists as potential human health hazards. *Environmental Health Perspectives* **14**:29–37.

Gori, G.B. 1980. The regulation of carcinogenic hazards. *Science* **208**:256–261.

Hawkes, C.L., and L.A. Norris. 1977. Chronic oral toxicity of 2,3,7,8-tetrachlorodibenzo-*p*-dioxin (TCDD) to rainbow trout. *Transactions of the American Fisheries Society* **106**:641–645.

Kouri, R.E., R.A. Salerno, and C.E. Whitmire. 1973. Relationships between aryl hydrocarbon hydroxylase inducibility and sensitivity to chemically induced subcutaneous sarcomas in various strains of mice. *Journal of the National Cancer Institute* **50**:263–268.

Miller, R.A., L.A. Norris, and B.R. Loper. 1979. The response of coho salmon and guppies to 2,3,7,8-tetrachlorodibenzo-*p*-dioxin (TCDD) in water. *Transactions of the American Fisheries Society* **108**:401–407.

Smith, R.J. 1980. Swifter action sought on food contamination. *Science* **207**:163.

Wallace, D. 1979. TCDD in the environment. *Transactions of the American Fisheries Society* **108**:103–109.

2
Physico-Chemical Concepts on the Fate of Organic Compounds

The concentration, behavior, and eventual fate of an organic compound in the aquatic environment are determined by a number of physico-chemical and biological processes. These processes include sorption-desorption, volatilization, and chemical and biological transformation. Solubility, vapor pressure, and the partition coefficient of a compound determine its concentration and residence time in water and hence the subsequent processes in that phase. The movement of an organic compound is largely dependent upon the physico-chemical interactions with other components of the aquatic environment. Such components include suspended solids, sediments, and biota.

Physico-Chemical Properties

Solubility

Although precise determination of solubility data is critical in evaluating the transformation process, such measurements still remain elusive for many compounds. In fact some of the data on aqueous solubility are no more than estimates. The problem is aggravated by the extremely low solubility of many environmentally significant contaminants. For example, the reported solubility of polychlorinated biphenyls (PCBs) varies by a factor of 2–4 depending on the procedures used (Haque and Schmedding, 1975; Wallhofer *et al.*, 1973). Several techniques are documented in the literature, but the one by Moriguchi (1975) seems to be simple and sound. This technique is based on factoring water solubility of an organic compound into two intrinsic components: free molecular volume and hydrophilic effect of polar groups.

Moriguchi chose Quayle's parachor (molecular volume) over five other additive parameters relating to molecular volume for predictive purposes (Quayle, 1953). Quayle's parachor value is calculated by considering the molecule as a sum of its functional groups. Each group is assigned a certain value, based on its empirical data. Thus, the molecular volume is calculated from the structure of the compound. The second component, hydrophilicity of polar groups, which accounts for solute-solvent and solvent-solvent interactions, is calculated for various functional groups from their empirical data (Moriguchi, 1975). Finally, using Quayle's parachor and the appropriate hydrophilic reference factor, aqueous solubility of an organic compound can be estimated (Appendix D-1).

Vapor Pressure

In simple terms, equilibrium vapor pressure can be interpreted as the solubility of the compound in air from the liquid phase. For some organic compounds, the solids possess a finite vapor pressure that should also be considered in the evaluation of its behavior in the environment. This is particularly important for compounds of low solubility. Vapor pressure at 25°C can be calculated directly where the constants are available at that temperature or by interpolation from other temperatures. The equation of Weast (1974) can be used for several priority pollutants for the calculation of vapor pressure (Appendix D-2).

In cases of insufficient data, tables in Dreisbach (1952) can be used from the knowledge of the boiling point at 760 torr and the chemical family to which the compound belongs. The tables developed the "Cox Chart" chemical families using Antoines equation, a modified version of equation D-2 (Appendix D). Charts are available for several organic compounds such as naphthalenes, halobenzenes with side chains, and phenols. The appropriate table is then referred to for an estimated vapor pressure.

The third method of calculating the vapor pressure is using the Clausius-Clapeyron equation (Appendix D-3). Calculations from this equation provide only a rough estimate of vapor pressure.

Partition Coefficient

Partition coefficient is a measure of the distribution of a given compound in two phases and expressed as a concentration ratio, assuming no interactions other than simple dissolution. In reality, the situation could be more complex as a result of dissociation/association of the molecule altering the speciation and stoichiometry. Partition coefficient values are valuable in describing the environmental behavior of the compound. Since partition coefficients are additive in nature, such values for a complex compound could be calculated from the values of the parent compound by adding on the values of the substituents (transfer constants). The organic compound is considered a sum of its functional groups, which cause a certain proportion of the partitioning between octanol/water phases. An index value called the "pi" value, which can be positive or negative, has been estimated for many of the common functional groups (Tute, 1971). Calculated and experimental values for some PCBs have shown good agreement. Equations have been developed for converting partition coefficient values from one set of solvents system to another (Leo *et al.*, 1971).

Physico-Chemical Processes

Sorption-Desorption

The term sorption used here covers both adsorption and absorption, which are difficult to distinguish in most situations. In general, the more hydrophobic the organic compound is, the more likely it is that it will be sorbed to the sediment. The solubility of an organic compound depends primarily upon the sorption-desorption characteristics of the sorbate (organic compound) in association with the sorbent (soil, sediment, or synthetic matrix such as resin-coated organic strips). The physico-chemical characteristics of (i) the sorbent such as surface area, nature of charge, charge density, presence of hydrophobic areas, and organic matter such as humic and fulvic acid, and (ii) the sorbate such as water solubility and the ionic form determine the extent and the strength of sorption.

Sorption can be expressed in terms of the equation,

$$C_s = K_p \, C_w^{\,1/n} \tag{1}$$

where C_s and C_w are the concentrations of the organic compound in solid and water phases respectively, K_p = partition coefficient for sorption, and $1/n$ = exponential factor. At environmentally significant concentrations that are low compared with the sorption capacities of the surface components, the term $1/n$ approximates to unity. It should be emphasized here that in the measurement of K_p, sufficient time must be allowed for the equilibration between phases to be established. This time could vary from a few minutes to several days depending on the organic compound. For neutral organic compounds, the sorption was shown to correspond to the organic content of the particulates (Kenaga and Goring, 1980). K_{oc} (K_p/fraction of organic carbon) was shown to correlate well with water solubility and K_{ow} (K for octanol/water mixture). However, this relationship between K_p and K_{oc} has limited predictability since many neutral organic compounds are also sorbed by materials with little or no organic content. Also, the field concentrations of organic compounds are often much lower than laboratory concentrations, and the span of the concentration in a given isotherm (for example, the pesticide concentration in soil) is generally less than an order of magnitude. Usually, laboratory concentration conditions were chosen to approximate the field conditions at the time of pesticide application and/or to meet analytical detection requirements. Thus, in many cases equation (1) has not been adequate to describe the extrapolation of sorption measurements to out-of-range (low concentration levels) field situations.

Recently, methods have been developed to estimate the equilibrium sorption behavior of hydrophobic pollutants applicable to environmental conditions (Karickhoff, 1981). This method is also based on the organic-carbon referenced sorption approach developed earlier by Goring (1967), Hamaker and Thompson (1972), Lambert (1968), Lambert *et al.* (1965), and Briggs (1973). In addition, it includes thermodynamic rationale to establish the relationship of sorption parameters to other physical properties of the sorbate, making the extrapolation to other sediments or structually similar sorbates valid. At low pollutant concentration (aqueous phase concentration less than half the solubility), sorption isotherms were linear, reversible, and characterized by a partition coefficient (Karickhoff, 1981). Partition coeffients normalized to organic carbon, K_{oc} ($K_{oc} = K_p$/fraction organic carbon) varied only twofold in a set of sediments and soils collected throughout the US. This can be compared with a 20- to 30-fold variation in K_p (without normalization for organic carbon).

From his findings as well as from those of Hassett *et al.* (1980) and Kenaga and Goring (1980), Karickhoff (1981) generalized that for neutral organic compounds of limited solubility ($<10^{-3}$ M) and not susceptible to speciation changes, sorption was "controlled" by organic carbon and amenable to quantification by K_{oc} format. In addition to organic carbon, sediment particle size governed the sorption of hydrophobic chemicals to natural sediments; the fine particles dominate the sorption and sand acting as diluent (10- to 40-fold reduction of K_{oc}). The K_{oc} for the whole sediment approximates that of the finest fraction. Karickhoff (1981) also derived equations for estimating K_{oc} from water solubility (including crystal energy) and octanol/water partition coefficients. The use of solubility data with no crystal energy correction was shown to introduce considerable error in the estimation of K_{oc} for organic solids containing polar groups (triazenes and carbamates) and for anomolously high melting compounds (β-BHC and anthracene). On the other hand, this corrected solubility equation apparently failed to extrapolate to high molecular weight chlorinated compounds (DDT, methoxychlor, hexachloro PCBs). The predictive capabilities of the equation were tested on literature sorption data for a wide variety of organic compounds and were found to agree within a factor of two.

Volatilization

The transport of a compound from the liquid to the vapor phase is called volatilization and it can be an important pathway for chemicals with high vapor pressures or low solubilities. Some early studies on the fate of chemicals misinterpreted volatilization losses as chemical or biological transformations. In the recent past, volatilization loss became recognized as a discrete process in the fate of organic compounds. Evaporation depends upon the equilibrium vapor pressure, diffusion (generally increasing inversely with molecular weight of the compound, and proportionally to turbulence), dispersion of emulsions, solubility, and temperature.

In general, the volatilization rate, R_v, is a first-order kinetic process (Appendix D-4; Liss and Slater, 1974; Mackay and Leinonen, 1975). For highly volatile compounds and for Henry's law constant $H_c > 3000$ torr M^{-1}, volatilization rate is determined by the diffusion through the liquid-phase boundary layer (Appendix D-5). In cases where $H_c < 10$ torr M^{-1}, the diffusion through the gas-phase boundary layer limits the volatilization rate. For conditions between 3000 and 10 torr M^{-1}, both liquid and gas phase are significant. In these cases, the mass transport coefficients of the chemical in the water column are estimated from representative values of mass transport coefficient for oxygen reaeration and water where liquid-phase resistance and gas-phase resistance are controlled, respectively.

Chemical Transformations

Chemical alteration of an organic compound in the environment could arise from one or more of the following reactions: (i) redox behavior, (ii) hydrolysis, (iii) halogenation-dehalogenation, and (iv) photochemical breakdown. The extent to which an organic compound breaks down to simple molecules in order to become part of natural biological processes will determine its persistence and toxicity. Studies have shown that some of these transformation processes can convert a compound into a derivative that may be substantially more hazardous and persistent. Examples are the photochemical degradation of hydrocarbons and nitrogen oxides to produce a smog that has more direct and active effect on the environment and its inhabitants. Thus, transformations magnify the environmental effects of organic compounds. Halogenation of aromatic compounds and aliphatic hydrocarbons are environmentally significant. Many active systems affecting such transformations occur in the biota owing to their catalytic enzymes and abundant bioenergy. Some of the transformation reactions are briefly described below.

Redox Behavior

A chemical reaction is one in which only neutral molecules and positively or negatively charged ions take part; electrons are not involved. An electrochemical reaction is one that involves, besides molecules and ions, negative electrons (e) arising from a metal or other substance. The reaction that liberates electrons is called an oxidation reaction, and the one that consumes electrons, a reduction reaction (Appendix D-6). pH and Pε are analogous, and the value of Pε will indicate the electrochemical direction of this environment; oxidation if this value is above the equilibrium potential, Pε° and reduction if the Pε is below Pε° value. Electrochemical thermodynamics, within limits, help to understand the complex transformation reactions and also predict certain species in the aqueous environment. It should be understood clearly that a thermodynamically feasible reaction need not take place unless it is kinetically fast. Hence, these equilibrium considerations should be used in conjunction with electrochemical kinetics to determine the affinity and the speed of a given redox reaction.

Many organic compounds can either accept or donate electrons, forming reduced or oxidized species. This is environmentally significant since the oxidized and reduced forms of an organic compound may have totally different biological and ecological properties. The rate of loss of a chemical by oxidation or reduction is generally a second-order kinetic reaction. For example, oxidation is expressed by equation (2):

$$-\frac{dc}{dt} = k_{ox}\,[ox]\,[C] \tag{2}$$

where k_{ox} = second-order rate constant for the oxidation of the chemical, C, and $[ox]$ and $[C]$ are the concentrations of oxidant and chemical, respectively. The use of k_{ox} in estimating oxidation half-lives of chemicals has been reviewed by Mill (1979) and Mill *et al.* (1979).

Hydrolysis

The ionic product (k_w) of water is $10^{-14} = ([H] + [OH^-])$ indicating low dissociation. However, the dissociated macro- and microsolutes usually interact with water, changing the concentration of H^+ and OH^- ions substantially. Fresh waters generally vary in pH from 6.0 to 8.0. However, lower pHs are encountered with leachates from mine wastes. A hydrolysis reaction is one where hydrogen, hydroxyl radicals, or the water molecule interacts with the organic compound depending on the pH and polarity of the site of attack on the molecule. Hydrogen ions lacking electrons are called electrophiles and essentially attack a site with a negative charge or lone pair of electrons or unsaturated compounds possessing a double bond. Typical examples are acid-catalyzed cleavage of ester linkage. On the other hand, nucleophiles, rich in electrons, interact with positive sites on the molecule being attacked.

The kinetics of the rate of hydrolysis of a chemical compound are described in Appendix D-7. The use of hydrolysis data in calculating the hydrolytic half-lives has been reviewed by Mabey and Mill (1978).

Halogenation-Dehalogenation

Halogenation of organic compounds occurs mostly under synthetic conditions or in drastic environments. Mild chlorination reactions are possible in natural waters in zones of mixing of different effluents or mixing of industrial with municipal effluents containing residual chlorine. Chlorine can be sorbed by algae and released with a time delay, and this could serve as a chlorine reservoir in natural waters.

Dehalogenation reactions occur in the environment and could be due to a combination of reactions such as hydrolysis and disproportionation reactions. The hydrolysis reactions can occur under neutral conditions with water nucleophile attack or under basic conditions, the OH^- ion being the nucleophile. The half-lives of some halogenated compounds at pH 7 and 25°C are given in Table 2.1. These conditions are relatively closer to the natural aquatic environment. Many halogenated compounds are susceptible to hydrolysis owing to charge separation between halogen atoms and carbon atoms. Chlorinated biphenyls are relatively inert to hydrolysis and consequent breakdown in the environment. Any breakdown of PCBs must be due to processes other than hydrolysis.

Table 2.1. Hydrolytic half-lives of some halogenated compounds.

Compound	Half-life ($t_{1/2}$)
CH_3F	30 years
CH_3Cl	339 days
CH_3Br	20 days
CH_3I	110 days
$CH_3CHClCH_3$	38 days
$CH_3CH_2CH_2Br$	26 days
$C(CH_3)_3Cl$	23 seconds
CH_2Cl_2	704 years
$CHCl_3$	3500 years
$CHBr_3$	686 years
CCl_4	7000 years
$C_6H_5CH_2Cl$	15 hours

Source: Mabey and Mill (1978).

Photochemical Breakdown Processes

Structural changes of a molecule induced by electromagnetic radiation in the near ultraviolet-visible light range (240–700 nm) are called photochemical reactions. However, ionizing radiation is not present in a concentrated form to inflict any molecular alterations. Photochemical reactions could take place either by (i) direct absorption by the molecule of an incident radiation leading to an excited state with subsequent deactivation reactions, or (ii) electron or energy transfer through an intermediate called a photosensitizer. In some cases, photochemical reactions are followed by secondary dark (thermal) reactions. Photochemical absorption can occur only when the electronic changes of the molecule correspond to the wavelength of the incident radiation. Absorption of light energy in terms of photons results in the excitation of an electron from a lower to a higher orbital. The excitation of an electron can have several possible transitions as described in Figure 2.1.

Fundamental spectral data for organic molecules that undergo such electronic transitions, wavelength of maximum response (λ_{max}), and their molar extinction coefficients (ε) (magnitude of the ability to absorb photons) are given in Table 2.2. The higher the value of λ_{max}, the lower is the energy difference in electronic transitions. Thus, the structure of an organic compound will determine whether or not a photochemical reaction takes place in the environment. Ultraviolet absorption is common with many aromatic and unsaturated compounds. Generally, an increase in the number of conjugated double bonds in the molecule will decrease the energy required for an electronic transition.

Thus, the reactions that are normally possible at the far ultraviolet region become feasible at the near ultraviolet-visible range. Figure 2.2 presents the energies of electromagnetic radiations at different wavelength regions and dissociation energies of some typical diatomic chemical bonds. Comparison

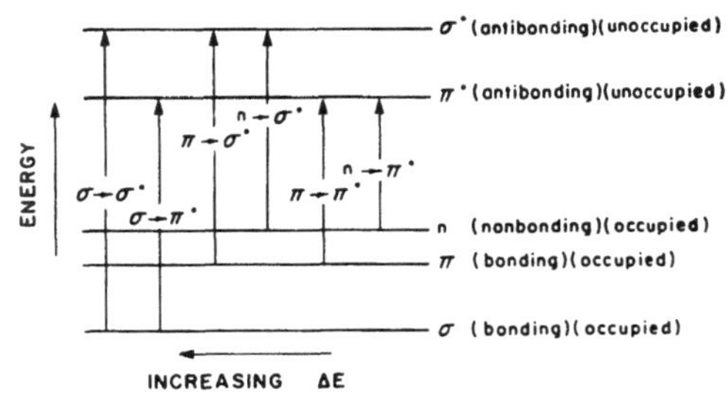

Figure 2.1. Transition of excited electrons from lower to higher orbitals. (From Tinsley, 1979).

Table 2.2. Spectral data of some chromophores.

Chromophore	Functional group	Electron transition	λ_{max}	ε_{max}
–O–	CH_3OH	$n \rightarrow \sigma^*$	1830	500
–S–	$C_6H_{13}SH$	$n \rightarrow \sigma^*$	2240	126
–N–	$(CH_3)_3N$	$n \rightarrow \sigma^*$	2270	900
–Cl:	CH_3Cl	$n \rightarrow \sigma^*$	1730	100
–Br:	CH_3Br	$n \rightarrow \sigma^*$	2040	200
–I:	CH_3I	$n \rightarrow \sigma^*$	2580	378
–C=C–	$H_2C{=}CH_2$	$\pi \rightarrow \pi^*$	1710	15,500
–C≡C–	HC≡CH	$\pi \rightarrow \pi^*$	1730	6000
>C=O	$(CH_3)_2CO$	$\pi \rightarrow \pi^*$	1890	900

Source: Tinsley (1979).

λ in A° unit, A° = 0.1 nm

$$\varepsilon = \frac{O.D.}{c \times d}$$

where O.D. = optical density
c = concentration in moles/L
d = length of optical cell, in mm.

of incident radiation energies with bond dissociation energies will provide an estimate of bond cleavage in a given wavelength region (Figure 2.2).

The excited organic molecule decays rapidly, returning either to the ground state after energy loss through collision and/or secondary radiations and/or chemical changes. The last category includes (i) ionization of the molecule resulting from ejection of an electron, (ii) molecular disproportionation yielding free radicals, (iii) molecular isomerization, and (iv) dark, thermal reactions involving free radicals and other molecules present in the environment. The rate of loss of a chemical ($-\,dc/dt$) by either direct or indirect photochemical reactions may be expressed by simple first-order kinetic expressions (Appendix D-8 and D-9).

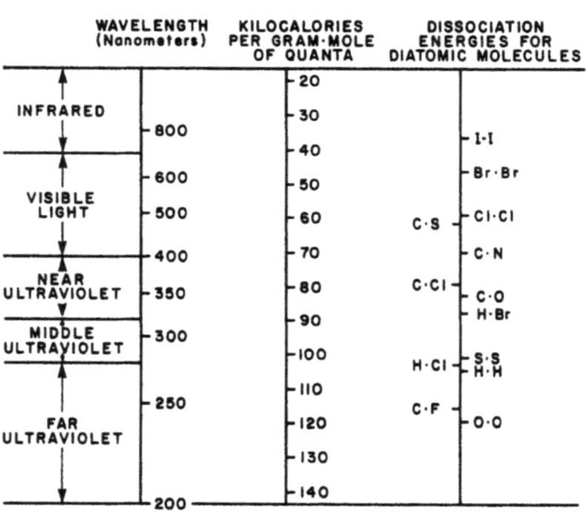

Figure 2.2. Comparison of radiation and bond energies. (From Tinsley, 1979.)

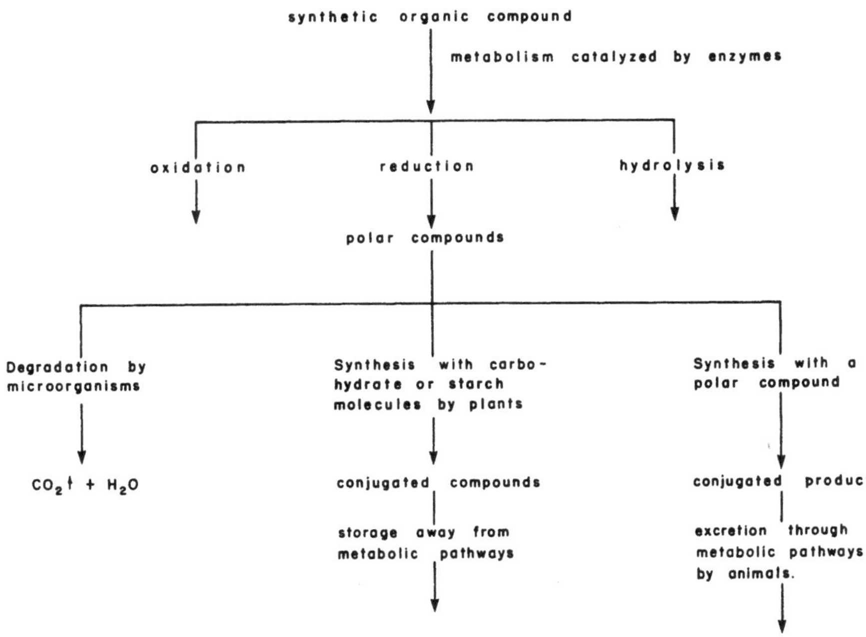

Figure 2.3. Possible pathways of enzymatic transformations of organic compounds.

Metabolic Transformations

Many microorganisms and biota in general develop resistance to most organic chemicals and transform them to compounds that are not toxic to themselves but may be toxic to the total environment. In general, the following enzyme-catalyzed reactions are possible in the metabolic transformation of organic compounds (Figure 2.3).

It is essential to determine the kinetics of these transformation reactions as a function of environmental variables to assess the half-life of the chemical under consideration. The rate for the biotransformation will be a function of the biomass and the chemical's concentration under given environmental conditions. When the organic compound is utilized as a carbon source, the growth rate of the organism is dependent upon the concentration of the former (Appendix D-10, D-11, D-12, and D-13). The half-life of the chemical under degradation ($t_{½}$ at a given cell concentration) can be calculated (Appendix D-14). In deriving this equation (Appendix D-14), it is assumed that the microbial community has already been acclimated to the chemical and that there is no lag time involved in the production of the necessary level of biodegrading organisms or mutants or the enzyme(s). However, when the chemical is newly introduced to the environment, the overall time required to reduce the chemical to 50% of its initial concentration will be given by:

$$T_{1/2} = t_o + t_{1/2}$$

where t_o = acclimation time and $t_{1/2}$ = half-life of transformation of the chemical.

Since the natural environment contains compounds of natural and anthropogenic origin, some biodegradation might require cofactors for microbial metabolism. Hence, cometabolism should be taken into account in the assessment of the environmental fate of a chemical. Half-life gives an estimate of the persistence of the organic compound in the environment (Appendix D-15). If the reactions are different from the first order, appropriate modifications must be made in the equation. Hence, the knowledge of kinetics and concentrations of the chemical are essential in the calculation of the half-lives of organic compounds in the environment.

References

Briggs, G.G. 1973. A simple relationship between soil adsorption of organic chemicals and their octanol/water partition coefficients. *Proceedings of the 7th British Insecticide and Fungicide Conference* **1**:83–86.

Dreisbach, R.R. 1952. *Pressure-volume-temperature relationship of organic compounds*. Handbook Publishers, Sandusky, Ohio, pp. 3–260.

Goring, C.A.I. 1967. Physical aspects of soil in relation to the action of soil fungicides. *Annual Review of Phytopathology* **5**:285–318.

Hamaker, J.W., and J.M. Thompson. 1972. Adsorption. *In*: C.A.I. Goring and J.W. Hamaker (Eds.), *Organic chemicals in the soil environment*, Vol. **1**, Dekker, New York. pp. 49–144.

Hassett, J.J., J.C. Means, W.L. Banwart, and S.G. Wood. 1980. *Sorption properties of sediments and energy related pollutants.* U.S. Environmental Protection Agency, Publication No. EPA-600/3-80-041, Athens, Georgia, 150 pp.

Haque, R., and D. Schmedding. 1975. A method of measuring the water solubility of hydrophobic chemicals: solubility of five polychlorinated biphenyls. *Bulletin of Environmental Contamination and Toxicology* **14**:13–18.

Karickhoff, S.W. 1981. Semi-empirical estimation of sorption of hydrophobic pollutants on natural sediments and soils. *Chemosphere* **10**:833–846.

Kenaga, E.E., and C.A.I. Goring. 1980. Relationship between water solubility, soil sorption, octanol-water partitioning, and concentration of chemicals in biota. *In*: J.G. Eaton, P.R. Parrish, and A.C. Hendricks (Eds.), *Proceedings of the 3rd Symposium on Aquatic Toxicology*, American Society for Testing and Materials, Philadelphia, pp. 78–115.

Lambert, S.M. 1968. Omega (Ω), a useful index of soil sorption equilibria. *Journal of Agricultural and Food Chemistry* **16**:340–343.

Lambert, S.M., P.E. Porter, and H. Schieferstein. 1965. Movement and sorption of chemicals applied to the soil. *Weeds* **13**:185–190.

Leo, A., C. Hansch, and D. Elkins. 1971. Partition coefficients and their uses. *Chemical Reviews* **71**:525–616.

Liss, P.S., and P.G. Slater. 1974. Flux of gases across the air-sea interface. *Nature* **247**:181–184.

Mabey, W., and T. Mill. 1978. Critical review of hydrolysis of organic compounds in water under environmental conditions. *Journal of Physical and Chemical Reference Data* **7**:383–415.

Mackay, D., and P.J. Leinonen. 1975. Rate of evaporation of low-solubility contaminants from water bodies to atmosphere. *Environmental Science and Technology* **9**:1178–1180.

Mill, T. 1979. *Structure reactivity correlations for environmental reactions*. U.S. Environmental Protection Agency, Publication No. EPA-560/11-79-012, Washington, D.C., 58 pp.

Mill, T., W.R. Mabey, and D.G. Hendry. 1979. *Test protocols for environmental processes: oxidation in water*. U.S. Environmental Protection Agency Draft Report, EPA Contract No. 68-03-2227, Washington, D.C.

Moriguchi, I. 1975. Quantitative structure-activity studies. I. Parameters relating to hydrophobicity. *Chemical and Pharmaceutical Bulletin (Tokyo)* **23**:247–257.

Quayle, O.R. 1953. The parachors of organic compounds. *Chemical Reviews* **53**:439–585.

Smith, J.H., D.C. Bomberger, Jr., and D.L. Haynes. 1980. Prediction of the volatilization rates of high-volatility chemicals from natural water bodies. *Environmental Science and Technology* **14**:1332–1337.

Tinsley, I.J. 1979. *Chemical concepts in pollutant behavior*. Wiley, New York, 265 pp.

Tute, M.S. 1971. Principles and practice of Hansch analysis: a guide to structure-activity correlation for the medicinal chemist. *Advances in Drug Research* **6**: 1–77.

Wallhofer, P.R., N. Koniger, and O. Hutzinger. 1973. *Analab Res. Notes* **13**:14.

Weast, R.C. (Ed.). 1974. *CRC Handbook of Chemistry and Physics*. 54th edition. CRC Press, Cleveland, Ohio, pp. D-162-D-188.

3
Aliphatic Hydrocarbons

Aliphatic hydrocarbons constitute a diverse group of organic compounds characterized by an open-chain structure and a variable number of single, double, and triple bonds. Saturated compounds are known as alkanes and fit the empirical formula C_nH_{2n+2}. Alkanes having four or more carbon atoms can exist in both straight-chain and branched-chain isomers. Compounds containing one double bond and having the formula C_nH_{2n} are known as alkenes. The first two members of the series, ethylene and propene, exist in only one form, whereas the next higher homolog, C_4H_8, has two straight-chain isomers and one branched-chain isomer. Aliphatics having two and three double bonds per molecule are termed alkadienes and alkatrienes, respectively, and those with one triple bond are referred to as alkynes.

Aliphatic hydrocarbons may undergo halogenation to form derivatives containing a variable number of chloride, bromide, fluoride, and iodide ions. Halogenation occurs naturally in the environment, producing toxic compounds such as the halomethanes. Industrial halogenation in various processes yields polyhalogenated derivatives that find wide application as solvents, degreasers, dry-cleaning agents, refrigerants, and organic syntheses agents.

Production, Uses, and Discharges

Production and Uses

Aliphatic hydrocarbons are produced at numerous plants throughout the USA, Europe, and Canada, and also find widespread application in many different processes. Consequently, industrially derived aliphatics are ubiquitous in the environment, originating from numerous point and nonpoint sources. In the USA, production of aliphatics steadily increased during the 1960s and 1970s but has recently decreased (Table 3.1). Notably large amounts of 1,2-dichloroethane, now exceeding 45 $\times$ 10^5 tons annually, have been manufactured since 1955, when it began to replace more toxic industrial solvents. Although annual production of chloroethane (1.5 $\times$ 10^5 tons) and 1,1,1-trichloroethane (2.8 $\times$ 10^5 tons) is also moderately high, 1,1-dichloroethane, 1,1,1,2-tetrachloroethane, pentachloroethane, and hexachloroethane do not appear to be commercially manufactured in the USA. In addition, direct production information for 1,1,2-trichloroethane and 1,1,2,2,-tetrachloroethane is not available. The heavy use of chlorinated ethanes reflects their low cost and properties that make them good degreasing agents, cutting fluids, fumigants, and solvents. Some are also used in the production of vinyl chloride, tetraethyl lead, plastics, and textiles and as intermediates in the synthesis of other organochlorine compounds.

Vinyl chloride (chloroethylene) has been used world-wide for more than 40 years in the manufacture of polymers and copolymers containing polyvinyl chloride, which is in turn the most important material in the manufacture of plastics. Production of vinyl chloride in the USA is second only to 1,2-dichloroethane, increasing from an annual average of 6.4 $\times$ 10^5 tons in the early 1960s to >30 $\times$ 10^5 tons in 1981 (Table 3.1). This accounts for approximately 25% of the world's total production. Vinyl chloride and PVC find numerous applications in nearly every branch of industry and commerce. Both compounds are used in building and construction products, the automotive industry, cables, piping, food packaging, medical supplies, and houshold and industrial implements. PVC and vinyl chloride copolymers are distributed and processed in several forms, including latex, organosols, plastisols, and dry resins.

Carbon tetrachloride production in the USA currently exceeds 3 $\times$ 10^5 tons annually and thus ranks third behind dichloroethane and vinyl chloride. Of this total, approximately 90% goes to the manufacture of fluorocarbons, which are used mainly as aerosol propellants. Since the demand for fluorocarbons is decreasing, it is expected that carbon tetrachloride use will also decline. Similarly, carbon tetrachloride was formerly employed as a deworming agent, grain fumigation agent, anaesthetic, and degreaser in the dry-cleaning industry but has been replaced with other compounds because of its toxicity. It is also used as an industrial and chemical solvent and a component of fire extinguisher solutions.

Tetrachloroethylene is employed primarily as a solvent in dry-cleaning shops and, to a lesser degree, as a degreaser in metal industries. Production in the USA has increased gradually during the last two decades and now exceeds 3 $\times$ 10^5 tons (Table 3.1). Although trichloroethylene is also occasionally employed as a dry-cleaning agent, its main use is as a degreaser in the metal industry. Other applications include use as an extractive solvent

Table 3.1. US production (metric tons × 10^4) of some aliphatic hydrocarbons.

	1960–65*	1966–70*	1971–75*	1976	1977	1978	1979	1980	1981	1982
Chloromethane	5.5	14.9	20.8	17.1	21.6	20.6	21.0	16.4	18.4	16.0
Dichloromethane	6.9	14.5	19.0	24.4	21.7	25.9	28.7	25.5	26.8	23.8
Chloroform	3.6	9.1	11.6	13.2	13.7	15.8	16.1	16.0	18.3	13.5
Carbon tetrachloride	21.8	36.5	45.9	38.9	36.7	33.4	32.4	32.1	32.9	—
Chloroethane	26.6	23.1	28.1	30.4	27.8	24.5	26.4	17.9	14.7	13.1
1,2-Dichloroethane	82.1	234.8	379.2	364.6	498.7	498.9	534.9	503.7	452.3	—
1,1,1-trichloroethane	—	14.0	21.9	28.6	28.8	29.2	32.5	31.4	27.8	25.0
Chlorodifluoromethane	—	—	—	7.7	8.1	9.3	9.6	10.3	11.4	—
Trichlorofluoromethane	5.6	10.4	13.6	11.6	9.7	8.8	7.6	10.3	7.4	18.5
Dichlorodifluoromethane	9.6	15.1	19.9	17.8	16.2	14.8	13.3	13.4	14.7	
Vinyl chloride	63.8	142.1	223.1	257.5	271.5	315.8	289.7	293.2	311.7	—
Trichloroethylene	13.9	24.5	18.8	14.3	13.5	13.6	14.5	12.1	11.7	8.1
Tetrachloroethylene	14.2	27.0	32.3	16.3	27.8	32.9	35.1	34.7	31.3	—
Bromoform	0.6	0.9	1.2	—	1.6	1.6	—	—	—	—
Ethylene dibromide	—	13.8	14.0	9.1	11.1	10.4	13.0	8.8	7.6	—

Sources: U.S. International Trade Commission(1960–1981); Chemical Marketing Reporter (1980–83).
*Annual average.
—, data not available from producers.

in foods (in extraction of caffeine from coffee) and as an inhalation anesthetic during some types of short-term surgery. Production is substantially less than tetrachloroethylene and has been gradually decreasing during the last 20 years (Table 3.1). A similar trend has occurred for the dichloroethylenes, with US figures now averaging 1.2×10^5 tons annually. The most important isomer (1,1-DCE) is employed as a chemical intermediate in the synthesis of methylchloroform and in the manufacture of polyvinylidene chloride copolymers. The impermeability of these latter compounds makes them useful as a barrier coating in the packaging industry, in sealing fuel storage tanks, ship tanks, railroad cars, and in coating steel pipes.

Halomethanes have numerous industrial applications (Table 3.2). One of the most important of the group, chloroform, is used mainly as a solvent and an intermediate in the production of various products. Chloromethane and tribromomethane are also employed as chemical intermediates, whereas dichloromethane is widely used as a solvent. Tribromomethane is a common fumigant, and fluoro-derivatives find general application as refrigerants and aerosol propellants. Recent restrictions placed on the use of fluorocarbons have resulted in a decrease in US production of fluoro-derivatives. During the last decade, production of trichlorofluoromethane declined from 1.3 to 0.7×10^5 tons while dichlorofluoromethane fell from 2.0 to 1.5×10^5 tons (Table 3.1). Production of other halomethanes has shown a consistent albeit small increase in recent years (Table 3.1). Total dichloromethane production now exceeds 2.5×10^5 tons compared with 1.5 and 1.8×10^5 tons for chloromethane and chloroform, respectively. It is estimated that bromomethane production will be only about 2.0×10^4 tons in 1982.

Canadian production of aliphatic hydrocarbons cannot be easily documented owing to incomplete records. Furthermore, there does not appear to

Table 3.2. Industrial uses of halomethanes.

Compound	Uses
Chloroform	Chemical solvent and intermediate in the production of refrigerants, plastics, and pharmaceuticals
Chloromethane	Intermediate in the production of silicone, gasoline antiknock, plastics, herbicides, and rubber
Dichloromethane	Solvent, paint remover, plastics manufacturing, and aerosol sprays
Bromomethane	Fumigant agent for soil, seed, feed, and space
Tribromomethane	Chemical intermediate
Bromodichloromethane	Chemical research reagent
Dichlorodifluoromethane and trichlorofluoromethane	Refrigerant and aerosol propellant

Source: Sittig (1980).

be any comprehensive accounting of imports and overall usage. Although Canadian production and usage of aliphatic compounds is undoubtedly lower than that of the USA, the difficulty involved in obtaining data can only hinder the development of comprehensive monitoring and impact assessment programs. It is known, however, that chloroethanes are used in moderately large amounts in Canada. In 1979, total usage was 10,400 metric tons compared with 12,200 tons in 1970. Chloromethane consumption for the same years was 10,360 and 2800 tons, respectively, whereas trichloroethylene use has remained low, <500 tons. Ethylene production reached a peak of 1.33×10^6 metric tons in 1981, declining to 1.07×10^6 in 1982.

Discharges

A wide range of aliphatic hydrocarbons occurs naturally in oil and gas deposits, and they are discharged to the environment during the course of spills and controlled emissions. Although variable, the principal components of crude petroleum are aliphatics within the *n*-alkane and iso-alkane groups. These usually range in carbon number from 1 to 40 and include compounds such as methane, ethane, *n*-hexane, iso-octane, and pristane. Some petroleums contain cycloalkanes (cyclopentane, cyclohexane, decalin) as their principal component. There have been several highly variable estimates of the total quantity of petroleum hydrocarbons entering the earth's oceans, ranging from 1.9 to 11.1×10^6 metric tons yr^{-1} (Connell and Miller, 1981). Of this total, 25–50% originates from land-based discharges such as refineries, waste oils, and sewage, and ~10% from atmospheric sources. Tankers not loading from the top (LOT) account for the majority of water-based discharges, although tankers using LOT, bilge discharge, and accidental spills also contribute significantly to the total load.

Chlorinated aliphatic compounds are frequently found in municipal wastewater and certain types of industrial discharges. Schwarzenbach *et al.* (1979) reported that the average concentration of tetrachloroethylene in the municipal effluent of Zurich (Switzerland) exceeded 1.5 $\mu g\ L^{-1}$, resulting in an average residue of 0.04 $\mu g\ L^{-1}$ in the water of Lake Zurich. Similarly, effluent from a specialty chemical plant (USA) contained dichloromethane at a concentration of 3–8 mg L^{-1} and, although not quantified, trichloroethylene, tetrachloroethylene, trichloroethane, and chloroform were also detected (Jungclaus *et al.*, 1978). In general, chlorination of liquid wastes results in the formation of halo-derivatives that are subsequently discharged to the environment. Kringstad *et al.* (1981) reported that chlorinated pulp mill effluent contained numerous compounds, including trichloroethylene, tetrachloroethylene, dichloromethane, chloroform, pentachloropropene, chlorinated methyl butenes, and brominated methyl butenes. Such compounds have also been found in excessively chlorinated municipal wastewater. Barrick (1982) reported that a sewage treatment facility in Seattle

(USA) discharged 475 metric tons of aliphatic hydrocarbons per year into Puget Sound.

The production of many organic compounds requires the chlorination of feedstock chemicals (Table 3.3). This yields untreated wastewater that contains significant amounts of chlorinated methanes, ethanes, propanes, ethylenes, and propylenes. Other related processes such as chlorohydrination and oxychlorination result in similar wastewater products. Furthermore, the use of vinyl chloride in the production of acrylic fibers and polyvinyl chloride resins yields chlorinated ethanes and ethylenes, whereas the production of epoxy resins results in the formation of dichloropropane and dichloropropylene through the use of epichlorohydrin (Wise and Fahrentholt, 1981).

Behavior in Natural Waters

Sorption

Chloromethanes have little or no affinity for surface interaction with suspended solids or sediments. In the case of chloromethane, its high vapor pressure, low aqueous solubility, and low log $P_{octanol}$ (logarithm of partition coefficient in octanol/water mixture) favor its partition primarily into air with low residence time in water. This accounts for its poor interaction with suspended solids or sediments. Laboratory experiments with dry bentonite clay and dry powdered dolimitic limestone showed about 20% sorption in 30 minutes (Dilling *et al.*, 1975). Sorption did not increase with time in these experiments and there was no selectivity in sorption among chloromethanes and chloroaliphatics. Peat moss was shown to sorb about 40% of chloromethanes (Dilling *et al.*, 1975). Thus, sediments with high organic detritus possess a relatively large sorptive capacity (McConnell *et al.*, 1975).

Volatilization

Evaporation is the primary route of loss of chloromethanes from water. At a concentration of 1 mg L^{-1} of chloromethane, the half-life in water was determined to be 27 minutes at 200 rpm and 25°C (Dilling *et al.*, 1975). As expected, the evaporation rate increased with increasing rate of stirring. With occasional stirring, the half-life of structurally similar compounds such as trichloromethane, trichloroethane, 1,1,1-trichloroethane, and 1,1,1-chloromethane was approximately three times greater than that under stirred conditions (Dilling *et al.*, 1975). It has to be emphasized here that extrapolation of laboratory results to natural situations (where the levels of chloromethanes are expected to be far less than 1 mg L^{-1}) is valid only for comparing relative rates of volatilization. These data show the preponder-

Table 3.3. Industrial products and processes that yield aliphatic hydrocarbons in untreated wastewater.

Industrial product	Industrial process	Chemical feedstock	Uncreated wastewater composition
Carbon tetrachloride	chlorination	methane, ethylene dichloride	chloromethanes, chloroethanes, chloroethylene
Chloroform	chlorination	methane, methyl chloride	chloromethanes, chloroethane, chloroethylenes
1,2-Dichloroethane	oxychlorination	ethylene, HCl	chloroethanes, chloroethylenes
Epichlorohydrin	chlorohydrination	allyl chloride	dichloropropanes, dichloropropylene
Ethylene amines	ammonation	1,2-dichloroethane, NH_3	chloroethanes, chloroethylenes
Ethylene diamine	ammonation	1,2-dichloroethane, NH_3	chloroethanes, chloroethylenes
Ethylene oxide	oxidation, chlorohydrination	ethylene	1,2-dichloroethane
Glycerine	hydrolysis	epichlorohydrin	dichloropropane, dichloropropylene

Methyl chloride	chlorination, hydrochlorination	methane, methanol	chloromethanes, chloroethanes, chloroethylenes
Methylene chloride	chlorination	methane, methyl chloride	chloromethanes, chloroethanes, chloroethylenes
Propylene oxide	chlorohydrination	propylene	dichloropropane dichloropropylene
Tetrachloroethylene	chlorination	1,2-dichlorethane	chloromethanes, -ethanes, -ethylenes, dichloropropane, dichloropropylene
Trichloroethylene	chlorination	1,2-dichloroethane	chloromethanes, chloroethanes, chloroethylenes
Vinyl chloride	dehydrochlorination	1,2-dichlorethane	chloromethanes, chloroethanes, chloroethylenes
Vinylidene chloride	dehydrochlorination	1,1,2-trichloroethane	chloromethane, chloroethanes, chloroethylenes

Source: Wise and Farhenthold (1981). © American Chemical Society. Reprinted with permission.

Table 3.4. Evaporation (minutes) of chloromethanes (1 mg L^{-1}) at 25°C and 200 rpm of stirring.

Compound	50% loss	90% loss
Chloromethane	27	91
Dichloromethane	21	60
Trichloromethane	21	62
Tetrachloromethane	29	97

Source: Dilling *et al.* (1975).

ance of the evaporation process over the other transport processes in a rough order of magnitude scale.

It seems that all chloromethanes evaporate more or less at a similar rate (Table 3.4). The absolute evaporation rates in natural waters could vary from these figures as a result of lower concentrations and highly fluctuating mixing conditions. In addition, other chloroaliphatics will also evaporate at the same relative rate, and this route will account for their transport from water to the atmosphere. Studies with closed and open aqueous systems under light at trichloromethane concentrations of 0.1–1 mg L^{-1} showed the predominance of evaporation over photolysis, oxidation, and hydrolysis (Jensen and Rosenberg, 1975). Fifty to sixty percent of trichloromethane was lost in the former case and in the latter, less than 5%. Assuming all reactions to be of first-order kinetics in nature, the evaporational loss is about an order of magnitude greater than the sum of other degradative processes. Aerial transport and reabsorption in water seems to play a major role in the wider distribution of these chloromethanes in upland waters (Pearson and McConnell, 1975).

Hydrolysis

Because of their high vapor pressure and low boiling points, chloromethanes will have a brief residence time in water resulting in transport to the atmosphere. Hydrolytic rate constants and half-lives of chloromethanes are therefore relatively long (Table 3.5).

Photolysis

Chemical oxidation is not significant in the fate of chloromethanes in natural waters, but once volatilized and present in the troposphere, chloromethanes are attacked by hydroxyl radicals via hydrogen abstraction. The rates of reaction and the half-life for chloromethanes are given in Table 3.6. The principal products of this photooxidation reaction are reported to be formyl

Table 3.5. Hydrolytic rate constants and half-life of some haloaliphatics in water.

Compound	Hydrolytic rate	Half-life ($t_{1/2}$)
Chloromethane[1]	1.9×10^{-8} sec^{-1}	417 days
Dichloromethane[1,2]	0.039 months^{-1}	18 months
Trichloromethane[2,3]	6.9×10^{-12} sec^{-1}	3500 years
	0.045 months^{-1}	15 months
Tetrachloromethane[3,4]	4.8×10^{-7} M^{-1} sec^{-1}	7000 years

Sources: [1]Versar (1979); [2]Dilling *et al.* (1975); [3]Radding *et al.* (1977); [4]Mabey and Mill (1978).

chloride (HCOCl), carbon monoxide (CO), hydrogen chloride (HCl), and phosgene ($COCl_2$) (Spence *et al.*, 1976). Phosgene was shown to hydrolyze readily into HCl and CO_2 (Morrison and Boyd, 1973). Assuming a troposphere-to-stratosphere turnover time of 30 years, the 0.37 year tropospheric lifetime for chloromethane corresponds to about 1% of the tropospheric chloromethane eventually reaching the stratosphere. For tetrachloromethane, this fraction will be about 90%. Chemical oxidation does not seem to play a role in the breakdown of tetrachloromethane in the environment (Table 3.6). Tropospheric stability and resistance to attack by hydroxyl radicals of tetrachloromethane has been established.

Photooxidation seems to be the predominant photolytic reaction in the fate of chloromethanes in the environment. In addition, any unreacted chloromethane reaching the stratosphere would undergo photodissociation. Generally, photodissociation in the stratosphere occurs in the ultraviolet region, 180–240 nm, and chloromethanes absorb from 174–200 nm (Robbins, 1976). Long-term studies on the degradation of dichloromethane at 1 mg L^{-1} concentration under light and dark conditions showed the insignificance of photolytic reactions in aqueous media. This could be due to the fact that

Table 3.6. Rates of photooxidation of chloromethanes.

Compound	Rate constant (cm^3 sec^{-1})	Half-life (years)*
Chloromethane[1]	8.5×10^{-14}	0.37
Dichloromethane[1]	1.04×10^{-13}	0.30
Trichloromethane[1,2]	11.9×10^{-14}	0.32
	16.8×10^{-14}	0.19

Sources: [1]Cox *et al.* (1976); [2]Yung *et al.* (1975).
*Reported as a lifetime for reduction to $1/e$ of original concentration.

chloromethanes have no chromophore to absorb light in the near ultraviolet or visible region (Jaffé and Orchin, 1962). The estimated photodecomposition half-life of dichloromethane is about 250 hours (Dilling *et al.*, 1976). On the other hand, trichloromethane has an absorption maximum at 175 nm (Lillian *et al.*, 1975) and thus might undergo photodissociation in the stratosphere (180–240 nm range). However, no direct photolysis will take place below the ozone layer (>290 nm of light energy). The high stability of tetrachloromethane in the troposphere might lead to one or both of the following: (i) wash down during wet precipitation (Pearson and McConnell, 1975) and (ii) diffusion to the stratosphere with eventual photodegradation to CCl_3 radicals and chlorine atoms. The CCl_3 radicals will then be oxidized to phosgene, which in turn will produce more chlorine atoms by photodissociation (Hanst, 1978; National Research Council, 1978). The released chlorine atoms are believed to cause rupturing of the ozone layer.

Biotransformation

Partitioning favorable to the air phase with low residence time in water minimizes any possible biotransformation of chloromethanes in the environment. Biodegradation/biotransformation of monochloromethane have not been reported. The rest of the chloromethanes are considered to be potentially biodegradable in biological sewage treatment (Thom and Agg, 1975). However, no microorganisms capable of biodegrading these compounds have been reported.

Chloromethanes from Humic Compounds

Chloromethane can be produced by the aqueous chlorination of organic compounds during waste treatment in alkaline conditions (Morris, 1975). The simplified reaction is given below:

$$H_3C{-}\overset{O}{\overset{\|}{C}}{-}R + 3HOCl \rightarrow Cl_3C{-}\overset{O}{\overset{\|}{C}}{-}R + 3H_2O \qquad (1)$$

$$Cl_3C{-}\overset{O}{\overset{\|}{C}}{-}R + H_2O \rightarrow CHCl_3 + RCOOH \qquad (2)$$

Humic and fulvic acids commonly present in natural waters have the required functional groups and structural arrangement for the haloform reaction. The extent of replacement of H atom on the CH_3 group by Cl atom determines the type of chloromethane formed.

A study of ten water treatment plants over a 10-month period concluded that only the total chlorine dosage was the determining factor for trihalomethane (THM) formation in water (Fast and McDonald, 1978). Multiple regression analysis of data from three plants in the Ottawa River showed that chlorine dosages and demands were dominant in determining chloroform levels (Otson *et al.*, 1981). Water temperature and variations in organic content are largely responsible for fluctuations in THM levels. The organic materials present in natural waters interact with chlorine to yield a variety of chlorinated products (Hileman, 1982; Norwood *et al.*, 1980). Humic substances are breakdown products of organic matter of natural origin. Humic substances are polyelectrolytes of varying molecular weight and can be classified into humic (acid-insoluble fraction) acids and fulvic (acid and alkali-soluble fraction) acids. Humic subtances can be soil-derived, leaching into natural waters, or of aquatic origin. The major THM precursors are usually aquatic humic materials. Molecular fractionation studies showed that the THM precursors are essentially of low molecular weight (<30,000) fulvic acid fractions (Oliver and Visser, 1980). Algae (Hoehn *et al.*, 1978), tannic acid (Youssefi *et al.*, 1978), and nitrogeneous compounds (Morris and Baum, 1978) have also been shown to produce THMs during chlorination of water.

Short-Chain Aliphatics

Bromomethanes. Hydrolysis (hydrolytic half-life = 20 days) and volatilization are the important processes in the fate of bromomethanes in water. Once volatilized, photooxidation yielding bromine atoms and inorganic bromides and diffusion to stratosphere with subsequent photodissociation will probably determine the fate of bromomethanes. At present, the importance of sorption and biotransformation processes in the fate of these compounds in the aquatic environment is not known.

Chloroethanes. Based on the results obtained on compounds analogous to chloroethanes, it could be concluded with reasonable confidence that volatilization is an important transport step in moving chloroethanes from the hydrosphere to the atmosphere. In addition, hydrolysis and photooxidation also significantly influence the fate of chloroethanes in the environment. The former largely reflects the high solubility and the relatively short hydrolytic half-life of monochloroethane.

No specific information is available on the sorption of chloroethanes in the aquatic environment. The experimental data available (Dilling *et al.*, 1975) on the analogs could be used to estimate the extent of sorption of chloroethanes. For 1,1-dichloroethane and 1,1,1-trichloroethane, the maximum sorption would be 22% at 750 mg L^{-1} of dry bentonite clay and 40% at 500 mg L^{-1} of peat moss.

Although low molecular weight chloroaliphatics seem to be resistant to microbial degradation (Pearson and McConnell, 1975; McConnell *et al.*,

1975), some evidence was found for the breakdown of dichloroethanes (1,2 and 1,1) in the tissues of fish and oysters. Studies with two strains of bacteria and three mixed populations of fungi showed no breakdown of vinyl chloride over a 5-week period at 20–120 mg L^{-1} concentration range (Hill *et al.*, 1976). Tri- and tetrachloroethanes were metabolized in mammals resulting in the formation of chloroacetic acids that are susceptible to further microbial degradation (Pearson and McConnell, 1975).

Other Haloaliphatics. The fate of these compounds in the aquatic environment is more or less similar to short-chain haloaliphatics. Volatilization and hydrolysis seem to be the important processes in that order. Hexachlorobutadiene (HCBD) is persistent in natural waters and does not volatilize rapidly owing to its low vapor pressure compared with hexachloromethane, which is somewhat structurally similar. Sorption is an active process for the depletion of HCBD from the water column (Leeuwangh *et al.*, 1975). The hydrolytic half-life of hexachlorocyclopentadiene (HCCPD) is 14 days at 25°C, while its photolytic half-life is 11 minutes, indicating that hydrolysis and near-surface photolysis are predominant reactions in the fate of HCCPD in the environment. Tetrachlorocyclopentadienone hydrate is likely to be the end product in the aquatic environment. Biotransformation was estimated to be about 1%. Hydrolysis is a slow process for chloroethers excepting bis(2-chloromethyl)ether (hydrolytic half-life 38 seconds) (Tou *et al.*, 1974). Volatilization with subsequent photooxidation seems to be the predominant fate process for other chloroethers. The high water solubility of 2-chloroethyl vinyl ether accounts for its relatively high persistence in natural waters leading to possible hydrolysis.

Residues

Air

As a result of emissions from industrial plants, automobiles, dumps, and other sources, aliphatic hydrocarbons are generally distributed in the atmosphere. Unsubstituted aliphatics are essentially ubiquitous, whereas the more toxic haloaliphatics are mainly found in urban and industrial areas. Batjer *et al.* (1980) reported that chloroform levels over cities in the FRG reached 220 ng m^{-3}, compared with maximum values of 125, 2, and 0.7 ng m^{-3} found for coastal, rural, and oceanic regions, respectively. Similarly, Simoneit and Mazurek (1981) found that fluorotrichloromethane and difluorodichloromethane were particularly common (0.04–0.8 ng m^{-3}) over cities, though chloroform, carbon tetrachloride, carbon tetrafluoride, and several other compounds were also present. In Milan (Italy), gaseous residues of trichloroethylene and tetrachloroethylene reached 75 and 85 μg

m^{-3}, respectively, during the day (Ziglio, 1981); however, levels generally decreased at night, following the reduction in industrial activity, and were greatest in the urban core. Residues in the air of dry-cleaning shops using tetrachloroethylene as a solvent can be much higher than those in the general atmosphere, exceeding 5 mg m^{-3} (Verberk and Scheffers, 1980).

During the mid-1970s, the American Public Health Association determined that vinyl chloride concentrations in air averaged 100–300 ppb within 1 km of several PVC and vinyl chloride plants in the USA (Sittig, 1980). Concentrations decreased gradually moving away from the plants, but residues still averaged 2–6 ppb at a distance of 5–8 km. Although specific data are not available, it is also likely that hexachlorocyclohexane and hexachlorobutadiene occur at relatively high levels in the vicinity of specific user industries. This may pose some threat to local residents but has not been confirmed with epidemiological evidence. Additives in gasoline have been directly implicated in the production of high levels of dibromoethane (~75 $\mu g\ m^{-3}$) and dichloroethane (2 $\mu g\ m^{-3}$) along motorways (Tsani-Bazaca *et al.*, 1981).

Water and Sediments

Aliphatics are generally distributed in surface waters, albeit at relatively low concentrations. Weaver *et al.* (1980) reported that average chloroform levels in the Kanawha River (USA) were $<5\ \mu g\ L^{-1}$ for much of the year, but occasionally rose to 45–85 $\mu g\ L^{-1}$, coincident with the release of industrial liquid wastes. Similarly, trichloroethylene levels in well water in Milan (Italy) reached 150–200 $\mu g\ L^{-1}$ in a heavily industrialized part of the city (Ziglio, 1981). The latter source also probably resulted in the contamination of ground water. In fact, Page (1981) demonstrated that the probability of finding haloaliphatics in ground water was generally comparable to that of surface waters. This does not mean that groundwater has become significantly contaminated over wide geographic areas, and there are certainly many sites where haloaliphatics remain at nondetectable levels. However, routine monitoring programs for haloaliphatics should include analysis of ground water, particularly in heavily industrialized areas.

Vinyl chloride is a common feature of liquid wastes discharged from PVC and VC plants. In a survey of 12 such plants during the early 1970s, concentrations varied from 0.05 to 20 mg L^{-1} with an average of 2.5 mg L^{-1} (U.S. Environmental Protection Agency, 1980). This in turn has resulted in the mild adulteration of some water supplies in the USA, with residues ranging from 0.3 to 5.7 $\mu g\ L^{-1}$. Such levels fall far below the proposed allowable limit for drinking water, 517 $\mu g\ L^{-1}$ (Sittig, 1980).

Chlorination from various municipal and industrial sources results in the formation of a wide range of substituted organic chemicals, such as the trihalomethanes. Since these compounds are either known or suspected carcinogens, they pose a potential, long-term risk to consumers. Consequently, guidelines and regulations limiting chloroform levels to <100 μg L^{-1} in US and <300 μg L^{-1} in Canadian drinking water have been proposed and adopted (Cotruvo, 1981). Overall, average concentrations of chloroform in drinking water generally range from 10 to 85 μg L^{-1} compared with levels of 15–28 and 8–11 μg L^{-1} for bromodichloromethane and chlorodibromomethane, respectively (Table 3.7). There is often a significant temporal variation in levels, which can be possibly related to alternate stagnation and flushing according to daily activities (Figure 3.1). In addition, air temperature, water temperature, and amount of humic acid also influence trihalomethane formation (Smith *et al.*, 1980). Consequently, monitoring the trihalomethanes should include sampling conducted over a 24-hour period.

Because of their volatility, most aliphatics occur at low or nondetectable levels in sediments. Consequently, there is generally little need to measure residues during monitoring and impact assessment programs except in cases of extreme pollution. Since hexachlorobutadiene has a relatively low vapor pressure, it has been found in sediments at a number of industrial sites. For example, residues of up to 920 μg kg^{-1} dry weight were reported for a landfill pond near the Mississippi River, whereas lower values (~200 μg kg^{-1}) were found in the Mississippi Delta region (U.S. Environmental Protection Agency, 1980).

Aquatic Plants, Invertebrates, and Fish

The majority of aliphatics do not significantly concentrate in fish or other aquatic species and thus pose little or no threat to fisheries and other forms of aquatic resource utilization. In some instances, residues are detected in tissues, particularly for populations inhabiting heavily polluted waters. In addition, high molecular weight compounds concentrate to a greater degree in tissues than low weight compounds (Figure 3.2). Dickson and Riley (1976) reported average chloroform concentrations of 51 μg kg^{-1} (wet weight) in the muscle of five species of fish from the Irish Sea (UK) compared with levels of 12, <1, and <1 μg kg^{-1} for trichloroethylene, tetrachloroethylene, and trichloroethane, respectively. Although variable, residues in the liver and gills of these fish were generally two to five times greater than those in the muscle. Unsubstituted aliphatics are also found in tissues, especially in areas where there have been oil spills or intentional release of hydrocarbon wastes. Total alkane levels in 48 species of benthic invertebrates from the Buccaneer oil field (Gulf of Mexico) averaged 2.7 mg kg^{-1} (Middleditch and Basile, 1980). Similarly, mussels *Mytilus edulis* inhabiting the Kiel Bight (FRG) carried total aliphatic residues ranging from 3 to 16 mg kg^{-1} (Ehrhardt and Heinemann, 1975).

Despite its widespread use and presence in water, vinyl chloride has not been reported from fish tissues. This presumably reflects its high volatility

Table 3.7. Average and range in concentration ($\mu g\ L^{-1}$) of halomethanes in water.

Source	Chloroform	Bromodichloromethane	Chlorodibromomethane	Bromoform
Potable water (113 cities, USA)[1]	43 (ND–271)	18 (ND–183)	8 (ND–190)	3 (ND–39)
Potable water (5 cities, USA)[2]	85 (1–301)	28 (4–73)	11 (3–32)	—
Potable water (London, UK)[3]	14 (ND–23)	17 (<1–28)	8 (<1–16)	—
Potable water (18 cities, FRG)[4]	4 (0.1–34)	—	—	4 (0.3–14)
Potable water (DDR)[5]	11 (2–18)	0.2 (0.1–0.3)	—	—
Water treatment plant (USA)[6]	92 (69–118)	9 (7–10)	3 (2–4)	—

Sources: [1]Cotruvo (1981); [2]Coleman *et al.* (1977); [3]Luong *et al.* (1980); [4]Gabel *et al.* (1981); [5]Koch and Tunger (1981); [6]Kasso and Wells (1981).
ND, not detected.

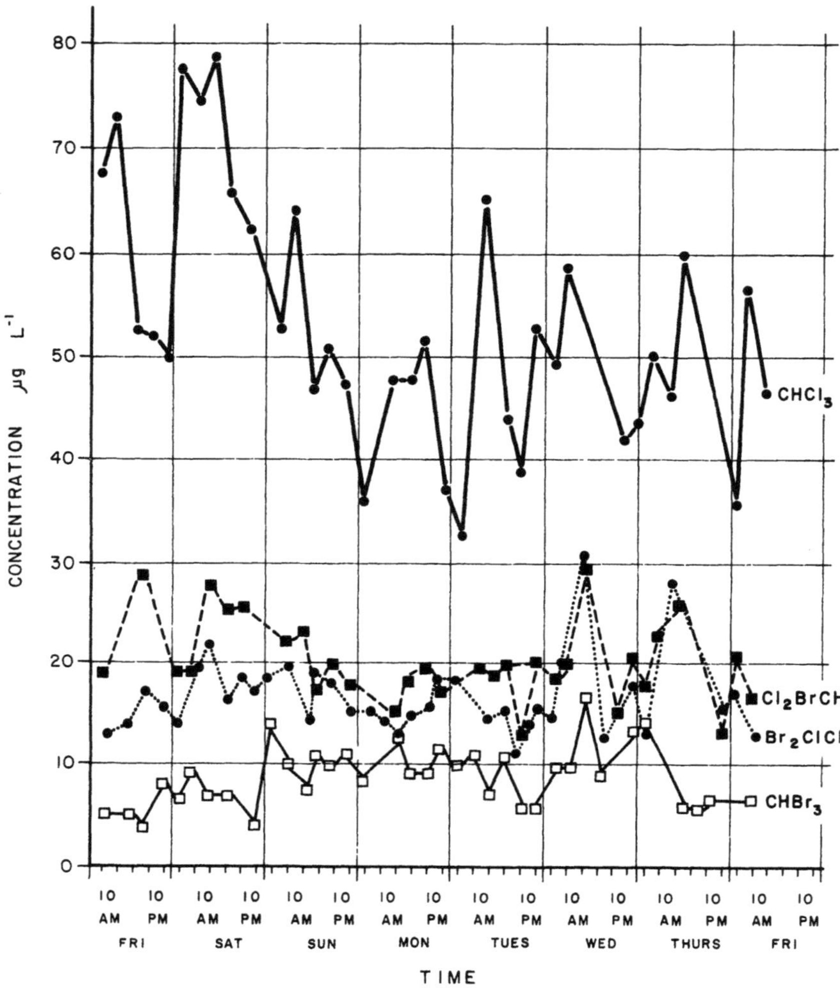

Figure 3.1. Concentrations of chloroform ($CHCl_3$), dichlorobromomethane (Cl_2BrCH), dibromochloromethane (Br_2ClCH), and bromoform ($CHBr_3$) in drinking water from the University of Texas. (From Smith *et al.*, 1980). © American Chemical Society. Reprinted with permission.

and relatively low log $P_{octanol}$ value. By contrast, hexachlorobutadiene is often found in fish tissues, albeit at generally low concentrations. Kuehl *et al.* (1981) reported that salmonids from Lake Ontario had whole body concentrations of ~2 $\mu g\ kg^{-1}$ wet weight. Although residues in fish from Lakes Superior, Huron, and Michigan were below detectable limits (1 $\mu g\ kg^{-1}$), concentrations of 445–3250 $\mu g\ kg^{-1}$ occurred in fish from two highly polluted rivers in Ohio.

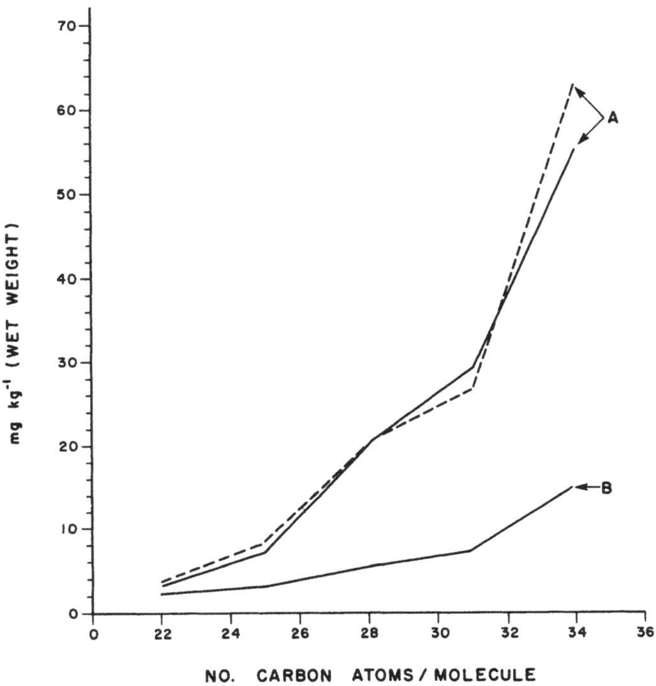

Figure 3.2. Normal alkane content of scarlet prawns *Plesiopenaeus edwardsianus* from Aruba, Dutch West Indies. A. Stations where oil was found. B. Stations where no oil was found. (From Thompson *et al.*, 1977.)

The half-life of most haloaliphatics in tissues is short, averaging <1 day (Table 3.8). This factor, combined with relatively slow uptake rates (Neely *et al.*, 1974), produce CFs that are generally <100 (Table 3.8). Comparably low (2–25) CFs have also been reported for fish inhabiting natural waters (Dickson and Riley, 1976). In most instances, there is a relatively small increase in CF with the degree of chlorination. These data imply that the discharge of aliphatics to natural waters could produce a reduction in aquatic resource utilization, though only for a short period of time.

Toxicity

Aquatic Plants, Invertebrates, and Fish

Most halogenated aliphatics are moderately to slightly toxic to aquatic algae, invertebrates, and fish (Table 3.9). Toxicity generally increases with the degree of halogenation, and consequently, the LC_{50}s for agents such as

Table 3.8. Concentration factors and half-life of haloaliphatics in bluegill sunfish.

Compound	Concentration factor	Half-life in tissues (days)
Chloroform	6	<1
Carbon tetrachloride	30	<1
1,2-dichloroethane	2	1–2
1,1,1-trichloroethane	9	<1
1,1,2,2-tetrachloroethane	8	<1
Pentachloroethane	67	<1
Hexachloroethane	139	<1
1,1,2-trichloroethylene	17	<1
Tetrachloroethylene	49	<1
Hexachlorocyclopentadiene*	29	—
Hexachlorobutadiene†	29	—

Sources: Barrows *et al.* (1980); Veith *et al.* (1979); U.S. Environmental Protection Agency (1980)
*Fathead minnow.
†Largemouth bass.

hexachlorobutadiene and hexachlorocyclopentadiene are relatively low (<0.2 mg L^{-1}). A comparable response occurs with increasing substitution of bromine for chlorine. Thus, the LC_{50} for carp eggs exposed to dibromochloromethane averaged 34 mg L^{-1}, compared with levels of 52, 67, and 97 mg L^{-1} for bromoform, bromodichloromethane, and chloroform, respectively (Mattice *et al.*, 1981). This trend has also been observed in marine algae (Erickson and Hawkins, 1980). Although the acute effects of chlorinated ethanes, propanes, and ethylenes to fish are approximately similar, there is a substantial difference in the hazard posed by different isomers of the same compound within these groups. For example, the LC_{50}s of 1,1-dichloropropane and 1,3-dichloropropane using bluegill sunfish were 98 and >520 mg L^{-1}, respectively (Table 3.9). Thus, initial monitoring and impact assessment programs should consider the analysis of a range of isomers within each aliphatic group.

Some species of larval and juvenile fish do not appear to be particularly sensitive to chlorinated aliphatics (Ward *et al.*, 1981; Mattice *et al.*, 1981). Although such findings may imply that the decision to discharge wastes does not have to consider concomitant effects on sensitive life stages, there is still relatively little information available on many important species. There is also some evidence indicating that immature stages of certain invertebrate species are more sensitive to aliphatics than adults. Stewart *et al.* (1979) reported that concentrations as low as 0.05 mg L^{-1} of bromoform and chloroform induced some mortality in oyster larvae. Overall, therefore, the question of enhanced sensitivity has not been adequately resolved and requires more management-oriented research data.

Table 3.9. Acute toxicity (96-h LC_{50}, mg L^{-1}) of halogenated aliphatics to algae, invertebrates, and fish.

		Invertebrates		Fish	
		Shrimp	Cladoceran		
		Mysidopsis	*Daphnia*	Bluegill	Sheepshead
	Algae‡	*bahia*[1]	*bahia*[1]	*magna*[2]*	minnow[4]
Chlorinated Ethanes					
1,2-dichloroethane	>433	113	220	430	130–230
1,1,1-trichloroethane	>670	31	>530	72	71
1,1,2-trichloroethane	60–260	—	18	40	—
1,1,2,2-tetrachloroethane	6–136	9	9.3	21	12
1,1,1,2-tetrachloroethane	—	—	24	20	—
pentachloroethane	58–121	0.4	63	7.2	116
hexachloroethane	8–90	0.9	8.1	1.0	2.4
Chlorinated Propanes					
1,1-dichloropropane	—	—	23	98	—
1,2-dichloropropane	—	—	52	280	139†
1,3-dichloropropane	1–93	0.8–10.3	280	>520	87
1,3-dichloropropene	—	—	6.2	6.1	1.8
Chlorinated Ethylenes					
1,1-dichloroethylene	>712	224	79	74	250
1,2-dichloroethylene	—	—	220	140	—
trichloroethylene	8	—	18	45	41†
tetrachloroethylene	>507	10	18	13	29–52

(continued)

Table 3.9 *(continued)*

		Invertebrates		Fish	
		Shrimp	**Cladoceran**		
	Algae‡	*Mysidopsis bahia*[1]	*Daphnia magna*[2]*	Bluegill sunfish[3]	Sheepshead minnow[4]
Others					
bromoform	12–114	24	46	29	18
carbon tetrachloride	—	—	35	27	43†
dichloromethane	>662	256	220	220	331
chloroform	—	—	29	13–22	16–22
hexachlorocyclopentadiene	—	0.03	0.04	0.13	0.06–0.18
hexachlorobutadiene	—	0.06	—	0.33	0.10

Sources: [1]U.S. Environmental Protection Agency (1980); [2]LeBlanc (1980); [3]Buccafusco *et al.* (1981); [4]Heitmuller *et al.* (1981)

*48h LC_{50}.

†Fathead minnow.

‡*Selenastrum capricornutum, Skeletonema costatum.*

Table 3.10. Known and suspected aliphatic mutagens and carcinogens found in drinking water.

Known/suspected mutagens	Known/suspected carcinogens
1,1,1-Trichloroethane	chloroform
Bromomethane	carbon tetrachloride
Chloromethane	1,1-dichloroethylene
Bromodichloromethane	1,1,2-trichloroethylene
Dichloromethane	1,1,2,2-tetrachloroethylene
Bromoform	1,1,2-trichloroethane
2-Chloropropane	dibromoethane
1,2-Dichloropropane	2-bromoethylpropane
1-Chloropropene	1,1,2,2-tetrachloroethane
1,1-Dichloroethane	hexachlorobutadiene
1,2-Dichloroethane	vinyl chloride
Chlorodibromomethane	
1,3-Dichloropropene	
Dibromomethane	
Hexachloroethane	

Sources: Kraybill (1980); Kraybill *et al.* (1978); Sittig (1980).

Human Health

Information to date indicates that essentially all of the haloaliphatics are known/suspected mutagens and that some are known carcinogens (Table 3.10). Trihalomethanes are ubiquitous in drinking water and are generally found at the highest concentration of any haloaliphatic. Although epidemiological evidence is inconclusive, it appears that chloroform alone presents a small but potentially important threat to the general population (Kraybill, 1980). Consequently, the maximum trihalomethane level in community water (USA) is set at 0.10 mg L^{-1} (Cotruvo, 1981). The lack of conclusive epidemiological evidence on carcinogenicity is due to a number of confounding factors such as population diversity and mobility, other exposure sources, socioeconomic and urbanization factors, and the usual 20–40 year latency period for many cancers. In addition, only limited data on THM levels in water are available, and in many instances these cover fewer than 5 years. This is due partly to the lack of adequate preservation and analytical methods during the early 1970s. Finally, there may be substantial variability in the amount of chloroform retained by different individuals (Cotruvo, 1981).

Other haloaliphatics (excluding trihalomethanes) may pose a threat to consumers, but there have been few attempts to correlate epidemiological evidence with the incidence of neoplasia. It has been suggested that since many of these compounds are either known or suspected carcinogens, their recommended concentration in water for maximum protection of human health should be zero (Sittig, 1980). Unlike trihalomethanes, the primary source of vinyl chloride, hexachlorobutadiene, chlorinated ethanes, ethylenes, and propanes in water is industrial waste. Such discharges can be regulated by controlling and treating the quality of the final effluent. This means that the target of zero in the water supply may be technologically feasible and is a desirable goal.

References

Barrick, R.C. 1982. Flux of aliphatic and polycyclic aromatic hydrocarbons to Central Puget Sound from Seattle (Westpoint) primary sewage effluent. *Environmental Science and Technology* **16**:682–692.

Barrows, M.E., S.R. Petrocelli, K.J. Macek, and J.J. Carroll. 1980. Bioconcentration and elimination of selected water pollutants by bluegill sunfish (*Lepomis macrochirus*). *In*: R. Haque (Ed.), *Dynamics, exposure, and hazard assessment of toxic chemicals*. Ann Arbor Science Publishers, Ann Arbor, Michigan, pp. 379–392.

Batjer, K., M. Cetinkaya, J. v. Duszeln, B. Gabel, U. Lahl, B. Stachel, and W. Thiemann. 1980. Chloroform emission into urban atmosphere. *Chemosphere* **9**:311–316.

Buccafusco, R.J., S.J. Ells, and G.A. LeBlanc. 1981. Acute toxicity of priority pollutants to bluegill (*Lepomis macrochirus*). *Bulletin of Environmental Contamination and Toxicology* **26**:446–452.

Chemical Marketing Reporter. 1980–83. Schnell Publishing Company, New York, New York.

Coleman, W.E., R.D. Lingg, R.G. Melton, and F.C. Kopfler. 1977. The occurrence of volatile organics in five drinking water supplies using gas chromatography/mass spectrometry. *In*: L.H. Keith (Ed.), *Identification and analysis of organic pollutants in water*. Ann Arbor Science Publishers, Ann Arbor, Michigan, pp. 305–327.

Connell, D.W., and G.J. Miller. 1981. Petroleum hydrocarbons in aquatic ecosystems—behavior and effects of sublethal concentrations: Part 1. *CRC Critical Reviews in Environmental Control* **11**:37–104.

Cotruvo, J.A. 1981. EPA policies to protect the health of consumers of drinking water in the United States. *The Science of the Total Environment* **18**:345–356.

Cox, R.A., R.G. Derwent, A.E.J. Eggleton, and J.E. Lovelock. 1976. Photochemical oxidation of halocarbons in the troposphere. *Atmospheric Environment* **10**:305–308.

Dickson, A.G., and J.P. Riley. 1976. The distribution of short-chain halogenated aliphatic hydrocarbons in some marine organisms. *Marine Pollution Bulletin* **7**:167–169.

Dilling, W.L., C.J. Bredeweg, and N.B. Tefertiller. 1976. Organic photochemistry. Simulated atmospheric rates of methylene chloride, 1,1,1-trichloroethane, trichloroethylene, tetrachloroethylene, and other compounds. *Environmental Science and Technology* **10**:351–356.

Dilling, W.L., N.B. Tefertiller, and G.J. Kallos. 1975. Evaporation rates and reactivities of methylene chloride, chloroform, 1,1,1-trichloroethane, trichloroethylene, tetrachloroethylene, and other chlorinated compounds in dilute aqueous solutions. *Environmental Science and Technology* **9**:833–838.

Ehrhardt, M., and J. Heinemann. 1975. Hydrocarbons in blue mussels from the Kiel Bight. *Environmental Pollution* **9**:263–282.

Erickson, W.J., and C.E. Hawkins. 1980. Effects of halogenated organic compounds on photosynthesis in estuarine phytoplankton. *Bulletin of Environmental Contamination and Toxicology* **24**:910–915.

Fast, D.A., and R.A. McDonald. 1978. *Halomethane study. Selected Saskatchewan Water Supplies*. Saskatchewan Department of the Environment, Water Pollution Control Branch, Report No. WPB 21, Regina, Saskatchewan, 142 pp.

Gabel, B., U. Lahl, K. Batjer, M. Cetinkaya, J. v. Duszeln, R. Kozichi, A. Podbielski, B. Stachel, and W. Thiemann. 1981. Volatile halogenated compounds (VOHal) in drinking waters of the FRG. *The Science of the Total Environment* **18**:363–366.

Hanst, P.L. 1978. Halogenated pollutants. Noxious trace gases in the air. Part II. *Chemistry* **51**:6–12.

Heitmuller, P.T., T.A. Hollister, and P.R. Parrish. 1981. Acute toxicity of 54 industrial chemicals to sheepshead minnows (*Cyprinodon variegatus*). *Bulletin of Environmental Contamination and Toxicology* **27**:596–604.

Hileman, B. 1982. The chlorination question. *Environmental Science and Technology* **16**:15A–18A.

Hill, J., H.P. Kollig, D.F. Paris, N.L. Wolfe, and R.G. Zepp. 1976. *Dynamic behavior of vinyl chloride in aquatic ecosystems*. U.S. Environmental Protection Agency, Publication No. EPA-600/3-76-001, Athens, Georgia, 64 pp.

Hoehn, R.C., C.W. Randall, R.P. Goode, and P.T.B. Shaffer. 1978. Chlorination and water treatment for minimizing trihalomethanes in drinking water. *In*: R.L. Jolley, H. Gorchev, and D.H. Hamilton (Eds.), *Water chlorination: environmental impact and health effects*, Vol. **2**. Ann Arbor Science Publishers, Ann Arbor, Michigan, pp. 519–535.

Jaffé, H.H., and M. Orchin. 1962. *Theory and applications of ultraviolet spectroscopy*. Wiley, New York, 624 pp.

Jensen, S., and R. Rosenberg. 1975. Degradability of some chlorinated aliphatic hydrocarbons in sea water and sterilized water. *Water Research* **9**:659–661.

Jungclaus, G.A., V. Lopez-Avila, and R.A. Hites. 1978. Organic compounds in an industrial wastewater: a case study of their environmental impact. *Environmental Science and Technology* **12**:88–96.

Kasso, W.B., and M.R. Wells. 1981. A survey of trihalomethanes in the drinking water system of Murfreesboro, Tennessee. *Bulletin of Environmental Contamination and Toxicology* **27**:295–302.

Koch, R., and A. Tunger. 1981. Kontamination von Wassern mit Halogenalkanen und -alkenen. *Acta Hydrochimica et Hydrobiologica* **9**:471–475.

Kraybill, H.F. 1980. Evaluation of public health aspects of carcinogenic/mutagenic biorefractories in drinking water. *Preventive Medicine* **9**:212–218.

Kraybill, H.F., C.T. Helmes, and C.C. Sigman. 1978. Biomedical aspects of biorefractories in water. *In*: O. Hutzinger, L.H. van Lelyveld, B.C.J. Zoeteman (Eds.), *Aquatic pollutants*. Pergamon Press, New York, pp 419–459.

Kringstad, K.P., P.O. Ljungquist, F. de Sousa, and L.M. Stromberg. 1981. Identification and mutagenic properties of some chlorinated aliphatic compounds in the spent liquor from kraft pulp chlorination. *Environmental Science and Technology* **15**:562–566.

Kuehl, D.W., K.L. Johnson, B.C. Butterworth, E.N. Leonard, and G.D. Veith. 1981. Quantification of octachlorostyrene and related compounds in Great Lakes fish by gas chromatography-mass spectrometry. *Journal of Great Lakes Research* **7**:330–335.

LeBlanc. G.A. 1980. Acute toxicity of priority pollutants to water flea (*Daphnia magna*). *Bulletin of Environmental Contamination and Toxicology* **24**: 684–691.

Leeuwangh, P., H. Bult, and L. Schneiders. 1975. Toxicity of hexachlorobutadiene in aquatic organisms. *In*: J.H. Koeman and J.J.T.W.A. Strik (Eds.), *Sublethal effects of toxic chemicals on aquatic animals*. Proceedings of the Swedish-Netherlands Symposium, Wageningen, The Netherlands, September 2–5, 1975, Elsevier/North Holland Press, New York. pp. 167–176.

Lillian, D., H.B. Singh, A. Appleby, L. Lobban, R. Arnts, R. Gumpert, R. Hague, J. Toomey, J. Kazazis, M. Antell, D. Hansen, and B. Scott. 1975. Atmospheric fates of halogenated compounds. *Environmental Science and Technology* **9**: 1042–1048.

Luong, T., C.J. Peters, R.J. Young, and R. Perry. 1980. Bromide and trihalomethanes in water supplies. *Environmental Technology Letters* **1**:299–310.

Mabey, W., and T. Mill. 1978. Critical review of hydrolysis of organic compounds in water under environmental conditions. *Journal of Physical and Chemical Reference Data* **7**:383–415.

Mattice, J.C., S.C. Tsai, M.B. Burch, and J.J. Beauchamp. 1981. Toxicity of trihalomethanes to common carp embryos. *Transactions of the American Fisheries Society* **110**:261–269.

McConnell, G., D.M. Ferguson, and C.R. Pearson. 1975. Chlorinated hydrocarbons and the environment. *Endeavor* **34**:13–18.

Merck Index. 1976. *An encyclopedia of chemicals and drugs*. 9th edition. M. Windholz (Ed.). Merck and Co., Rahway, New Jersey, 1313 pp.

Middleditch, B.S., and B. Basile. 1980. Alkanes in benthic organisms from the Buccaneer oil field. *Bulletin of Environmental Contamination and Toxicology* **24**:945–952.

Morris, J.C. 1975. *Formation of halogenated organics by chlorination of water supplies*. U.S. Environmental Protection Agency, Publication No. EPA-600/1-75-002, Washington, D.C., 54 pp.

Morris, J.C., and B. Baum. 1978. Precursors and mechanisms of haloform formation in the chlorination of water supplies. *In*: R.L. Jolley, H. Gorchev, and D.H. Hamilton (Eds.), *Water chlorination: environmental impact and health effects*, Vol. **2**. Ann Arbor Science Publishers, Ann Arbor, Michigan, pp. 29–48.

Morrison, R.T., and R.W. Boyd. 1973. *Organic chemistry*. 3rd edition. Allyn and Bacon Inc., Boston, Massachusetts, 1285 pp.

National Research Council. 1978. *Nonfluorinated halomethanes in the environment*. National Academy of Sciences, Washington, D.C., 297 pp.

Neely, W.B., D.R. Branson, and G.E. Blau. 1974. Partition coefficient to measure bioconcentration potential of organic chemicals in fish. *Environmental Science and Technology* **8**:1113–1115.

Norwood, D.L., J.D. Johnson, R.F. Christman, J.R. Hass, and M.J. Bobenrieth. 1980. Reactions of chlorine with selected aromatic models of aquatic humic material. *Environmental Science and Technology* **14**:187–190.

Oliver, B.G., and S.A. Visser. 1980. Chloroform production from the chlorination of aquatic humic material: the effect of molecular weight, environment and season. *Water Research* **14**:1137–1141.

Otson, R., D.T. Williams, P.D. Bothwell, and T.K. Quon. 1981. Comparison of trihalomethane levels and other water quality parameters for three treatment plants on the Ottawa River. *Environmental Science and Technology* **15**:1075–1080.

Page, W.G. 1981. Comparison of groundwater and surface water for patterns and levels of contamination by toxic substances. *Environmental Science and Technology* **15**:1475–1481.

Pearson, C.R., and G. McConnell. 1975. Chlorinated C_1 and C_2 hydrocarbons in the marine environment. *Proceedings of the Royal Society of London, Series B* **189**:305–332.

Radding, S.B., D.H. Liu, H.L. Johnson, and T. Mill. 1977. *Review of the environmental fate of selected chemicals*. U.S. Environmental Protection Agency, Publication No. EPA-560/5-77-003, Washington, D.C., 147 pp.

Robbins, D.C. 1976. Photodissociation of methyl chloride and methyl bromide in the atmosphere. *Geophysical Research Letters* **3**:213–216.

Schwarzenbach, R.P., E. Molnar-Kubica, W. Giger, and S.G. Wakeham. 1979. Distribution, residence time, and fluxes of tetrachloroethylene and 1,4-dichlorobenzene in Lake Zurich, Switzerland. *Environmental Science and Technology* **13**:1367–1373.

Simoneit, B.R.T., and M.A. Mazurek. 1981. Air pollution: the organic components. *CRC Critical Reviews in Environmental Control* **11**:219–276.

Sittig, M. 1980. *Priority toxic pollutants. Health impacts and allowable limits.* Noyes Data Corporation, New Jersey, 370 pp.

Smith, V.L., I. Cech, J.H. Brown, and G.F. Bogdan. 1980. Temporal variations in trihalomethane content of drinking water. *Environmental Science and Technology* **14**:190–196.

Spence, J.W., P.L. Hanst, and B.W. Gay, Jr. 1976. Atmospheric oxidation of methyl chloride, methylene chloride, and chloroform. *Journal of the Air Pollution Control Association* **26**:994–996.

Stewart, M.E., W.J. Blogoslawski, R.Y. Hsu, and G.R. Helz. 1979. By-products of oxidative biocides: toxicity to oyster larvae. *Marine Pollution Bulletin* **10**: 166–169.

Thom, N.S., and A.R. Agg. 1975. The breakdown of synthetic organic compounds in biological processes. *Proceedings of the Royal Society of London, Series B* **189**:347–357.

Thompson, Jr., H.C., R.N. Farragut, and M.H. Thompson. 1977. Relationship of scarlet prawns (*Plesiopenaeus edwardsianus*) to a benthic oil deposit off the northwest coast of Aruba, Dutch West Indies. *Environmental Pollution* **13**: 239–253.

Tou, J.C., L.B. Westover, and L.F. Sonnabend. 1974. Kinetic studies of bis (chloromethyl)ether hydrolysis by mass spectrometry. *Journal of Physical Chemistry* **78**:1096–1098.

Tsani-Bazaca, T., A.E. McIntyre, J.N. Lester, and R. Perry. 1981. Concentrations and correlations of 1,2-dibromoethane, 1,2-dichloroethane, benzene and toluene in vehicle exhaust and ambient air. *Environmental Technology Letters* **2**:303–316.

U.S. Environmental Protection Agency. 1980. *Ambient Water Quality Criteria Reports*. Office of Water Regulations and Standards, Washington, D.C.

U.S. International Trade Commission. 1960–1981. *Synthetic organic chemicals, U.S. production and sales*. U.S. Government Printing Office, Washington, D.C.

Veith, G.D., D.L. DeFoe, and B.V. Bergstedt. 1979. Measuring and estimating the bioconcentration factor of chemicals in fish. *Journal of the Fisheries Research Board of Canada* **36**:1040–1048.

Verberk, M.M., and T.M.L. Scheffers. 1980. Tetrachloroethylene in exhaled air of residents near dry cleaning shops. *Environmental Research* **21**:432–437.

Versar. 1979. *Water related environmental fate of 129 priority pollutants*. Vol. **II**. U.S. Environmental Protection Agency, Publication No. EPA-440/4-79-029b, Washington, D.C.

Ward, G.S., P.R. Parrish, and R.A. Rigby. 1981. Early life stage toxicity tests with a saltwater fish: effects of eight chemicals on survival, growth, and development of sheepshead minnows. (*Cyprinodon variegatus*). *Journal of Toxicology and Environmental Health* **8**:225–240.

Weaver, L., R. Boes, and G. Moore. 1980. An organics detection system for the Ohio river. *Progress in Water Technology* **13**:227–236.

Wise, Jr., H.E., and P.D. Fahrenthold. 1981. Predicting priority pollutants from petrochemical processes. *Environmental Science and Technology* **15**: 1292–1304.

Youssefi, M., S.T. Zenchelsky, and S.D. Faust. 1978. Chlorination of naturally occurring organic compounds in water. *Journal of Environmental Science and Health A* **13**:629–637.

Yung, Y.L., M.B. McElroy, and S.C. Wofsy. 1975. Atmospheric halocarbons: a discussion with emphasis on chloroform. *Geophysical Research Letters* **2**: 397–399.

Ziglio, G. 1981. Human exposure to environmental trichloroethylene and tetrachloroethylene: preliminary data on population groups of Milan, Italy. *Bulletin of Environmental Contamination and Toxicology* **26**:131–136.

4
Aromatic Hydrocarbons—Monocyclics

Monocyclic aromatic compounds consist of a basic benzene ring with six carbon atoms, six hydrogen atoms, and three double bonds. Substitution of the hydrogen atoms is common and yields chlorobenzenes, nitrobenzenes, ethyl benzene, toluene, and other derivatives. Although the carbon atom is bonded to three other atoms rather than four, the ring is not considered unsaturated and its stability results from a resonating electron structure. Thus, the larger the number of alternative arrangements of the electrons, the greater the stability of the molecule. This is exemplified by the increase in stability of chlorobenzenes with increasing halogenation. Aromatic compounds owe their name to the fact that many of the early derivatives of the series possessed a characteristic odor or were obtained from odoriferous material.

Production, Uses, and Discharges

Production and Uses

Monocyclic aromatic hydrocarbons have many industrial and agricultural applications and are therefore widely distributed in the environment. In the USA, total production has steadily increased during the 1960s and 1970s but has shown some decline in the last two years. Some of the more commonly produced aromatics include benzene, toluene, xylene, ethylbenzene, and chlorobenzene. Annual benzene production in the USA

Table 4.1. US production of some aromatic hydrocarbons.

	Benzene ($\times 10^7$ gallons)	Toluene ($\times 10^7$ gallons)	Xylene ($\times 10^7$ gallons)	Monochloro-benzene	1,2-Dichloro-benzene	1,4-Dichloro-benzene	Alkyl benzenes	2,4+2,6-Dini-trotoluene	Ethyl benzene	Nitro benzene
				$\times 10^4$ metric tons						
1960–1965*	62.6	39.1	31.9	25.0	1.9	3.2	28.3	—	106.1	9.7
1966–1970*	104.8	70.2	44.9	25.0	2.7	2.9	29.4	12.6	184.7	19.1
1971–1975*	125.8	87.5	53.2	17.2	2.7	2.9	23.6	18.6	246.9	20.2
1976	142.5	99.9	72.2	14.9	2.2	1.7	—	18.0	261.7	18.6
1977	143.5	101.8	81.1	14.8	2.1	3.0	23.6	9.5†	376.9	25.0
1978	148.8	105.4	84.5	13.4	1.9	1.9	23.8	29.8	380.3	26.1
1979	167.2	101.0	97.2	14.7	2.6	3.8	28.4	31.2	383.1	43.2
1980	200.7	101.7	90.8	12.8	2.2	3.4	40.6	—	346.5	27.7
1981	133.9	85.6	88.2	12.9	2.3	3.3	24.2	22.8†	354.3	40.8
1982	107.0	105.2	73.0	—	—	—	—	—	299.7	—

Sources: U.S. International Trade Commission (1960–1981); Chemical Marketing Reporter (1980–83).
*Annual average.
†2,4-dinitrotoluene only.
—, data not available.

averaged 62.6×10^7 gallons during the period 1960–1965, increasing to $>200 \times 10^7$ gallons by 1980 (Table 4.1). Benzene has many applications, including: (i) an intermediate in the synthesis of pharmaceuticals and other chemicals, such as styrene, detergents, pesticides, and cyclohexane, (ii) a degreasing and cleaning agent, (iii) an antiknock fuel additive, and (iv) a solvent for industrial extraction. It is also used as a thinner for lacquers, as a solvent in the rubber industry, and in the preparation of inks.

Production of toluene in the USA has always been less than that of benzene, now reaching 85×10^7 gallons annually (Table 4.1). Approximately 70% of this total goes to the synthesis of benzene while an additional 15% is employed in the manufacture of other chemicals. The remaining production is used as a solvent for paints and as a gasoline additive.

The chlorinated benzenes are a diverse group of industrially derived chemicals with wide practical application (Table 4.2). Monochlorobenzene is used primarily in the derivation of organic chemicals, some of which (e.g., DDT) are no longer made in large quantities. This has resulted in a decline of approximately 50% in total US production over the last 20 years. Usage of 1,2- and 1,4-dichlorobenzene has been variable in recent times but has shown a modest overall increase in production (Table 4.1). Although both compounds have a number of applications (Table 4.2), the 1,3-isomer does not appear to be used in significant quantities. Production of 1,2,4-trichlorobenzene (TCB) and 1,2,4,5-tetrachlorobenzene (TeCB) is substan-

Table 4.2. Uses of chlorinated benzenes.

Compound	Uses
Monochlorobenzene	synthesis of nitrochlorobenzenes (50%), solvent uses (20%), phenol production (10%), and DDT manufacture
1,2-Dichlorobenzene	synthesis of toluene diisocyanate, dyestuffs, herbicides, and degreasers
1,4-Dichlorobenzene	air deodorant and insecticide (90%)
1,2,4-Trichlorobenzene	dye carrier (46%), herbicide intermediate (28%), heat transfer medium, dielectric fluid in transformers, degreaser, and lubricant
1,2,4,5-Tetrachlorobenzene	synthesis of the defoliant 2,4,5-trichlorophenoxy acetic acid (56%), synthesis of 2,4,5-trichlorophenol (33%), and fungicide (11%)
Pentachlorobenzene	intermediate in the manufacture of specialty chemicals
Hexachlorobenzene	fungicide, dye manufacture, wood preservative, intermediate in organic synthesis, additive in pyrotechnic compositions, and porosity controller in the production of electrodes

Source: Sittig (1980).

Table 4.3. Canadian consumption ($\times 10^4$ metric tons) of some aromatic hydrocarbons

	Benzene	Toluene	1,4-Dichlorobenzene	Xylene
1960–1965	7.5	1.7	0.12	—
1966–1970	8.8	2.7	0.16	—
1971–1975	6.1	2.9	0.15	—
1976	—	6.1	0.11	—
1977	10.8	8.0	0.09	—
1978	20.2	8.6	0.11	—
1979	34.4	8.1	0.15	—
1980	56.0	47.0	—	47.9
1981	56.6	48.0	—	40.1
1982	49.7	47.7	—	39.9

Source: Statistics Canada (1960–1982).
—, data not available.

tially less than that of mono- and dichlorobenzene. In 1975, annual US output of 1,2,4-TCB was 1.3×10^4 metric tons, compared with 0.8×10^4 metric tons for 1,2,4,5-TeCB (Sittig, 1980). Since the latter compound is used in the synthesis of 2,4,5-T, it is likely that total output will decrease if 2,4,5-T use is further limited. Other trichlorobenzene isomers, as well as pentachlorobenzene, are not manufactured or used in significant quantities.

Direct information on the production of aromatics in Canada is not available. However, total usage can be estimated from consumption data (Table 4.3). As in the USA, the use of benzene has increased in recent years from 10.8×10^4 metric tons in 1977 to 56.6×10^4 tons in 1981. This is comparable to the increase recorded for toluene ($8–48 \times 10^4$ tons) during the same time period. By contrast, use of 1,4-dichlorobenzene has remained essentially steady during the last 20 years. No data could be found on the production/consumption of other aromatic compounds in Canada. This probably means that consumption of each chemical is low and not individually categorized.

Discharges

Monocyclic aromatic compounds originate from numerous sources such as industrial effluents, municipal effluents, seepage from dumps and landfill sites, oil spills, and combustion. It has been estimated that 1.9–11.1 million metric tons of petroleum oil enter the world's oceans annually (Connell and Miller, 1981). Assuming an average aromatic hydrocarbon content of 15% (Petrakis *et al.*, 1980), total deposition of these compounds amounts to 0.3–1.7 million metric tons per year. It therefore comes as no surprise to find high concentrations of monocyclics in surface waters near oil spill sites. For example, the blow-out of a well in the Gulf of Mexico resulted in the following sea-water residues: benzene 17,600 mg L^{-1}, toluene 7600 mg L^{-1}, ethyl benzene 1000 mg L^{-1}, and xylene 22,500 mg L^{-1} (Table 4.4). Although overall deposition into fresh waters is substantially less than the amount recorded for oceans, many oil/gas fields discharge considerable amounts of hydrocarbons into relatively small receiving waters. For example, the largest tar sands plant in northern Alberta (Canada) releases ~165,000 kg of oil and grease into the Athabasca River per year, equivalent to approximately 25,000 kg of aromatics.

User and producing industries are primary sources of monocyclic aromatic compounds (Table 4.4). Benzene residues exceeding 100 $\mu g\ L^{-1}$ have been reported for a number of plants, whereas dichlorobenzene and trichlorobenzene have occurred at substantially higher levels (>500 $\mu g\ L^{-1}$) in some areas. Municipal effluents are another major source of monocyclic compounds (Table 4.4). Although comprehensive data on total discharges are generally lacking, Eganhouse and Kaplan (1982) reported an overall aromatic discharge of ~3,500 metric tons per year from the Los Angeles sewage treatment facilities. Such high levels are probably typical of major cities and reflect the discharge of aromatic compounds from small industries and other users into the sewage systems. In addition, large quantities of benzene and toluene are washed from roadways during periods of heavy rainfall and eventually find their way into municipal storm sewers.

The production of many organic compounds requires the use of a variety of industrial processes and feedstock chemicals (Wise and Fahrenthold, 1981). This results in wastewater that contains a range of aromatics including benzene, toluene, ethylbenzene, chlorinated benzenes, and nitrobenzenes. It is also known that the production of most resins (acrylic, epoxy, alkyl, polypropylene, and phenolic polyester) requires the use of monomers, which again leads to discharge of benzene, toluene, and ethyl benzene. The same may be said about the production of polycarbonates, polyester, and styrene.

Behavior in Natural Waters

Several monocyclic compounds are sparingly soluble in water and thus generally have a low residence time under natural conditions. Halogen substitution (either mono or di) of the benzene ring in the *ortho* and *meta* positions increases the span of the liquid phase of the compound by lowering the melting point and elevating the boiling point, whereas *para* substitution seems to have the opposite effect; the melting point is elevated from 5.5°C (benzene) to 53.1°C (*p*-dichlorobenzene). The latter response, markedly elevated in hexachlorobenzene, is also reflected in the vapor pressures of the

Table 4.4. Concentration of some aromatic compounds in water and effluents.

Compound	Range (μg L^{-1})	Source
Benzene	12×10^3–$17,600 \times 10^3$	Gulf of Mexico, well blow-out
	<1–179	Effluent from consumer industry (USA)
	0.1–1.0	Potable water, 10 cities (USA)
Toluene	3×10^3–7600×10^3	Gulf of Mexico, well blow-out
	0.1–19	Potable water, 5 cities (USA)
	max. 11	Potable water, New Orleans
Monochlorobenzene	0.1–5.6	River water receiving textile industry waste (USA)
	90–530	Effluent from dye manufacturing plant (USA)
	max. 27.0	Municipal effluents, 2 cities (USA)
1,2-Dichlorobenzene	<1	Potable water, 4 cities (USA)
	1.0	Ground water (Miami)
	15–690	Effluent from consumption plants
	<0.01–440	Municipal effluent, 6 cities (USA)
	0.006–0.022	Municipal effluent, 4 cities (Canada)
1,3-Dichlorobenzene	<0.5	Potable water, 4 cities (USA)
	0.5	Ground water (Miami)
	0.007–0.013	Municipal effluents, 4 cities (Canada)
1,4-Dichlorobenzene	<0.5	Potable water, 4 cities (USA)
	0.5	Ground water (Miami)
	0.48–0.92	Municipal effluents, 4 cities (Canada)
	0.5–230	Municipal effluents, 6 cities (USA)
	1.5–2.9	Municipal effluents (Zurich)
	90–380	Effluent from dye-manufacturing plant (USA)
1,2,3-Trichlorobenzene	21–46	Creek receiving municipal effluent (North Carolina)
	0.002–0.003	Municipal effluent, 4 cities (Canada)
1,2,4-Trichlorobenzene	<0.01–275	Municipal wastewater, 5 cities (USA)
	0.005–0.018	Municipal effluent, 4 cities (Canada)
	12–500	Effluent from consumer industries

Table 4.4 *(continued)*

Compound	Range (μg L^{-1})	Source
	0.007	Los Angeles River (USA)
1,3,5-Trichlorobenzene	<0.01–0.9	Municipal wastewater, 6 cities (USA)
	26.0	Industrial waste discharge
	0.006	Los Angeles River (USA)
	<0.0005	Municipal effluents, 4 cities (Canada)
Alklybenzenes	100–230	Effluent, tire manufacturing plant (USA)
	2–1000	Gulf of Mexico, well blow-out

Sources: U.S. Environmental Protection Agency (1980); Brooks *et al.* (1981); Schwarzenbach *et al.* (1979); Jungclaus *edt al.* (1976); Games and Hites (1977); Oliver and Nicol (1982).

corresponding compounds. Solubility decreases with increasing chlorination of the benzene ring, and the opposite is true for the log *P* (octanol/water mixture) values. Chlorination of the benzene ring yields 12 different compounds: 1 mono-, 3 di-, 3 tri-, 3 tetra-, 1 penta-, and 1 hexachlorobenzene. Most of them are colorless liquids with a pleasant odor. Chlorination increases the solvent power, viscosity, and chemical reactivity of the benzene ring. All of the chlorobenzenes are thermally stable.

Sorption

Sediments and suspended solids are complex mixtures of aluminosilicate minerals (clays), metal oxides, and humic materials. The proportions of these components will vary widely from one area to the other depending on geological formation and weathering. The latter will also determine the particle size distribution. Many monocyclic aromatic compounds, which are nonpolar and sparingly soluble or insoluble in water, will be sorbed strongly to the organic component of the suspended solids and sediments. The relatively high log *P* (octanol/water mixture) values of chlorobenzenes and other monocyclic aromatic compounds indicate that, at environmental concentration levels, they will be sorbed strongly by sedimentary organic material. If the fraction of the chemical sorbed is large, sorption tends to buffer the concentration of the organic compound present in the aqueous phase. This in turn will likely slow down other transport and/or transformation processes of monocyclic aromatics in the aquatic environment. Thus, the sorption process should be considered as a discrete process with its effects on the rate of other processes such as volatilization in estimating the fate of an organic compound in natural waters.

Laboratory studies have shown that sediments accumulated (332 $\mu g\ kg^{-1}$) hexachlorobenzene from water (8.3 $\mu g\ L^{-1}$) with a concentration factor of 40 (Laseter *et al.*, 1976). Depuration was relatively long, exceeding that of the biota. Laska *et al.* (1976) showed that the concentration of hexachlorobenzene in the vicinity of an industrialized zone bordering the Mississippi River between Baton Rouge and New Orleans was 80 times greater in the soil adjacent to the river than in the water. This was due to the suspended and dissolved load in the river water. No specific environmental sorption data are available on other monocyclic aromatic compounds.

For predicting the sorption of monocyclic aromatics, one can use the linear regression relationship developed by Mill (1980) as given below:

$$\log K_{oc} = -0.782 \log [C] - 0.27$$

where $[C]$ = concentration in moles liter $^{-1}$.

This equation was developed using the data of Karickhoff *et al.* (1979), Kenaga and Goring (1980) and Smith *et al.* (1978) and rescaling to one coordinate set. Figure 4.1 presents the combined data plotted as log K_{oc} versus log solubility (of the chemical) in moles liter^{-1} and the regression line. One can easily estimate K_{oc} (sorption constant corrected for the organic

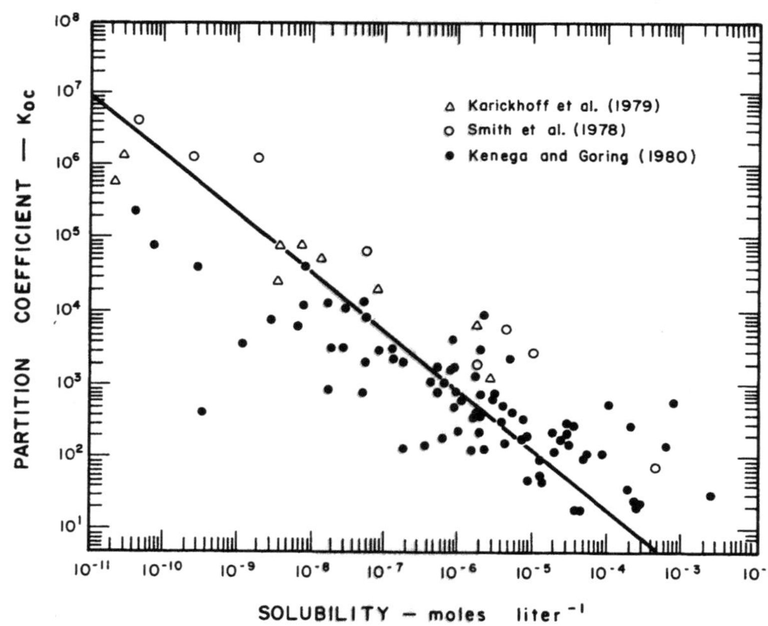

Figure 4.1. Correlation of aqueous solubility from partition coefficient (sediment/suspended solids). (From Mill, 1980.)

content) from the knowledge of accurate aqueous solubility data and use of the above equation. This estimate is reliable to a power of ten for most nonpolar chemicals, which is sufficiently accurate for screening purposes in most cases.

Volatilization

Based on laboratory studies, volatilization is the primary pathway in the fate of most monocyclic aromatic compounds. For benzene, the rate of volatilization is dependent on temperature only when the system is vapor phase controlled. The extrapolation of laboratory data to field conditions should take into consideration the variation in mixing as a function of depth of the water body and water velocity. The common assumptions made in the estimation of the rate of volatilization and half-lives of organic compounds are as follows: (i) the chemical is in the dissolved phase as a simple compound as opposed to the possible complex interaction with micro- and macrosolutes and surfaces, (ii) equilibrium between liquid and vapor phases, (iii) interfacial aqueous concentration is the same as the bulk water, assuming thorough mixing, and (iv) negligible effect of other chemicals present in the evaporation process. The validity of the above assumptions has to be checked prior to extrapolation to natural situations. Compounds with high boiling points and correspondingly low vapor pressures at ambient temperatures might be expected to volatilize slowly from water. However, it is argued that the sparingly soluble organic compounds often have high activity coefficients in the aqueous phase leading to unexpectedly high equilibrium partial vapor pressures. This, in turn, increases greatly the rate of volatilization (Mackay and Wolkoff, 1973). Calculations according to Mackay and Leinonen (1975) predicted for hexachlorobenzene (aqueous solubility $\sim 10\ \mu g\ L^{-1}$) a half-life of ~8 hours for volatilization from a water column of 1 m depth. Table 4.5 lists the half-lives for volatilization of some typical aromatic monocyclic compounds.

Oxidation and Hydrolysis

Many aromatic monocyclic compounds are resistant to aqueous oxidation. The only compound known to be susceptible to oxidation in water is 2,4-dinitrotoluene, where the CH_3 group is converted to form a hydroperoxide. This reaction may be important in highly aerated waters, but the extent to which it could occur in natural waters is difficult to estimate.

Owing to the extreme difficulty of nucleophilic substitution of aryl halides, hydrolysis will not likely occur in natural waters. Nitro compounds also resist hydrolytic scission. Thus, in general, hydrolysis under ambient environmental conditions is highly improbable.

Table 4.5. Half-lives for volatilization of some typical aromatic monocyclic compounds

Compound	Half-life ($t_{1/2}$)
Benzene[1]	4.81 hours
Chlorobenzene[1,2]	9.0 hours
	10–11 hours
1,2-Dichlorobenzene[1,2]	<9.0 hours
	8–9 hours
1,3-Dichlorobenzene[1]	~10 hours
1,4-Dichlorobenzene[1,2]	<9 hours
	11–12 hours
Hexachlorobenzene[3]	~8 hours
Ethyl benzene[1]	5–6 hours
Nitrobenzene[1]	~200 hours
Toluene[1]	5.18 hours
2,4-Dinitrotoluene[1]	~hundreds of days
2,6-Dinitrotoluene[1]	~hundreds of days

Sources: [1]Versar (1979); [2]Garrison and Hill (1972); [3]Mackay and Leinonen (1975).

Photolysis

Laboratory studies indicate that direct photolysis of most aromatic monocyclic compounds is not likely to occur unless a substantial shift in the wavelength of the incident light below 280 nm is caused by environmental factors. Detailed studies on the photodegradation of hexachlorobenzene showed that (i) the process is extremely slow and no breakdown products were identified and that (ii) the reaction is unlikely to be accelerated by the presence of known chemical sensitizers (Plimmer and Klingebiel, 1976). In natural waters, however, it is possible that the photodegradation of hexachlorobenzene could be photosensitized by humic compounds. Laboratory experiments carried out in smog chambers (Laity *et al.*, 1973) showed that *m*-xylene and 1,3,5-trimethyl benzene have half-lives of less than 4 hours. Using this value and the table of relative reactivities (Laity *et al.*, 1973), the photolytic half-life of ethylbenzene was calculated to be approximately 15 hours. However, extrapolation of laboratory data to the natural airshed should take into account meteorological variables.

Nitro compounds may be photoreduced by suitable hydrogen donors under sunlight. For example, nitrobenzene was reduced by toluene to yield aniline, 4-amino phenol, azoxy benzene, and benzoic acid. Alkyl aromatics that can form benzyl-type radicals were efficient hydrogen donors for this reaction. The sorption of nitrobenzene to humus with increased residence time and reaction sites will likely make this pathway play a significant role in the fate of nitrobenzene. Half-life of toluene calculated from smog chamber experiments on analogous compounds was 15 hours with benzaldehyde as the principal breakdown product (Laity *et al.*, 1973). For 2,4-dinitrotoluene and analogous compounds with moderate solubility and low volatility, vapor phase photolysis may not be significant. Photolysis in solution may be highly probable with a variety of possible breakdown products depending on the conditions and presence of other compounds. The nitro compounds are photochromic, colorless in the absence of light and becoming colored upon exposure to light. The environmental significance of this phenomenon is the possible photolytic breakdown of the nitrogroup under sunlight to yield a hydroxylamino, nitroso, or amino group with concomitant oxidation of the CH_3 group to an alcohol, aldehyde, or carboxylic acid. Specific studies to substantiate this possibility have not been carried out.

Biotransformations

Several soil and aquatic microorganisms can utilize some aromatic monocyclic compounds as a carbon source. Table 4.6 lists the biological breakdown products of some typical aromatic monocyclic compounds.

Isolation of specific microbes for biodegradation should not necessarily mean that other microorganisms are incapable of breaking down the organic compounds. Enrichment procedures using growth media tend to be selective toward faster growing species. In the natural environment, a given organic compound may in fact be readily metabolized in spite of the inability to identify a specific microbial species using it as a carbon source. On the other hand, isolation of a specific microbial species capable of utilizing an organic compound as a nutrient source need not guarantee that such a process will take place in ambient environmental conditions. Generally, the substrate concentration used in pure culture studies is considerably higher than normally encountered in the environment. Consequently, the enzymes involved in the degradation may not be induced under the environmental conditions.

Residues

Air

Benzene, toluene, and their derivatives enter the atmosphere by various mechanisms such as controlled emissions from consumer industries, volatilization from waste dumps and landfill sites, intentional spraying and dusting, and from automobile exhaust. Since such compounds are moderately soluble in water, they are probably washed out of the atmosphere with rainfall, deposited in surface waters, and then evaporated back into the atmosphere. Although this recycling process may be a significant source of aromatic

Table 4.6. Biotransformation of aromatic monocyclic compounds.

Compound	Organism	Mechanism and products
Benzene[1,2,3]	soil bacteria	oxidative degradation yielding catechols
	oil-degrading bacteria (only in presence of dodecane or naphthalene and at low concentration)	utilization as carbon source
	mammals	oxidation to arene oxides and transidihydrodiols
Chlorobenzene[4,5]	*Pseudomonas putida* (only in presence of another aromatic hydrocarbon source)	oxidative degradation yielding 3-chlorocatechol
	mixed cultures of aerobes	complete volatilization
1,2-Dichlorobenzene[6]	biological sewage treatment	nondegradable
1,4-Trichlorobenzene[7]	mixed cultures	nondegradable
Hexachlorobenzene[8]	Atlantic salmon	direct excretion
	algae (*Oedogonium cardiacum*)	85% of HCB intact
	snail (*Physa* sp.)	91% intact
	Daphnia magna	87% intact
	mosquito larva	58% intact
	mosquitofish	27% intact
	snail	84% intact
	cladoceran	67% intact
	mosquito larva	65% intact
	fish	64% intact (pentachlorophenol found in alga, mosquito larva, and in water)

Sources: [1]Gibson (1976); [2]Claus and Walker (1964); [3]Walker and Colwell (1975); [4]Gibson *et al.* (1968b); [5]Garrison and Hill (1972); [6]Thom and Agg (1975); [7]Ware and West (1977); [8]Zitko (1977).

compounds to aquatic ecosystems, particularly in heavily industrial parts of the world, no data are currently available to confirm this point.

Benzene can usually be detected in the atmosphere over cities. Residues generally range from 4 to 40 $\mu g\ m^{-3}$ but may exceed 16,000 $\mu g\ m^{-3}$ in the vicinity of solvent reclamation plants (U.S. Environmental Protection Agency, 1980). Toluene is also a common component of urban atmospheres and originates primarily from automobile exhaust and solvent-using industries. Average residues vary from 24 to 40 $\mu g\ m^{-3}$ over major metropolitan areas and therefore exceed those (4–8 $\mu g\ m^{-3}$) reported for ethyl benzene (Grob and Grob, 1974; Pilar and Graydon, 1973). It has been suggested that this latter compound accounts for approximately 10% of the total aromatics detected in air and roughly 1% of the total organics detected. Although monochlorobenzene and dichlorobenzene occur at moderately high concentrations in urban atmospheres (1–5 $\mu g\ m^{-3}$), little or no information is available on residues of other substituted aromatics. It is likely that significant quantities of hexachlorobenzene can be emitted from agricultural sources, but no information is available on this point.

Water

Under natural conditions, elevated residues in surface waters are primarily restricted to site-specific discharge points (Table 4.4). Two major sources of substituted benzenes are municipal effluents and discharges from consumer-industries (Table 4.4). Oliver and Nicol (1982) demonstrated that the concentrations of chlorobenzenes (Cl_2–Cl_6) in the upper reaches of the Niagara River (Canada/USA) ranged from <0.01 to 1.0 ng L^{-1} but increased to 1–126 ng L^{-1} downstream of the City of Niagara Falls. This was due to the discharge of effluents from chemical plants into the river and resulted in an annual loading of ~10 metric tons of chlorobenzenes into Lake Ontario. Higher levels have been reported for other industrialized parts of the world. For example, Meijers and Van der Leer (1976) reported that 1,2- and 1,4-dichlorobenzene reached a maximum combined concentration of 12.5 $\mu g\ L^{-1}$ in the River Waal (Netherlands) compared with 0.9 $\mu g\ L^{-1}$ in the River Maas. Similarly, maximum monochlorobenzene and trichlorobenzene residues in the Delaware River (USA) were 7.0 and 1.0 $\mu g\ L^{-1}$, respectively (Sheldon and Hites, 1978), whereas Weaver *et al.* (1980) reported that monochlorobenzene in the Ohio River (USA) ranged from <0.1 to 10 $\mu g\ L^{-1}$. These compounds and related agents generally occur at much lower concentrations and are often nondetectable in treated tap water (Grob and Grob, 1974; Stottmeister and Engewald, 1981). Consequently, chlorination of water supplies has not been implicated in the large-scale formation of halogenated benzene compounds, as is the case with some haloaliphatics. Hexachlorobenzene is widely distributed in water, owing to its use in agriculture, relatively long half-life (~8 hours), and high sorption affinity to suspended solids. Residues are now generally low, ranging from <0.001 to 0.010 $\mu g\ L^{-1}$, but in earlier years residues exceeded 0.2 $\mu g\ L^{-1}$ in some surface waters (Puccetti and Leoni, 1980).

Sediments

Benzene, toluene, and some of their derivatives are both moderately volatile and soluble in water. Consequently, large-scale sorption to sediments does not occur, which in turn results in low/nondetectable residues of these compounds in most lakes, rivers, and coastal areas. However, since increasing substitution of the basic benzene ring causes a decrease in the solubility and an increase in partition coefficient, compounds such as dichlorobenzene, hexachlorobenzene, and dinitrotoluene have been recorded at relatively high concentrations in sediments, particularly near specific industries. Some of the highest residues on record are those for the lower Great Lakes region, particularly the Niagara River and parts of Lake Ontario. Both areas receive seepage from chemical dumps and direct industrial discharges. Elder *et al.* (1981), working on tributary streams of the Niagara River, recorded maximum residues of 30–200 mg kg^{-1} for Cl_3–Cl_6 benzene compounds, and 6–20 mg kg^{-1} for Cl_1–Cl_3 naphthalene derivatives. Such levels have been implicated in the adulteration of fish tissues in Lake Ontario (Niimi, 1979). Nelson and Hites (1980) similarly reported that nitro-derivatives, including nitrobenzene, 2,4-dinitrotoluene, and 2,6-dinitrotoluene, reached concentrations of 8–50 mg kg^{-1} in soil samples taken adjacent to tributaries of the Niagara River and Lake Erie.

The concentrations of chlorinated benzenes in the sediments of Lake Ontario are substantially greater than those recorded for the other Great Lakes. Hexachlorobenzene, 1,4-dichlorobenzene, and 1,2,4-trichlorobenzene occur in particularly high concentrations averaging 94–97 μg kg^{-1}. This is primarily due to the input of wastewater from chemical plants at Niagara Falls (Oliver and Nicol, 1982). Although the same three compounds are also prevalent in Lakes Erie and Huron, their levels range from only 2 to 16 μg kg^{-1}. This can be compared with residues of 0.2–5 μg kg^{-1} in Lake Superior.

The presence of substituted aromatics in sediments has been related to the discharge of municipal wastewaters. Young *et al.* (1978), working with effluents from southern California, found that total chlorobenzene levels in wastewater averaged 0.11 μg L^{-1} with a maximum of 0.80 μg L^{-1}. This resulted in sediment residues of 0.36 mg kg^{-1} in the California Bight, though there was little or no accumulation of chlorobenzenes in fish tissues. In many other industrial/agricultural areas, chlorobenzenes occur at either low or nondetectable levels in sediments, reflecting a relatively low input rate and the effect of physico-chemical conditions on accumulation in sea water.

Aquatic Plants, Invertebrates, and Fish

All monocyclic aromatic compounds can be sorbed from the water but, since the half-life of unsubstituted compounds is short, benzene and toluene are seldom detected in plant tissues. Increasing substitution leads to increased stability in both plants and animals. Consequently, although data are limited for most substances, a concentration factor of 730 was reported for the alga

Oedogonium cardiacum exposed to hexachlorobenzene for 33 days (Isensee *et al.*, 1976). Invertebrate species similarly sorb monocyclic compounds directly from the water, but food may also be a primary route of entry, particularly for the less soluble derivatives such as hexachlorobenzene. Isensee *et al.* (1976) reported that exposure of gastropods and cladocerans to this latter compound resulted in concentration factors of 910 to 1500, whereas the corresponding values for grass shrimp *Palaemonetes pugio* and pink shrimp *Penaeus duorarum* were 2000 to 4100.

Monocyclic aromatic compounds are widespread contaminants of fish tissues. Highly substituted compounds, such as Cl_4–Cl_6 chlorinated benzenes, are particularly common and may account for more than half of the total monocyclic burden in fish (Table 4.7). Although dichlorobenzene and trichlorobenzene generally occur at lower concentrations, residues exceeding 20 $\mu g\ kg^{-1}$ wet weight have been reported for some sites in Yugoslavia, presumably a reflection of large-scale industrial discharge into receiving waters (Table 4.7). Benzene and toluene also usually occur at low levels in fish and, in many instances, are undetectable. This is due to the abbreviated half-life and volatility of these compounds.

The concentration factor of chlorinated benzenes increases with the degree of substitution, reaching 7800–18,500 in hexachlorobenzene (Table 4.8). Such values are 50–90% lower than those reported for DDT, its derivatives, and PCBs, but exceed those of many common pesticides such as lindane. The concentration factors of bromo-derivatives also increase with the degree of substitution and, although data are limited, such compounds appear to concentrate at least as much as the chloro-derivatives (Table 4.8). The half-life of dichlorobenzenes in tissues is relatively short, <1 day, whereas the corresponding values for pentachlorobenzene and hexachlorobenzene range from 6 to 7 days (Table 4.8). Although the half-life of bromo-derivatives in tissues probably increases with the degree of substitution, there are few data to confirm this point.

Toxicity

Aquatic Plants, Invertebrates, and Fish

Toxicity of aromatic hydrocarbons to aquatic plants is extremely variable, depending on species, compound, and environmental conditions (Table 4.9). Unsubstituted compounds such as benzene and toluene generally exhibit low toxicity, regardless of species. Although chlorobenzene is also only slightly toxic, increasing substitution leads to enhanced toxicity to many algae. Thus the 96-h LC_{50} of pentachlorobenzene is 40–170 times lower than that of

Table 4.7. Average concentration ($\mu g\ kg^{-1}$ wet weight) of chlorobenzenes in fish.

	1,2-Dichloro-benzene	1,4-Dichloro-benzene	1,3,5-Trichloro-benzene	1,2,4-Trichloro-benzene	1,2,3-Tricholoro-benzene	1,2,4,5-Tetrachloro-benzene	1,2,3,4,-Tetrachloro-benzene	Pentachloro-benzene	Hexachloro-benzene
Lake Superior									
Lake trout[1,2,*]	0.3	<1	<1	1.2	<1	<5	<5	<5	6.5
Lake Michigan									
Lake trout[1,*]	—	—	<1	2	<1	<5	<5	<5	6.3
Lake Ontario									
Lake trout[1,2,*]	1	4	<1	2.4	<1	2.9	4.5	9.1	117
Brown trout[1,*]	—	—	<1	2.8	1.1	2.5	5.0	11.0	115
Ashtabula River, Ohio									
Pike[1,*]	—	—	11	7.5	<1	96	20	157	1140
Fields Brook, Ohio									
Carp, bass[2,*]	—	—	31	14	1.9	193	43	352	2210
Sucker[2*]	—	—	75	124	20	1150	205	789	1530
Drava River,									
Yugoslavia, nase[3,†]	1140	450	5	5	48	12	2	75	160
Gulf of Triest, Yugoslavia,									
pilchard[3,†]	220	30	15	7	130	18	3	9	26

Sources: [1]Kuehl *et al.* (1981); [2]Oliver and Nicol (1982); [3]Jan and Malnersic (1980).
*Whole fish.
†Fat.
—, not measured.

chlorobenzene. Other derivatives that appear to be particularly toxic to aquatic plants include dinitrotoluene and 2,4,6-trinitrotoluene.

The acute toxicity of aromatics to invertebrates generally parallels that outlined for aquatic plants (Table 4.9). Thus, highly substituted compounds such as pentachlorobenzene are much more toxic to both marine and freshwater species than chlorobenzene. Similarly, sensitivity to dinitrotoluene is substantially greater than that to nitrotoluene, whereas most unsubstituted compounds are not acutely toxic to invertebrates. Symptoms of sublethal/chronic intoxication of monocyclic aromatic compounds are numerous and generally nonspecific. They include a reduction in growth, frequency of molting, fecundity, egg survival, and ability to osmoregulate. There may also be a depression followed by an increase in oxygen consumption, and depletion of glycogen.

Fish appear to be particularly sensitive to tetrachlorobenzene and pentachlorobenzene (Table 4.9). Although dinitrotoluene and 2,4,6-trinitrotoluene are also toxic, unsubstituted compounds such as benzene and toluene, as well as ethylbenzene, do not present an immediate threat to fish survival, except at high concentrations. Induction of chronic/sublethal

Table 4.8. Concentration factors and half-life of aromatic compounds in muscle tissue of fish

	Concentration factor	Half-life in tissues (days)
1,2-Dichlorobenzene[1]	89	<1
1,3-Dichlorobenzene[1]	66	<1
1,4-Dichlorobenzene[1,2,3]	60–220	0.5
1,2,4-Trichlorobenzene[1]	182	1–3
1,2,3,5-Tetrachlorobenzene[1]	1800	2–4
Pentachlorobenzene[1]	3400	>7
Hexachlorobenzene[2,3,4]	7800–18,500	6–7
2,5-Dibromotoluene[5]	470	3–4
1,2,3-Tribromobenzene[5]	1100	4–5
1,3,5-Tribromobenzene[5]	1130	4
1,2,4,5-Tetrabromobenzene[5]	1400	4–5
Toluene[3]	—	0.5
Benzene[6]	450	—

Sources: [1]Barrows *et al.* (1980); [2]Neely *et al.* (1974); [3]Neely (1979); [4]Veith *et al.* (1979); [5]Zitko (1979); [6]Woodward *et al.* (1981).

—, not measured.

Table 4.9. Acute toxicity (96-h LC_{50}, mg L^{-1}) of monocyclic aromatic hydrocarbons to algae, invertebrates, and fish.

	Algae		Invertebrates		Fish	
Compound	*Selenastrum capricornutum*[1,2]	*Skeletonema costatum*[1,2]	Cladoceran *Daphnia magna*[1,3] (48-h LC_{50})	Shrimp *Mysidopsis bahia*[1,2]	Bluegill[1,4]	Sheepshead minnow[5]
Benzene	525*	—	200	39†	22.5	33‡
Toluene	>433	>433	310	56	13	280–480
Chlorobenzene	228	342	86	16.4	16	10
1,2-Dichlorobenzene	95	44	2.4	2.0	5.6	9.7
1,3-Dichlorobenzene	64	52	28	2.9	5.0	7.8
1,4-Dichlorobenzene	97	57	11	2.0	4.3	7.4
1,2,4-Trichlorobenzene	36	8.9	—	0.5	3.4	21
1,2,3,5-Tetrachlorobenzene	17	0.7–7.1	9.7	0.3	6.4	37
1,2,4,5-Tetrachlorobenzene	50	7.3	—	1.5	1.6	0.8
Pentachlorobenzene	6.7	2.1	5.3	0.2	0.25	0.8
Ethyl benzene	>438	>438	75	88	150	280
Nitrobenzene	44	9.6	27	6.7	43	59
Dinitrotoluene	1.5	0.4	0.7	0.6	0.3	—
2,4,6-Trinitrotoluene	5	50	—	—	—	2.6‡

Sources: [1]U.S. Environmental Protection Agency (1980); [2]Kauss and Hutchinson (1975); [3]LeBlanc (1980); [4]Buccafusco *et al.* (1981); [5]Heitmuller *et al.* (1981).

**Chlorella vulgaris.*

†Grass shrimp *Palaemonetes pugio.*

‡Fathead minnow.

—, not measured.

intoxication has been reported for most compounds. Struhsaker (1977) exposed female Pacific herring to benzene at a concentration of 0.7 mg L^{-1} for 48 hours just prior to spawning. This resulted in a reduction in the survival of embryos at hatching and of continuously exposed larvae. Similarly, striped bass showed a temporary loss in weight following treatment with benzene at concentrations of 6 mg L^{-1} for 168 hours (Korn *et al.*, 1976). In both of these studies, the concentrations required to induce symptoms of intoxication represented 3–20% of the acute LC_{50} level, a range that has been reported for several other compounds. For example, Smock *et al.* (1976) found that a behavioral response of fathead minnows to 2,4,6-trinitrotoluene occurred at 0.46 mg L^{-1} compared with the acute level of 2.58 mg L^{-1}. Exposure of embryos and larvae of the same species to 1,3- and 1,4-dichlorobenzene yielded chronic intoxication levels of 0.8–2.0 mg L^{-1} (U.S. Environmental Protection Agency, 1980). Thus the acute-chornic ratio for both compounds was 5.2.

Human Health

Monocyclic aromatic hydrocarbons are highly soluble in lipids and commonly occur in human fat and milk. Morita *et al.* (1975) found that dichlorobenzene in the fat of deceased persons from Tokyo averaged 1700 $\mu g\ kg^{-1}$ compared with 19 and 210 $\mu g\ kg^{-1}$ for tetrachlorobenzene and hexachlorobenzene, respectively. Comparable or slightly higher levels have been reported from Canada, New Zealand, and Italy (IARC, 1979). Similarly, hexachlorobenzene concentrations in human milk generally range from 0.01 to 0.10 mg L^{-1}, though average residues of 1.24 mg L^{-1} have been reported from Austria (IARC, 1979). Because of their short biological half-life, benzene, toluene, and ethyl benzene are not commonly found in human tissues. Although the same may be said about nitrobenzene, this compound can be absorbed through the skin at rates as high as 2 mg $cm^{-2}\ h^{-1}$. Thus, there are many reports of poisoning from absorption of the nitrobenzene in shoe dyes and laundry marking ink, particularly during the first half of this century. In more recent times, the main route of exposure is inhalation at plants producing or using nitrobenzene.

Many monocyclic aromatic compounds, including chlorobenzene, dichlorobenzene, trichlorobenzene, tetrachlorobenzene, pentachlorobenzene, ethyl benzene, toluene, and nitrobenzene are not carcinogenic. By contrast, oral administration of hexachlorobenzene to hamsters produced liver tumors, liver haemangiotheliomas, and thyroid adenomas (IARC, 1979). Although hexachlorobenzene is also fetotoxic and produces some teratogenic effects,it is not mutagenic to yeast. Other compounds that are either promoters or known/suspected carcinogens include benzene, 2,6-dinitrotoluene, 2,4,6-

trinitrotoluene, azobenzene, and dodecylbenzene (Kraybill, 1980; Chu *et al.*, 1981).

References

Barrows, M.E., S.R. Petrocelli, K.J. Macek, and J.J. Carroll. 1980. Bioconcentration and elimination of selected water pollutants by bluegill sunfish (*Lepomis macrochirus*). *In*: R. Haque (Ed.), *Dynamics, exposure and hazard assessment of toxic chemicals*. Ann Arbor Science Publishers, Ann Arbor, Michigan, pp. 379–392.

Brooks, J.M., D.A. Wiesenburg, R.A. Burke, Jr., and M.C. Kennicutt. 1981. Gaseous and volatile hydrocarbon inputs from a subsurface oil spill in the Gulf of Mexico. *Environmental Science and Technology* **15**:951–959.

Buccafusco, R.J., S.J. Ells, and G.A. LeBlanc. 1981. Acute toxicity of priority pollutants to bluegill (*Lepomis macrochirus*). *Bulletin of Environmental Contamination and Toxicology* **26**:446–452.

Chemical Marketing Reporter. 1980–83. Schnell Publishing Company, New York, New York.

Chu, K.C., C. Cueto, Jr., and J.M. Ward. 1981. Factors in the evaluation of 200 National Cancer Institute carcinogen bioassays. *Journal of Toxicology and Environmental Health* **8**:251–280.

Claus, D., and N. Walker. 1964. The decomposition of toluene by soil bacteria. *The Journal of General Microbiology* **36**:107–122.

Connell, D.W., and G.J. Miller. 1981. Petroleum hydrocarbons in aquatic ecosystems—behavior and effects of sublethal concentrations: Part 1. *CRC Critical Reviews in Environmental Control* **11**:37–104.

Eganhouse, R.P., and I.R. Kaplan. 1982. Extractable organic matter in municipal wastewaters. 1. Petroleum hydrocarbons: temporal variations and mass emission rates to the ocean. *Environmental Science and Technology* **16**:180–186.

Elder, V.A., B.L. Proctor, and R.A. Hites. 1981. Organic compounds found near dump sites in Niagara Falls, New York. *Environmental Science and Technology* **15**:1237–1243.

Games, L.M., and R.A. Hites. 1977. Composition, treatment efficiency, and environmental significance of dye manufacturing plants effluents. *Analytical Chemistry* **49**:1433–1440.

Garrison, A.W., and D.W. Hill. 1972. Organic pollutants from mill persist in downstream waters. *American Dyestuff Report* **61**:21–25.

Gibson, D.T. 1976. Initial reactions in the bacterial degradation of aromatic hydrocarbons. *Zbl. Bakt. Hyg., I. Abt.* Orig. B. **162**:157–168.

Gibson, D.T., J.R. Koch, and R.E. Kallio. 1968a. Oxidative degradation of aromatic hydrocarbons by microorganisms. I. Enzymatic formation of catechol from benzene. *Biochemistry* **7**:2653–2662.

Gibson, D.T., J.R. Koch, C.L. Shuld, and R.E. Kallio. 1968b. Oxidative degradation of aromatic hydrocarbons by microorganisms. II. Metabolism of halogenated aromatic hydrocarbons. *Biochemistry* **7**:3795–3802.

Grob, K., and G. Grob. 1974. Organic substances in potable water and in its precursor. Part II. Applications in the area of Zurich. *Journal of Chromatography* **90**:303–313.

Heitmuller, P.T., T.A. Hollister, and P.R. Parrish. 1981. Acute toxicity of 54 industrial chemicals to sheepshead minnows (*Cyprinodon variegatus*). *Bulletin of Environmental Contamination and Toxicology* **27**:596–604.

International Agency for Research on Cancer. 1979. *Some halogenated hydrocarbons*. IARC monographs on the evaluation of the carcinogenic risk. Vol. **20**, 609 pp.

Isensee, A.R., E.R. Holden, E.A. Woolson, and G.E. Jones. 1976. Soil persistence and aquatic bioaccumulation potential of hexachlorobenzene (HCB). *Journal of Agricultural and Food Chemistry* **24**:1210–1214.

Jan, J., and S. Malnersic. 1980. Chlorinated benzene residues in fish in Slovenia (Yugoslavia). *Bulletin of Environmental Contamination and Toxicology* **24**: 824–827.

Jungclaus, G.A., L.M. Games, and R.A. Hites. 1976. Identification of trace organic compounds in tire manufacturing plant wastewaters. *Analytical Chemistry* **48**:1894–1896.

Karickhoff, S.W., D.S. Brown, and T.A. Scott. 1979. Sorption of hydrophobic pollutants on natural sediments. *Water Research* **13**:241–248.

Kauss, P.B., and T.C. Hutchinson. 1975. The effects of water-soluble petroleum components on the growth of *Chlorella vulgaris* Beijerinck. *Environmental Pollution* **9**:157–174.

Kenaga, E.E., and C.A.I. Goring. 1980. Relationship between water solubility, soil sorption, octanol-water partitioning, and concentration of chemicals in biota. *In*: J.G. Eaton, P.R. Parrish, and A.C. Hendricks (Eds.), *Proceedings of the 3rd Symposium on Aquatic Toxicology*, American Society for Testing and Materials, Philadelphia, 1980, pp. 78–115.

Korn, S., J.W. Struhsaker, and P. Benville. 1976. Effects of benzene on growth, fat content, and caloric content of striped bass, *Morone saxatilis*. *Fishery Bulletin* **74**:694–698.

Kraybill, H.F. 1980. Evaluation of public health aspects of carcinogenic/mutagenic biorefractories in drinking water. *Preventive Medicine* **9**:212–218.

Kuehl, D.W., K.L. Johnson, B.C. Butterworth, E.N. Leonard, and G.D. Veith. 1981. Quantification of octachlorostyrene and related compounds in Great Lakes fish by gas chromatography-mass spectrometry. *Journal of Great Lakes Research* **7**:330–335.

Laity, J.L., I.G. Burstain, and B.R. Appel. 1973. Photochemical smog and the atmospheric reactions of solvents. *In*: R.W. Tess (Ed.), *Solvents theory and practice*. Advances in Chemistry Series 124, American Chemical Society, Washington, D.C., pp. 95–112.

Laseter, J.L., C.K. Bartell, A.L. Laska, D.G. Holmquist, D.B. Condie, J.W. Brown, and R.L. Evans. 1976. *An ecological study of hexachlorobenzene*. U.S. Environmental Protection Agency, Publication No. EPA-560/6-76-009, Washington, D.C., 62 pp.

Laska, A.L., C.K. Bartell, and J.L. Laseter. 1976. Distribution of hexachlorobenzene and hexachlorobutadiene in water, soil, and selected aquatic organisms along the lower Mississippi River, Louisiana. *Bulletin of Environmental Contamination and Toxicology* **15**:535–542.

LeBlanc, G.A. 1980. Acute toxicity of priority pollutants to water flea (*Daphnia magna*). *Bulletin of Environmental Contamination and Toxicology* **24**: 684–691.

Mackay, D., and P.J. Leinonen. 1975. Rate of evaporation of low solubility contaminants from water bodies to atmosphere. *Environmental Science and Technology* **9**:1178–1180.

Mackay, D., and A.W. Wolkoff. 1973. Rate of evaporation of low-solubility contaminants from water bodies to atmosphere. *Environmental Science and Technology* **7**:611–614.

Meijers, A.P., and R.C. Van der Leer. 1976. The occurrence of organic micropollutants in the river Rhine and the river Maas in 1974. *Water Research* **10**: 597–604.

Mill, T. 1980. Data needed to predict the environmental fate of organic chemicals. *In*: R. Haque (Ed.), *Dynamics, exposure and hazard assessment of toxic chemicals*. Ann Arbor Science Publishers, Ann Arbor, Michigan, pp. 297–322.

Morita, M., S. Mimura, G. Ohi, H. Yagyu, and T. Nishizawa. 1975. A systematic determination of chlorinated benzenes in human adipose tissue. *Environmental Pollution* **9**:175–179.

Neely, W.B. 1979. A preliminary assessment of the environmental exposure to be expected from the addiction of a chemical to a simulated aquatic ecosystem. *International Journal of Environmental Studies* **13**:101–108.

Neely, W.B., D.R. Branson, and G.E. Blau. 1974. Partition coefficient to measure bioconcentration potential of organic chemicals in fish. *Environmental Science and Technology* **8**:1113–1115.

Nelson, C.R., and R.A. Hites. 1980. Aromatic amines in and near the Buffalo River. *Environmental Science and Technology* **14**:1147–1149.

Niimi, J.J. 1979. Hexachlorobenzene (HCB) levels in Lake Ontario salmonids. *Bulletin of Environmental Contamination and Toxicology* **23**:20–24.

Oliver, B.G., and K.D. Nicol. 1982. Chlorobenzenes in sediments, water and selected fish from Lakes Superior, Huron, Erie, and Ontario. *Environmental Science and Technology* **16**:532–536.

Petrakis, L., D.M. Jewell, and W.F. Benusa. 1980. Analytical chemistry of petroleum. An overview of practices in petroleum industry laboratories with emphasis on biodegradation. *In*: L. Petrakis and F.T. Weiss (Eds.), *Petroleum in the marine environment*. American Chemical Society, Washington, D.C., pp. 23–53.

Pilar, S., and W.F. Graydon. 1973. Benzene and toluene distribution in Toronto atmosphere. *Environmental Science and Technology* **7**:628–631.

Plimmer, J.R., and U.I. Klingbiel. 1976. Photolysis of hexachlorobenzene. *Journal of Agricultural and Food Chemistry* **24**:721–723.

Puccetti, G., and V. Leoni. 1980. PCB and HCB in the sediments and waters of the Tiber estuary. *Marine Pollution Bulletin* **11**:22–25.

Schwarzenbach, R.P., E. Molnar-Kubica, W. Giger, and S.G. Wakeham. 1979. Distribution, residence time, and fluxes of tetrachloroethylene and 1,4-dichlorobenzene in Lake Zurich, Switzerland. *Environmental Science and Technology* **13**:1367–1373.

Sheldon, L.S., and R.A. Hites. 1978. Organic compounds in the Delaware River. *Environmental Science and Technology* **12**:1188–1194.

Sittig, M. 1980. *Priority toxic pollutants. Health impacts and allowable limits.* Noyes Data Corporation, New Jersey, 370 pp.

Smith, J.H. W.R. Mabey, N. Bohonos, B.R. Holt, S.S. Lee, T.-W. Chou, D.C. Bomberger, and T. Mill. 1978. *Environmental pathways of selected chemicals in freshwater systems*. Parts I and II. U.S. Environmental Protection Agency, Publication Nos. 600/7-77-113 and 600/7-78-074.

Smock, L.A., D.L. Stoneburner, and J.R. Clark. 1976. The toxic effects of trinitrotoluene (TNT) and its primary degradation products on two species of algae and the fathead minnow. *Water Research* **10**:537–543.

Statistics Canada. 1960–1982. *Manufacturers of Industrial Chemicals*. Manufacturing and Primary Industries Division, Catalogue 46-219, Ministry of Supply and Services, Ottawa, Canada.

Stottmeister, E., and W. Engewald. 1981. Some aspects of the isolation and concentration of volatile organic micropollutants from water. *Acta Hydrochimica et Hydrobiologica* **9**:479–494.

Struhsaker, J.W. 1977. Effects of benzene (a toxic component of petroleum) on spawning Pacific herring, *Clupea harengus pallasi*. *Fishery Bulletin* **75**:43–49.

Thom, N.S., and A.R. Agg. 1975. The breakdown of synthetic organic compounds in biological processes. *Proceedings of the Royal Society of London, Series B* **189**:347–357.

U.S. Environmental Protection Agency. 1980. *Ambient water quality criteria reports*. Office of Water Regulations and Standards, Washington, D.C.

U.S. International Trade Commission. 1960–1981. *Synthetic organic chemicals. U.S. production and sales*. U.S. Government Printing Office, Washington, D.C.

Veith, G.D., D.W. Kuehl, E.N. Leonard, F.A. Puglisi, and A.E. Lemke. 1979. Polychlorinated biphenyls and other organic chemical residues in fish from major watersheds of the United States. 1976. *Pesticides Monitoring Journal* **13**:1–11.

Versar. 1979. *Water related environmental fate of 129 priority pollutants*. Vol. **II**. U.S. Environmental Protection Agency, Publication No. EPA-440/4-79-029b, Washington, D.C.

Walker, J.D., and R.R. Colwell. 1975. Degradation of hydrocarbons and mixed hydrocarbon substrate by microorganisms from Chesapeake Bay. *Progress in Water Technology* **7**:783–791.

Ware, S.A., and W.L. West. 1977. *Investigation of selected potential environmental contaminants: halogenated benzenes*. U.S. Environmental Protection Agency, Publication No. EPA-560/2-77-004, Washington, D.C., 283 pp.

Weaver, L., R. Boes, and G. Moore. 1980. An organics detection system for the Ohio River. *Progress in Water Technology* **13**:227–236.

Wise, Jr., H.E., and P.D. Fahrenthold. 1981. Predicting priority pollutants from petrochemical processes. *Environmental Science and Technology* **15**: 1292–1304.

Woodward, D.F., P.M. Mehrle, Jr., and W.L. Mauck. 1981. Accumulation and sublethal effects of a Wyoming crude oil in cutthroat trout. *Transactions of the American Fisheries Society* **110**:437–445.

Young, D.R., T.-K Jan, and T.C. Heesen. 1978. Cycling of trace metal and chlorinated hydrocarbon wastes in the southern California bight. *In*: M.L. Wiley (Ed.), *Estuarine interactions*. Academic Press, New York, pp. 481–496.

Zitko, V. 1977. *Uptake and excretion of chlorinated and brominated hydrocarbons by fish*. Fisheries and Marine Service Technical Report No. 737, Biological Station, St. Andrews, New Brunswick, 14 pp.

Zitko, V. 1979. The fate of highly brominated aromatic hydrocarbons in fish. *In*: M.A.Q. Khan, J.J. Lech, and J.J. Menn (Eds.), *Pesticide and xenobiotic metabolism in aquatic organisms*. American Chemical Society Symposium series No. 99, American Chemical Society, Washington, D.C., pp. 177–182.

5
Aromatic Hydrocarbons—Polycyclics

Polycyclic aromatic hydrocarbons (PAH), also known as polynuclear aromatic hydrocarbons, are fused compounds built on benzene rings. When a pair of carbon atoms is shared, then the two sharing aromatic rings are considered fused. The resulting structure is a molecule where all carbon and hydrogen atoms lie in one plane. Fusion imparts chemical properties in between those of benzene, a highly aromatic compound, and those of olefinic hydrocarbons. The lowest PAH is naphthalene ($C_{10}H_8$), consisting of two fused rings, whereas the ultimate member is graphite, an allotropic form of elemental carbon. The environmentally significant PAH range between naphthalene and coronene ($C_{24}H_{12}$). In this range, there is a large number of PAH differing in the number and positions of aromatic rings with varying number, position, and eventual chemistry of substituents on the basic ring system. Physical and chemical properties of PAH vary approximately in a regular trend with molecular weight. Susceptibility to redox reactions increases with increasing molecular weight, whereas aqueous solubility and vapor pressure decrease almost logarithmically with increasing molecular weight (Neff, 1979). Thus, PAH differ in their environmental behavior and interactions with the biological systems.

PAH originate from both natural and anthropogenic sources and are generally distributed in plant and animal tissues, surface waters, sediments, soils, and air. Many PAH can be found in smoked food, cigarette smoke, vegetable oils, and margarine, as well as surface waters and fish. Although PAH are not acutely toxic to most forms of life, several compounds are either known or suspected carcinogens. The addition of alkyl substituents generally enhances the carcinogenic potency of PAH, whereas hydrogenation and methylation cause a decrease in potency. Halogenation of PAH during industrial processes generally enhances their persistence and acute toxicity in surface waters.

Production, Uses, and Discharges

Production and Uses

Although many polycyclic hydrocarbons have been identified, only a few are produced and used commercially. These generally include the simpler compounds, such as naphthalene, acenaphthene, fluorene, and phenanthrene. In the USA, commercial production of petroleum naphthalene was 4.9×10^4 metric tons in 1976, compared with 4.7 and 6.5×10^4 metric tons for 1980 and 1981, respectively (U.S. International Trade Commission, 1960–81). This represented 8–10% of total crude naphthalene production. In Canada, the corresponding consumption was $0.04–0.07 \times 10^4$ metric tons for the same time period (Statistics Canada, 1960–80). Naphthalene is used in the manufacture of chemicals such as solvents, lubricants, and dyes. It is also employed as a moth repellent, insecticide, vermicide, anthelmintic, and intestinal antiseptic, and as a feedstock chemical for the synthesis of phthalic anhydride.

Other polycyclic aromatic hydrocarbons are manufactured in only small amounts. Acenaphthene finds limited use as an intermediate in dyestuff, plastics, and pesticide synthesis. Phenanthrene is also used as an intermediate, whereas 4-ring and 5-ring compounds (chrysene, pyrene, perylene, benzopyrene, dibenzanthracene, and benzanthracene) have few industrial uses. Thus, engine exhaust and other forms of combustion are some of the main sources of PAH in the environment.

Mixtures of tri- and tetrachloronaphthalenes comprise the bulk of the market use as the paper impregnant in automobile capacitors; secondary uses of mono- and dichloronaphthalenes include oil additives for engine cleaning and fabric dyeing. Annual production of all chlorinated naphthalenes in the USA has probably been <270 metric tons in recent years (Sittig, 1980).

Discharges

There are numerous sources of PAH in surface water, including municipal and industrial effluent, atmospheric fallout, fly-ash precipitation, road runoff, and leaching from contaminated soils. Overall, forest fires and combustion of coal are the most important sources of PAH in the atmosphere. Although automobiles, particularly those with diesel engines, formerly produced large quantities of PAH, recent environmental restrictions have significantly reduced their emissions. This will, however, probably be balanced by the increase in use of coal, shale oil, and tar sands oil as energy sources predicted for the 1980s. In many instances, domestic sewage and urban run-off contain higher PAH residues than effluents from user industries. This is primarily due to the erosion of automobile combustion products from roads, particularly during periods of heavy rainfall. Barrick (1982) found that a sewage treatment plant in Seattle (USA) discharged ~1 metric ton of PAH per year into Puget Sound.

Crude oil contains high levels of PAH, but the relative concentration of each compound depends largely on the type and source of oil. For example, Graf and Winter (1968) reported that the average concentrations of benzo[a]pyrene in Persian Gulf, Libyan, and Venezuelan crude were 0.04, 1.3, and 1.6 mg L^{-1}, respectively. This variability in PAH content is also found in refined petroleum products. Fluoranthene, pyrene, and benz[a] anthracene may be particularly common in gasoline compared with diesel fuel and heating oil. Part of the variability in content is induced by different distillation methods and levels of sophistication of the refinery. In general, PAH concentrate in the higher boiling point distillates and solid residues. Refining and petrochemical plant waste oils have been estimated to contribute ~200,000 metric tons of petroleum products and crude oil to the world's oceans annually. Although corresponding values are not available for fresh-water inputs, there are several reports of refinery effluents contributing to PAH contamination in lakes and rivers. For example, benzo[a]pyrene residues averaged 12 $\mu g\ L^{-1}$ in the effluent of a shale-oil plant (FRG), 1 $\mu g\ L^{-1}$ in the receiving waters 3.5 km downstream of the plant, and 0.1 $\mu g\ L^{-1}$ upstream of the plant (Pucknat, 1981).

Behavior in Natural Waters

Based on their physical and chemical properties, PAH can be grouped into two classes: (i) lower molecular weight 2–3 ring aromatics such as naphthalenes, fluorenes, phenanthrenes, and anthracenes, which are volatile and relatively toxic to aquatic organisms, and (ii) higher molecular weight 4- to 7-ring aromatics (chrysene-coronene), which are not acutely toxic but have been proven carcinogenic. Although PAH are only slightly soluble in water owing to their high molecular weight and low polarity, addition of anionic detergents such as potassium laurate can increase the PAH solubility by 10^4 times by forming micelles (Klevens, 1950).

Sorption

Because of their low aqueous solubility and vapor pressure, PAH are sorbed by particulate matter on entry into natural waters and deposited in bottom sediments, accumulating to high concentrations. Thus, water generally contains low PAH concentrations, reflecting unfavorable partitioning to the aqueous phase whereas the content in aquatic organisms is several orders of magnitude higher than that in water but less than in sediments. Low molecular weight PAH are depleted from the water body through volatilization, microbial oxidation, and sedimentation. Higher molecular weight PAH are removed essentially by photooxidation and sedimentation. About 50% of B[a]P and other high molecular weight PAH entering natural waters appears in the sediment (Neff, 1979). The sorptive affinity of PAH is exploited in their removal in waste treatment processes such as coagulation and flocculation, sedimentation, and filtration with sand or activated carbon. Conventional sewage treatment processes as investigated by Borneff and Kunte (1967) removed up to 90% of PAH. Removal of PAH during the primary sedimentation stage varied from 20 to 80%. Further treatment with synthetic flocculants, followed by filtration through activated charcoal, yields a water with PAH concentration similar to that of ground water (Reichert *et al.*, 1971). Removal by both rapid sand filters and activated carbon filters is a result of the high sorption affinity of PAH for surfaces. With powdered carbon, the sorption equilibrium of PAH can be reached in less than 5 minutes with continual stirring (Reichert *et al.*, 1971).

Volatilization

Volatilization can be a significant transport process for 2-ring PAH in the aquatic environment. For example, up to 50% of naphthalene was lost from a marine oil spill depending on water temperature, wind speed, and wave action (Lee, 1975). Based on laboratory studies, Southworth (1979) showed that the volatilization rate of several PAH decreased with a decrease in vapor pressure. The rate of change was inversely related to the number of aromatic rings and highly dependent upon the mixing rates within both water and air columns. Lower PAH such as naphthalene and anthracene were more sensitive to mixing within the water column than the higher PAH (Southworth, 1979). In fact, the volatilization half-life for naphthalene increased 7.5 times compared to 1.4 for benzo[a]pyrene with a ten-fold increase in stream velocity. This implied that PAH with 4 or more rings will have insignificant volatilizational loss under all environmental conditions and that lower PAH such as naphthalene will evaporate substantially from a clear, rapidly flowing shallow stream (Southworth, 1979). It was also shown that the volatilization half-lives for benzo[a]pyrene and benz[a]anthracene in well-stirred aqueous solution were 22 and 89 hours, respectively. These values are higher than their photolytic half-lives ($t_{1/2} = \sim 1-2$ hours) (Smith *et al.*, 1978). Anthracene had a calculated volatilization half-life of 300 hours in a 1 meter deep quiescent water body and about 18 hours under well-stirred conditions (Southworth, 1977). These calculated values assume an instantaneous mixing throughout the water column and presence of concentration gradient only at the surface. Such assumptions may not be totally valid under natural environmental conditions.

Photolysis

Atmospheric oxidants or sulfur oxides oxidize PAH to compounds including quinones, which are carcinogenic. PAH vary in their sensitivity to photooxidation. For example, 60% of B[a]P in soot particles was degraded under light in 40 minutes, whereas several other PAH showed little or no photooxidation (Thomas *et al.*, 1968). On exposure of equimolar concentrations of B[a]P and benz[a]anthracene to a simulated atmosphere of ozone and ultraviolet irradiation, more than half of B[a]P was degraded in 30 minutes compared with only 20% degradation of benz[a]anthracene.

The two major oxidizing species in urban air are the HO· radical and ozone; the former is formed by a cycle of reactions involving NO_2, H_2O, CO, and simple organic compounds, and the latter is formed from the oxygen atom and molecular oxygen (Radding *et al.*, 1976). Reactions involving singlet oxygen are unimportant in the atmosphere since its residence is limited owing to quenching by other interactions. Photooxidation by singlet oxygen seems to be the dominant process for the breakdown of PAH and other organics in water (Zafiriou, 1977). Under ozone and light, half-lives of several PAH vary between a few minutes to a few hours. Alkyl PAH appear to be more susceptible to photooxidation than the parent PAH (Radding *et al.*, 1976). Photolysis products are either (i) endoperoxides that undergo secondary reactions to yield a variety of products or (ii) diones (Figure 5.1).

PAH sorbed to particulates are more susceptible to photooxidation than in solution. Also, the oxidative pathway is different, yielding quinones. For example, anthracene sorbed to alumina or silica gel is photooxidized to 1,4-dihydroxy-9,10-anthraquinone. The photodegradation of surface-sorbed B[a]P is dependent on oxygen concentration (half-life decreasing with increasing oxygen levels), temperature (exponential increase with temperature) and extent of solar radiation in the water body. The rate of photodegradation of B[a]P will decrease with depth owing to (i) decrease in light intensity through absorption and scattering by water and suspended solids and (ii) decreases in temperature and dissolved oxygen. Thus, photooxidation of PAH will be negligible in bottom sediments.

Biotransformation

Microorganisms present in soil, sewage, and sea water are capable of degrading PAH. The reactions involve the introduction of two hydroxyl groups through a dihydrodiol intermediate into the aromatic nucleus

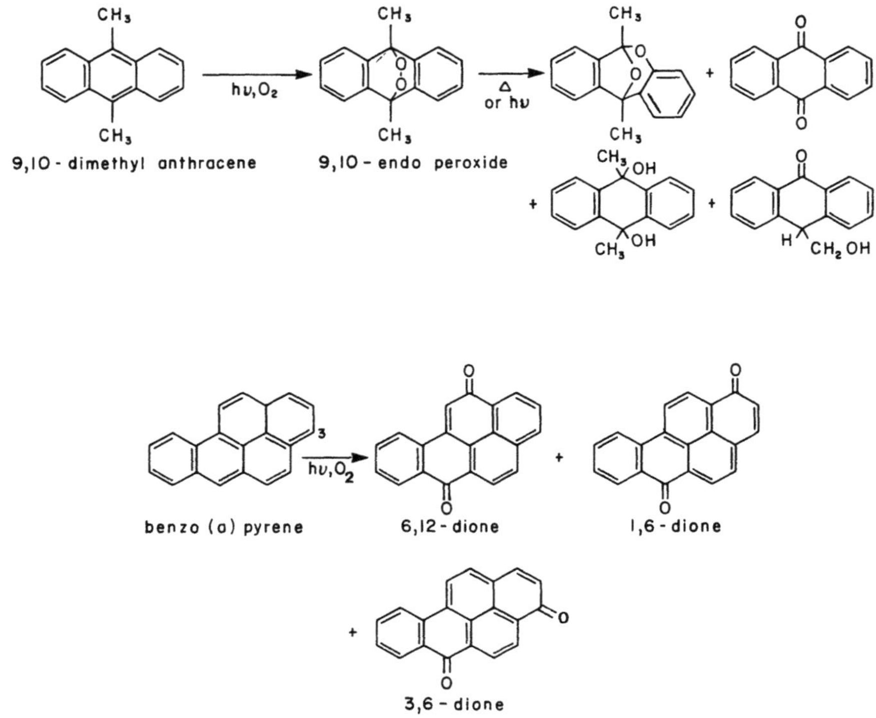

Figure 5.1. Photolysis products of 9,10-dimethyl anthracene and benzo[a]pyrene. (From National Academy of Sciences, 1972.)

(Gibson, 1976, 1977). The hydroxyl groups could be *ortho* or *para* to each other and undergo subsequent enzymatic cleavage. The bacterial pathway produces the *cis*-dihydrodiol intermediate, whereas the mammalian microsomal system yields the *trans*-isomer through an arene oxide intermediate. This arene oxide or its immediate oxidation products appear to be responsible for the carcinogenicity or mutagenicity of PAH (Heidelberger, 1976). Lower molecular weight PAH can be degraded completely to CO_2 and H_2O (for example, naphthalene by *Pseudomonas putida*) (Jerina *et al.*, 1971), whereas higher molecular weight PAH form various phenolic and acidic metabolites. A wild-type bacterium, *Beijerinckia* sp., isolated from polluted fresh-water streams, metabolized biphenyl, B[a]P, phenanthrene, anthracene, and benz[a]anthracene (Gibson *et al.*, 1975). Several *Pseudomonas* species degraded fluoranthene and B(a)P rapidly during the stationary phase of the growth cycle (Barnsley, 1975).

In mammals, an enzyme system called Mixed Function Oxygenase (MFO) located in the liver and sometimes in several other organs of vertebrates and invertebrates, is responsible for metabolizing several organic compounds, including PAH. Although this system is effective in detoxifying xenobiotics, PAH are transformed into mutagenic and carcinogenic intermediates such as arene oxides. These intermediates could be further broken down to less toxic products by various enzymatic and nonenzymatic reactions (Neff, 1979). The MFO system functions as an electron transport system in the presence of NADPH (nicotinamide adenine dinucleotide phosphate) with its reducing power, molecular oxygen, and the substrate, PAH. The system catalyzes the transfer of one oxygen atom to H_2O and one to PAH. Methods have been developed to identify the MFO system in nonmammalian species including many aquatic organisms (Neff, 1979); many species were found to metabolize PAH into polar metabolites. Table 5.1 lists the identified PAH metabolites by aquatic organisms. Production of dieldrin is also a good indicator of MFO activity in fresh-water invertebrates (Khan *et al.*, 1974). The metabolites formed by marine and fresh-water organisms are analogous to those formed in mammalian metabolism of PAH.

Table 5.1. Identified PAH metabolites by some typical aquatic organisms on exposure to PAH.

Species	Substrate(s)	Metabolites
Annelida *Nereis succinea* *Nereis virens*	benz[a]anthracene	5,6-dihydroxy-5,6-dihydrobenz[a]anthracene
Crassostrea virginica (oysters)	benzo[a]pyrene	9,10-dihydroxy-9,10-dihydrobenzo[a]pyrene 4,5 (or 7,8)-dihydroxy-4,5 (or 7,8)-dihydrobenzo[a]pyrene 1,6 (or 3,6)-diketobenzo[a]-pyrene 3-hydroxy-benzo[a]-pyrene
Brown trout	benzo[a]pyrene	Dihydroxy-dihydro-benzo[a]pyrene; 3,6-diketo-benzo[a]pyrene; 3-hydroxy-benzo[a]pyrene
Coho salmon	naphthalene	1-naphthol; 1-naphthyl-glucuronate; 1-naphthyl-glucoside; mercapturic acid; 1-naphthyl sulphate; 1,2-dihyroxy-1,2-dihydro-naphthalene

Source: Versar (1979).

Residues

Air

Polycyclic aromatic hydrocarbons are ubiquitous in the atmosphere. Maximum concentrations often occur in urban areas, resulting primarily from engine exhaust and other forms of combustion. These processes produce compounds that may be carried across continents and oceans, particularly in highly industrialized areas of the northern hemisphere (Simoneit and Mazurek, 1981). High molecular weight PAH, such as fluoranthene, pyrene, chrysene, benz[a]anthracene, benzofluoranthenes, benzo[a]pyrene, and benzo[e]pyrene are commonly encountered in urban atmospheres (Simoneit and Mazurek, 1981). Low molecular weight PAH are also widely distributed and include unsubstituted and alkyl naphathalene, phenanthrene, acenaphthene, and fluorene. PAH react in the upper atmosphere with NO_x to form nitro-derivatives, which are mutagenic.

Many aromatics such as benzo[a]pyrene and other PAH can be produced in small quantities by plants, bacteria, algae, and other microorganisms. This partially accounts for the presence of these compounds in remote regions of the world. Although data are limited and variable, atmospheric deposition is considered an important source of hydrocarbons in natural waters. Connell and Miller (1981a) indicated that 10–81% of input of PAH to the world's oceans was from atmospheric sources. Much of the variability in these data can be related to methods of analysis, differences in residence times, and environmental conditions, such as rainfall. Eisenreich *et al.* (1981) showed that total PAH deposition into the Great Lakes was high, amounting to 38–163 metric tons per year, compared with 13–56 and 7–32 metric tons per year for pesticides and phthalates, respectively.

Water

PAH are commonly found in natural waters, particularly in industrialized areas of the world. Andelman and Snodgrass (1974) found total PAH concentrations of 0.7–1.5 $\mu g\ L^{-1}$ in the Rhine River, compared with 0.1–3.1 $\mu g\ L^{-1}$ for other German rivers. Much higher levels (>50 $\mu g\ L^{-1}$) may be found near oil wells and other hydrocarbon sources, whereas in the open ocean and unpolluted lakes, residues are often <1 $\mu g\ L^{-1}$ (Connell and Miller, 1981a; Wiesenburg *et al.*, 1981). Benzo[a]pyrene is one of the most common PAH and, because it is carcinogenic, has been assayed in many monitoring programs. Russian workers recorded B[a]P levels of 0.05–3.5 $\mu g\ L^{-1}$ immediately downstream of an oil field, and 0.07–1.06 $\mu g\ L^{-1}$ a further 22 km downstream (Harrison *et al.*, 1975). Concentrations of 1 $\mu g\ L^{-1}$ have also been found below a shale processing plant, whereas 3.5–4.0 $\mu g\ L^{-1}$ of B[a]P occurred in the Clipperton Atoll Lagoon in the northeast Pacific

Table 5.2. Concentration (μg L^{-1}) of PAH in river waters.

	Thames (UK)	Danube (FRG)	Aach (FRG)
Fluoranthene	0.23(0.14–0.36)	0.09(0.06–0.11)	0.58(0.38–0.76)
Benzo[b]fluoranthene	0.08(0.04–0.12)	0.02(0.01–0.02)	0.24(0.08–0.36)
Benzo[j]fluoranthene	—	0.02(0.01–0.02)	0.32(0.14–0.42)
Benzo[k]fluoranthene	—	<0.01	0.14(0.13–0.17)
Benz[a]anthracene	—	0.01(<0.01–0.01)	0.20(0.10–0.39)
Benzo[a]pyrene	0.21(0.13–0.35)	<0.01	0.02(0.01–0.04)
Indeno[1,2,3-cd]pyrene	0.12(0.05–0.21)	<0.01	0.17(0.12–0.22)
Perylene	0.08(0.04–0.13)	—	—
Benzo[g,h,i]perylene	0.11(0.06–0.16)	0.01	0.07(0.04–0.11)

Source: Harrison *et al.* (1975).
—, not determined.

Ocean (Andelman and Snodgrass, 1974). These latter residues were probably produced through the metabolism of large algal and bacterial populations.

Apart from B[a]P, many other PAH are commonly found in natural waters and, in many instances, their concentrations exceed those reported for B(a)P (Table 5.2). Consequently, monitoring and impact assessment studies should consider these compounds and not necessarily be limited to the analysis of B[a]P. It is also important to monitor seasonal changes in residues, which may be extreme in some cases. Gschwend *et al.* (1982) reported that concentrations of naphthalene in Vineyard Sound (USA) ranged from <1 to 35 ng L^{-1} over a period of 16 months (Figure 5.2). Comparable variability may occur in just a few hours in rivers that are subject to heavy run-off during rain storms (Figure 5.3).

Sediments

Unsubstituted aromatics are generally distributed in the surficial sediments of lakes, rivers, and coastal marine areas. High levels are usually associated with heavy industrial activity and high population densities. Some of the more frequently encountered compounds are benzo[a]pyrene, fluoranthene, pyrene, and anthracene, though a wide range of other aromatics are also associated with sediments. Heit *et al.* (1981), working on two remote lakes (USA), demonstrated that PAH residues in recent surficial sediments (0–4 cm) were much higher than those in older, deeper cores. Benzo[a] pyrene averaged 0.13–0.69 mg kg^{-1} at the surface and only 0.001 mg kg^{-1} below 50 cm; the corresponding values for fluoranthene, pyrene, and anthracene were 0.85–0.004, 0.63–0.003, and 0.03–0.002 mg kg^{-1}, respectively. This gave

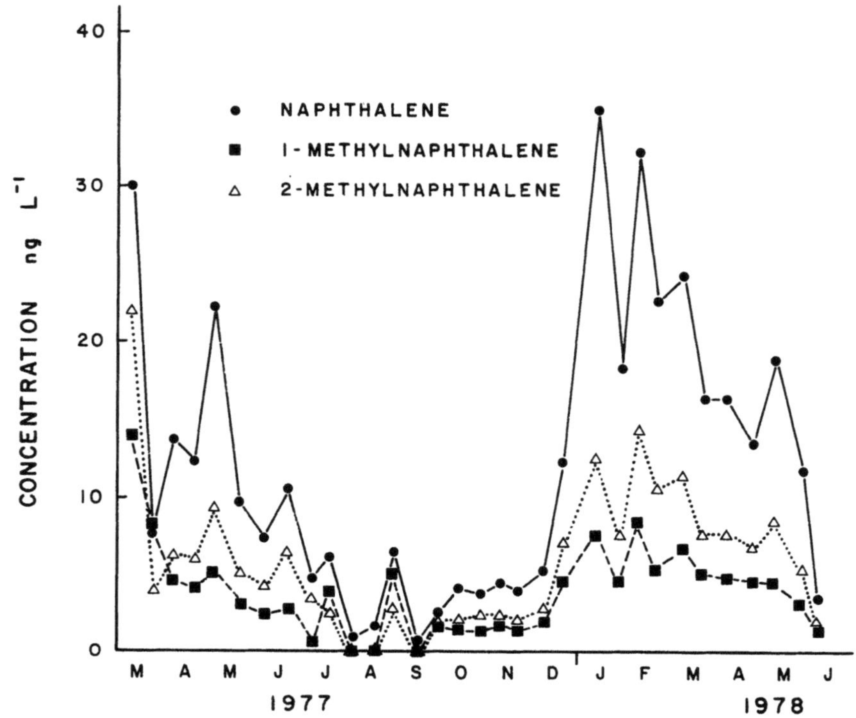

Figure 5.2. Seasonal changes in the concentrations of naphthalene, 1-methylnaphthalene, and 2-methylnaphthalene in Vinyard Sound, Massachusetts. (From Gschwend *et al.*, 1982.) © American Chemical Society. Reprinted with permission.

cultural enrichment factors in sediments that were approximately ten times greater than those recorded for metals. Davies *et al.* (1981) showed that very high levels of PAH can be found in sediments around drilling platforms. In their study of the North Sea fields (UK), anthracene + phenanthrene ranged up to 13.1 mg kg^{-1}, whereas naphthalenes, acenaphthene + fluorene, and total PAH reached 7.2, 0.7, and 33.5 mg kg^{-1}, respectively.

Aquatic Plants, Invertebrates, and Fish

A wide range of polycyclic aromatics occur in plant tissues albeit at relatively low concentrations. Obana *et al.* (1981a) showed that benzo[e] pyrene and pyrene were the predominant PAH in seaweed collected from the Osaka port area (Japan), averaging 72 and 60 $\mu g\ kg^{-1}$ dry weight, respectively (Table 5-3). These compounds were also among the most common in sediments and oysters collected from the harbor. Equally high

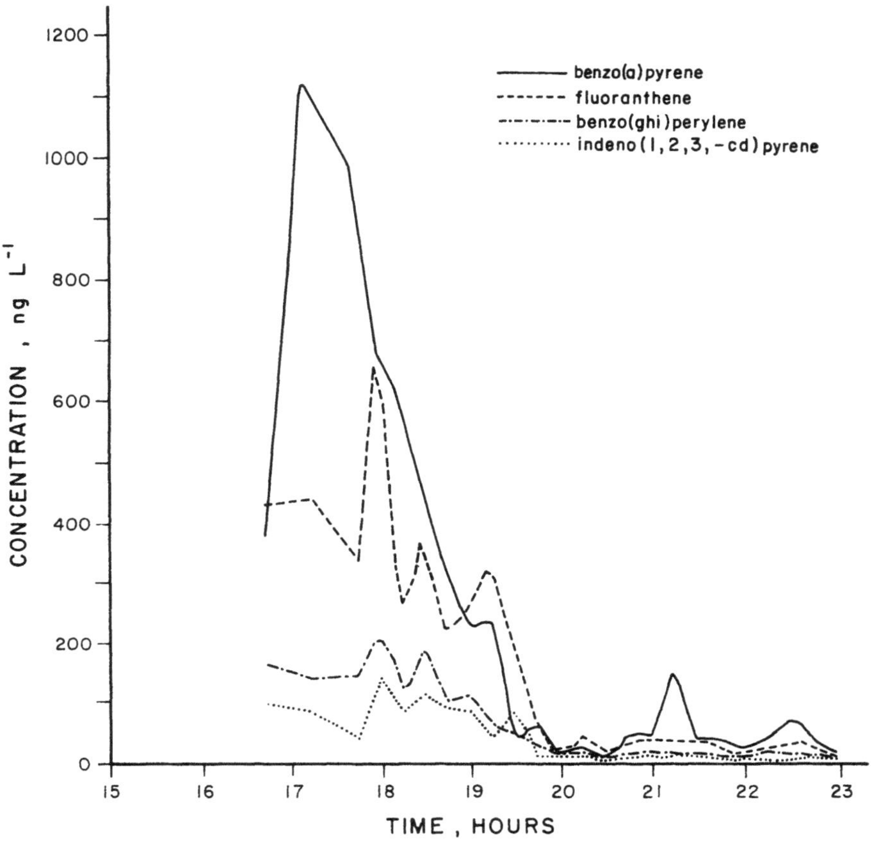

Figure 5.3. Hourly changes in the concentrations of some polycyclic aromatic hydrocarbons in the Sendelbach River (FRG) following a rainstorm. (From Herrmann, 1981.)

Table 5.3. Average and range in concentration ($\mu g\ kg^{-1}$) of polycyclic aromatic hydrocarbons in sediments, seaweed, and oysters collected from Osaka Harbor (Japan).

	Sediment (wet weight)	Seaweed (dry weight)	Oysters (wet weight)
Anthracene	42(15–88)	4(1.6–12)	1(<0.3–4.2)
Pyrene	338(150–630)	60(12–260)	25(7–52)
Benz[a]anthracene	87(38–230)	13(1.6–69)	4(1.5–10)
Benzo[e]pyrene	332(160–650)	72(0.9–410)	11(3–11)
Benzo[b]fluoranthene	372(220–640)	17(1.9–68)	7(3–20)
Benzo[k]fluoranthene	154(96–260)	23(0.9–120)	1(<0.1–5)
Benzo[a]pyrene	188(100–320)	16(0.6–81)	1(0.3–2.6)
Benzo[g,h,i]perylene	325(69–520)	24(0.6–130)	1.5(<0.2–2.8)
Dibenz[a,h]anthracene	29(19–40)	2(<0.1–10)	<0.1

Source: Obana *et al.* (1981a).

concentrations of PAH, including B(a)P, have been detected in remote parts of the world such as Greenland. This partially reflects world-wide atmospheric dissemination and anthropogenic emissions and implies that monitoring programs using plants must be designed to distinguish regional inputs from those of other sources. Relatively little is known about the half-life and bioconcentration of PAH in plant tissues. The alga *Oedogonium cardiacum* was exposed to benzo[a]pyrene for 3 days in a model ecosystem (U.S. Environmental Protection Agency, 1980). A concentration factor of 5300 was obtained, compared with 82,000–134,000 for two invertebrate species and 930 for the mosquito fish. Walsh *et al.* (1977) similarly treated the alga *Chlorococcum* sp. with a mixture of chlorinated naphthalene compounds for 24 hours and obtained a concentration factor of 25–140.

Southworth *et al.* (1978) reported that elimination rates in the cladoceran *Daphnia pulex* increased in the order perylene>benz[a]anthracene= methylanthracene > pyrene > phenanthrene > anthracene > naphthalene. Hence, naphthalene often occurs in relatively low concentrations in natural invertebrate populations. Anderson *et al.* (1976) demonstrated that methyl-derivatives of naphthalene were retained in invertebrates longer than the parent compound owing to a decrease in solubility and rate of volatilization. Varanasi and Malins (1977) reported bioconcentration factors of 2, 8, 17, and 27 for clams *Rangia cuneata* exposed to naphthalene, methylnaphthalene, dimethylnaphthalene, and trimethylnaphthalene, respectively.

Detectable levels of PAH have been recorded for invertebrates inhabiting remote parts of the world such as the antarctic and South Georgia Islands (Platt and Mackie, 1981). This is probably the result of world-wide dissemination of petrogeneic and combustion products, though production from natural sources cannot be discounted (Clarke and Law, 1981). In populated areas, several PAH are commonly found in invertebrate tissues. Obana *et al.* (1981a) reported that the main PAH in oysters from Osaka Bay (pyrene, benzo[e]pyrene, benzo[b]fluoranthene) were comparable to those in sediments (Table 5.3). Sediment residues were similarly correlated with the presence of PAH in oligochaetes and insect larvae inhabiting a tributary stream of the lower Great Lakes (Eadie *et al.*, 1982). Both species contained considerable amounts of pyrene, benzo[e]pyrene, benzo[a]pyrene, and fluoranthene, averaging 50 $\mu g\ kg^{-1}$ (oligochaetes) and 130 $\mu g\ kg^{-1}$ (insects).

Although many PAH have been found in both marine and fresh-water fish, residues are generally low except at site-specific discharge points. Consequently, PAH do not present a threat to most fisheries resources. Based on experimental data, it is known that the concentration factors for naphthalene are low, ranging from 30–150, with a half-life of 1–2 days. Similarly, concentration factors of 150, 75, 390, and 930 have been reported for fluorene, phenanthrene, acenaphthalene, and benzo[a]pyrene, respectively, and the corresponding half-life from <1 to 2 days (Woodward *et al.*, 1981; U.S. Environmental Protection Agency, 1980). No steady-state data are available for many of the remaining PAH. However, the equation log $CF = (0.85 \log P) - 0.70$ can be used to estimate the steady-state concentration factor for fish that contain approximately 7.6% lipid from the octanol/water partition coefficient (P). An adjustment factor of $3.0/7.6 = 0.395$ is used to adjust the CF from the 7.6% lipid to 3.0% lipid that represents the edible muscle portion of fish (Table 5.4).

Uptake of PAH from water is rapid and increases with exposure concentration; consequently, the deposition of petroleum-based products in water leads to rapid albeit short-term accumulation in fish tissues. Woodward *et al.* (1981) demonstrated that the main aromatic components in crude oil from Wyoming (naphthalene and benzene) were also prevalent in the tissues of cutthroat trout exposed to this oil for 90 days. Muscle residues reached 8.5 and 2.9 mg kg^{-1} for naphthalene and benzene, yielding CFs of 150 and 450, respectively. Naphthalene and methyl-naphthalene derivatives are also frequently encountered in natural populations but at relatively low concentrations. Nulton and Johnson (1981) found that both compounds seldom exceeded 50 $\mu g\ kg^{-1}$ in fish from the Gulf of Mexico, despite the presence of drilling platforms. Methyl-naphthalene was more common than the parent compound, because of its long half-life in tissues.

Table 5.4. Calculated concentration factors (CF) of polycyclic aromatic hydrocarbons based upon the octanol/water partition coefficient.

Chemical	Log P	Estimated steady state CF	Weighted average CF
Acenaphthalene	3.74	301	119
Dibenzofuran	4.12	634	250
Fluorene	4.18	713	282
Anthracene	4.45	1210	478
Phenanthrene	4.46	1230	486
Pyrene	4.88	2800	1100
Fluoranthene	4.90	2920	1150
1-Methylphenanthrene	5.00	3550	1400
Chrysene	5.61	11,700	4620
Benz[a]anthracene	5.61	11,700	4620
Dibenz[a,h]acridine	5.73	14,800	5850
Benzo[a]pyrene	6.06	28,200	11,100
Benzo[k]fluoranthene	6.06	28,200	11,100
Benzo[b]fluoranthene	6.06	28,200	11,100
Benzo[g,h,i]perylene	6.51	68,200	26,900
Indeno[1,2,3-cd]pyrene	6.51	68,200	26,900
Dibenz[a,h]anthracene	6.77	113,000	44,600
3-Methylcholanthrene	6.97	168,000	66,400

Source: U.S. Environmental Protection Agency (1980).

Toxicity

Aquatic Plants and Invertebrates

The toxicity of polycyclic aromatic hydrocarbons to aquatic plants is highly variable, depending on species, compound, and environmental conditions (Table 5.5). Some of the least toxic compounds include octachloronaphthalene and fluoranthene, whereas 1-chloronaphthalene is relatively toxic. Exposure of algae to PAH generally results in a rapid loss of chlorophyll and other changes in cell mineral composition, such as a decrease in manganese and potassium (Hutchinson *et al.*, 1981). There may also be a decline in protein levels and a concomitant increase in carbohydrates and lipids. Although these changes often result in mortality, many cultures are able to recover when placed in uncontaminated media. This ability to recover is partly due to the high volatility of many aromatics, which leads to a rapid decrease in the level of exposure in both culture and natural waters (Kauss and Hutchinson, 1975). In addition, polycyclics can be broken down to water-soluble derivatives that are then excreted by the plant.

The acute toxicity of PAH to invertebrates generally parallels that outlined for aquatic plants (Table 5.5). Several compounds are essentially nontoxic to invertebrates, whereas the LC_{50} of 1-chloronaphthalene is 0.4–1.6 mg L^{-1} depending on species. Symptoms of sublethal/chronic intoxication are numerous and include reduced growth and molting rate, depressed fecundity, and behavioral disorders such as locomotor impairment and abnormal burrow construction (Connell and Miller, 1981b). There may also be an initial depression followed by an increase in respiration rate and an impairment in ability to osmoregulate. Depletion of glycogen, development of gonadal tumors, generalized increase in leukocyctosis, and vacuolization of diverticula in the stomach and intestine have been observed in molluscs and presumably occurs in other species. Since these changes are nonspecific, they cannot be used as the sole indicator of aromatic hydrocarbon intoxication. However, the amount of work done in these areas is small, lagging far behind that for heavy metals.

Fish

Most PAH probably do not present a threat to fish survival, except at high concentrations (Table 5.5). Exposure to chronic/sublethal levels of aromatics has been implicated in a reduction in growth and fecundity of several species (Moles *et al.*, 1981). There may also be behavioral abnormalities such as loss of equilibrium, avoidance of mildly contaminated water, and increase in spontaneous activity. Since many aromatics are mutagens/teratogens, developmental abnormalities are often reported for treated fish.

Table 5.5. Acute toxicity (96-h LC_{50}; mg L^{-1}) of polycyclic aromatic hydrocarbons to algae, invertebrates, and fish

	Algae		Invertebrates		Fish	
Compound	*Selenastrum capricornutm*	*Skeletonema costatum*	*Daphnia magna*	*Mysidopsis bahia*	Bluegill	Sheepshead minnow
Naphthalene	33 (48 h)	—	8.6	—	4.9–8.9*	7.9*
1-Chloronaphthalene	1.0	1.2	1.6	0.4	2.3	2.4
Octachloronaphthalene	>500	>500	>530	>500	>600	>560
Acenaphthene	0.5	0.5	41	1.0	1.7	2.2
Fluoranthene	55	45	325	—	4.0	>560
Fluorene	—	—	1.0	—	—	—
Phenanthrene	—	—	0.3–0.6	—	—	—

Sources: U.S. Environmental Protection Agency (1980); Kauss and Hutchinson (1975); LeBlanc (1980); Buccafusco *et al.* (1981); Heitmuller *et al.* (1981); DeGraeve *et al.* (1982).

—, no data.

*Fathead minnow.

These include development of liver tumors, gill hyperplasia, kidney lesions, scoliosis, twinning, and nuclear pyenosis (Hose *et al.*, 1982). Benzo[a] pyrene is also known to cause chromosomal aberrations and induce spontaneous mutations in fish cell cultures (Kocan *et al.*, 1981; Hooftman and Vink, 1981). Treatment of fish with aromatics generally induces an increase in the number of white blood cells and calcium levels in blood and a decrease in the number of red blood cells, hemoglobin, and protein (Goel and Garg, 1980).

As with invertebrates, relatively little is known about factors influencing the toxicity of aromatics to fish. Synergism/antagonism has not been studied in detail, though it is known that water hardness has little effect on toxicity of at least some compounds (Pickering and Henderson, 1968). Temperature appears to have an inconsistent and unpredictable effect on survival, possibly reflecting the inadequate data base collected to date (Malins and Hodgins, 1981). It is known, however, that the larvae and fry of several species are much more susceptible to aromatics than adults. Similarly, Pacific salmon are more sensitive to hydrocarbons upon entering the sea than fresh water owing to osmotic stress (Malins and Hodgins, 1981).

Human Health

Polycyclic aromatic hydrocarbons and their substituted derivatives are highly soluble in lipids and therefore commonly found in human adipose tissue, milk, and liver. Obana *et al.* (1981b) reported average residues of 0.3 and 1.4 $\mu g\ kg^{-1}$ wet weight for anthracene and pyrene in the fat of ten deceased persons from Japan, whereas the concentration in liver averaged 0.2 and 0.4 $\mu g\ kg^{-1}$, respectively. Several other PAH, such as benzo[b]fluoranthene and benzo[a]pyrene were also found in both tissues, but at lower concentrations, $<0.05\ \mu g\ kg^{-1}$. The hepatobiliary system and gastrointestinal tract are the primary routes of elimination of PAH and their metabolites, although noncarcinogenic PAH such as anthracene, naphthalene, and phenanthrene are also eliminated in the urine. The half-life of PAH in mice and presumably humans is highly variable. Heidelberger and Jones (1948) reported that disappearance rates of radioactivity from sites of subcutaneous injections in mice were 12 weeks for dibenz[a,h]anthracene, 3.5 weeks for 3-methylcholanthrene, and 12 days for benzo[a]pyrene. The relative carcinogenicity of each compound was directly proportional to the half-life of the compounds.

The presence of residues in humans has not been directly implicated in the induction of cancer. However, many polycyclic aromatic compounds are either known or suspected carcinogens and contribute to the total carcinogenic burden in people (Table 5.6). Consequently, exposure guidelines have been developed for many aromatics, particularly those occurring in food, potable water, and the work place.

Table 5.6. Carcinogenicity of some polycyclic aromatic hydrocarbons.

Promoters and Known/Suspected Carcinogens	
Benzo[a]pyrene	Benzo(g,h,i)perylene
Benzo[e]pyrene	Benzo(a)fluorene
Benz[a]anthracene	Chrysene
Benzo[a]fluoranthene	Fluorene
Benzo[b]fluoranthene	Perylene
Benzo[j]fluoranthene	Pyrene
Benzo[k]fluoranthene	
Not Carcinogenic	
Acenaphthene	Naphthalene
Anthracene	Phenanthrene
Fluoranthene	

Sources: Kraybill *et al.* (1978); Kraybill (1980); Chu *et al.* (1981); Sittig (1980).

Apart from their carcinogenic properties, PAH are commonly implicated in damage to the hematoporetic and lymphoid systems. Degeneration of the spleen, thymus, and mesenteric lymph nodes, and inhibition in the development of bone marrow has also been observed. There may be damage to oocytes and increase in ovarian aryl hydrocarbon hydroxylase activity. Although many PAH are teratogenic in experimental animals, there is no evidence of similar effects in humans.

References

Andelman, J.B., and J.E. Snodgrass. 1974. Incidence and significance of polynuclear aromatic hydrocarbons in the water environment. *CRC Critical Reviews in Environmental Control* **4**:69–83.

Anderson, J.W., L.J. Moore, J.W. Blaylock, D.L. Woodruff, and S.L. Kiesser. 1976. Bioavailability of sediment-sorbed naphthalenes to the sipunculid worm, *Phascolosoma agassizii. In*: D.A. Wolfe (Ed.), *Proceedings of a symposium on fate and effects of petroleum hydrocarbons in marine ecosystems and organisms*, November 10–12, 1976, Seattle, Washington. pp. 276–285.

Barnsley, E.A. 1975. The bacterial degradation of fluoranthene and benzo(a)pyrene. *Canadian Journal of Microbiology* **21**:1004–1008.

Barrick, R.C. 1982. Flux of aliphatic and polycyclic aromatic hydrocarbons to Central Puget Sound from Seattle (Westpoint) primary sewage effluent. *Environmental Science and Technology* **16**:682–692.

Borneff, J., and H. Kunte. 1967. Carcinogenic substances in water and soil. Part XIX. The effect of sewage purification on polycyclic aromatic hydrocarbons. *Archiv für Hygiene und Bakteriologie* **151**:202–210.

Buccafusco, R.J., S.J. Ells, and G.A. LeBlanc. 1981. Acute toxicity of priority pollutants to bluegill (*Lepomis macrochirus*). *Bulletin of Environmental Contamination and Toxicology* **26**:446–452.

Chu, K.C., C. Cueto, Jr., and J.M. Ward. 1981. Factors in the evaluation of 200 National Cancer Institute carcinogen bioassays. *Journal of Toxicology and Environmental Health* **8**:251–280.

Clarke, A., and R. Law. 1981. Aliphatic and aromatic hydrocarbons in benthic invertebrates from two sites in Antarctica. *Marine Pollution Bulletin* **12**:10–14.

Connell, D.W., and G.J. Miller. 1981a. Petroleum hydrocarbons in aquatic ecosystems—behavior and effects of sublethal concentrations: Part 1. *CRC Critical Reviews in Environmental Control* **11**:37–104.

Connell, D.W., and G.J. Miller. 1981b. Petroleum hydrocarbons in aquatic ecosystems—behavior and effects of sublethal concentrations: Part 2. *CRC Critical Reviews in Environmental Control* **11**:105–162.

Davies, J.M., R. Hardy, and A.D. McIntyre. 1981. Environmental effects of North Sea oil operations. *Marine Pollution Bulletin* **12**:412–416.

DeGraeve, G.M., R.G. Elder, D.C. Woods, and H.L. Bergman. 1982. Effects of naphthalene and benzene on fathead minnows and rainbow trout. *Archives of Environmental Contamination and Toxicology* **11**:487–490.

Eadie, B.J., W. Faust, W.S. Gardner, and T. Nalepa. 1982. Polycyclic aromatic hydrocarbons in sediments and associated benthos in Lake Erie. Chemosphere **11**:185–191.

Eisenreich, S.J., B.B. Looney, and J.D. Thornton. 1981. Airborne organic contaminants in the Great Lakes ecosystem. *Environmental Science and Technology* **15**:30–38.

Gibson, D.T. 1976. Microbial degradation of carcinogenic hydrocarbons and related compounds. *In*: *Sources, effects and sinks of hydrocarbons in the aquatic environment*. American Institute of Biological Sciences, Washington, D.C., pp. 224–238.

Gibson, D.T. 1977. Biodegradation of aromatic petroleum hydrocarbons. *In*: D.A. Wolfe (Ed.), *Fate and effects of petroleum hydrocarbons in marine ecosystems and organisms*. Pergamon Press, New York, pp. 36–46.

Gibson, D.T., V. Mahadevan, D.M. Jerina, H. Yagi, and H.J.C. Yeh. 1975. Oxidation of the carcinogens benzo(a)pyrene and benzo(a) anthracene to dihydrodiols by a bacterium. *Science* **189**:295–297.

Goel, K.A., and V. Garg. 1980. 2,3′,4-triaminoazobenzene-induced hematobiochemical anomalies in fish (*Channa punctatus*). *Bulletin of Environmental Contamination and Toxicology* **25**:136–141.

Graf, V.W., and C. Winter. 1968. 3,4-benzpyren im Erdol. *Archiv fuer Hygiene and Bakteriologie* **152**:289–293.

Gschwend, P.M., O.C. Zafiriou, R.F.C. Mantoura, R.P. Schwarzenbach, and R.B. Gagosian. 1982. Volatile organic compounds at a coastal site. 1. Seasonal variations. *Environmental Science and Technology* **16**:31–38.

Harrison, R.M., R. Perry, and R.A. Wellings. 1975. Polynuclear aromatic hydrocarbons in raw, potable and waste waters. *Water Research* **9**:331–346.

Heidelberger, C. 1976. Studies on the mechanisms of carcinogenesis by polycyclic aromatic hydrocarbons and their derivatives. *In*: R. Freudenthal and P.W. Jones (Eds.), *Carcinogenesis—a comprehensive survey. Volume I. Polynuclear aromatic hydrocarbons: chemistry, metabolism, and carcinogenesis*. Raven Press, New York, pp. 1–8.

Heidelberger, C., and H.B. Jones. 1948. The distribution of radioactivity in the mouse following administration of dibenzanthracene labelled in the 9 and 10 positions with carbon^{14}. *Cancer* **1**:252–260.

Heit, M., Y. Tan., C. Klusek, and J.C. Burke. 1981. Anthropogenic trace elements and polycyclic aromatic hydrocarbon levels in sediment cores from two lakes in the Adirondack acid lake region. *Water, Air and Soil Pollution* **15**:441–464.

Heitmuller, P.T., T.A. Hollister, and P.R. Parrish. 1981. Acute toxicity of 54 industrial chemicals to sheepshead minnows (*Cyprinodon variegatus*). *Bulletin of Environmental Contamination and Toxicology* **27**:596–604.

Herrmann, R. 1981. Transport of polycyclic aromatic hydrocarbons through a partly urbanized river basin. *Water, Air and Soil Pollution* **16**:445–468.

Hooftman, R.N., and G.J. Vink. 1981. Cytogenetic effects on the eastern mud-minnow, *Umbra pygmaea*, exposed to ethyl methanesulfonate, benzo(a)pyrene, and river water. *Ecotoxicology and Environmental Safety* **5**:261–269.

Hose, J.E., J.B. Hannah, D. DiJulio, M.L. Landolt, B.S. Miller, W.T. Iwaoka, and S.P. Felton, 1982. Effects of benzo(a)pyrene on early development of flatfish. *Archives of Environmental Contamination and Toxicology* **11**:167–171.

Hutchinson, T.C., J.A. Hellebust, and C. Soto. 1981. Effect of naphthalene and aqueous crude oil extracts on the green flagellate *Chlamydomonas angulosa*. IV. Decreases in cellular manganese and potassium. *Canadian Journal of Botany* **59**:742–749.

Jerina, D.M., J.W. Daly, A.M. Jeffrey, and D.T. Gibson. 1971. *Cis*-1,2-dihydroxy-1,2-dihydronaphthalene: a bacterial metabolite from naphthalene. *Archives of Biochemistry and Biophysics* **142**:394–396.

Kauss, P.B., and T.C. Hutchinson. 1975. The effects of water-soluble petroleum components on the growth of *Chlorella vulgaris* Beijerinck. *Environmental Pollution* **9**:157–174.

Khan, M.A.Q., R.H. Stanton, and G. Reddy. 1974. Detoxication of foreign chemicals by invertebrates. *In*: M.A.Q. Khan and J.P. Bederka (Eds.), *Survival in toxic environments*. Academic Press, New York, pp. 177–201.

Klevens, H.B. 1950. Solubilization of polycyclic hydrocarbons. *J. Phys. Colloid. Chem.* **54**:283–298.

Kocan, R.M., M.L. Landolt, J. Bond, and E.P. Benditt. 1981. *In vitro* effect of some mutagens/carcinogens on cultured fish cells. *Archives of Environmental Contamination and Toxicology* **10**:663–671.

Kraybill, H.F. 1980. Evaluation of public health aspects of carcinogenic/mutagenic biorefractories in drinking water. *Preventive Medicine* **9**:212–218.

Kraybill, H.F., C.T. Helmes, and C.C. Sigman. 1978. Biomedical aspects of biorefractories in water. *In*: O. Hutzinger, L.H. Van Lelyveld, B.C.J. Zoeteman (Eds.), *Aquatic pollutants*. Pergamon Press, New York, pp. 419–459.

LeBlanc, G.A. 1980. Acute toxicity of priority pollutants to water flea (*Daphnia magna*). *Bulletin of Environmental Contamination and Toxicology* **24**: 684–691.

Lee, R.F. 1975. Fate of petroleum hydrocarbons in marine zooplankton. *In*: Sources, effects and sinks of hydrocarbons in the aquatic environment. American Institute of Biological Sciences, Washington, D.C., pp. 549–554.

Malins, D.C., and H.O. Hodgins. 1981. Petroleum and marine fishes: a review of uptake, disposition, and effects. *Environmental Science and Technology* **15**: 1272–1280.

Moles, A., S. Bates, S.D. Rice, and S. Korn. 1981. Reduced growth of coho salmon fry exposed to two petroleum components, toluene and naphthalene, in fresh water. *Transactions of the American Fisheries Society* **110**:430–436.

National Academy of Sciences. 1972. *Particulate polycyclic organic matter*. NAS, Washington, D.C., 361 pp.

Neff, J.M. 1979. *Polycyclic aromatic hydrocarbons in the aquatic environment. Sources, fates and biological effects*. Applied Science Publishers, England, 262 pp.

Nulton, C.P., and D.E. Johnson. 1981. Aromatic hydrocarbons in marine tissues from the central Gulf of Mexico. *Journal of Environmental Science and Health* **A16**:271–288.

Obana, H., S. Hori, and T. Kashimoto. 1981a. Determination of polycyclic aromatic hydrocarbons in marine samples by high-performance liquid chromatography. *Bulletin of Environmental Contamination and Toxicology* **26**:613–620.

Obana, H., S. Hori, T. Kashimoto, and N. Kunita. 1981b. Polycyclic aromatic hydrocarbons in human fat and liver. *Bulletin of Environmental Contamination and Toxicology* **27**:23–27.

Pickering, Q.H., and C. Henderson. 1968. Acute toxicity of some important petrochemicals to fish. *Journal Water Pollution Control Federation* **38**: 1419–1429.

Platt, H.M., and P.R. Mackie. 1981. Sources of Antarctic hydrocarbons. *Marine Pollution Bulletin* **12**:407–409.

Pucknat, A.W. 1981. *Health impacts of polynuclear aromatic hydrocarbons*. Environmental Health Review No. 5., Noyes Data Corporation, Park Ridge, New Jersey, 271 pp.

Radding, S.B., T. Mill, C.W. Gould, D.H. Liu, H.L. Johnson, D.C. Bomberger, and C.B. Fojo. 1976. *The environmental fate of selected polynuclear aromatic hydrocarbons*. U.S. Environmental Protection Agency, Publication No. EPA-560/5-75-009, Washington, D.C., 122 pp.

Reichert, J., H. Kunte, K. Engelhardt, and J. Borneff. 1971. Carcinogenic substances occurring in water and soil—XXVII: further studies on the elimination from waste water of carcinogenic polycyclic aromatic hydrocarbons. *Archiv fuer Hygiene und Bakteriologie* **155**:18–40.

Simoneit, B.R.T., and M.A. Mazurek. 1981. Air pollution: the organic components. *CRC Critical Reviews in Environmental Control* **11**:219–276.

Sittig, M. 1980. *Priority toxic pollutants. Health impacts and allowable limits*. Noyes Data Corporation, New Jersey, 370 pp.

Smith, J.H., W.R. Mabey, N. Bohonos, B.R. Holt, S.S. Lee, T.-W. Chou, D.C. Bomberger, and T. Mill. 1978. *Environmental pathways of selected chemicals in freshwater systems; Part II: Laboratory studies*. U.S. Environmental Protection Agency, Publication No. EPA-600/7-78-074, Athens, Georgia, 432 pp.

Southworth, G.R. 1977. *Transport and transformation of anthracene in natural waters: process rate studies*. U.S. Department of Energy, Oak Ridge National Laboratory, Oak Ridge, Tennessee, 26 pp.

Southworth, G.R. 1979. The role of volatilization in removing polycyclic aromatic hydrocarbons from aquatic environments. *Bulletin of Environmental Contamination and Toxicology* **21**:507–514.

Southworth, G.R., J.J. Beauchamp, and P.K. Schmieder. 1978. Bioaccumulation potential of polycyclic aromatic hydrocarbons in *Daphnia pulex*. *Water Research* **12**:973–977.

Statistics Canada. 1960–1980. *Miscellaneous Chemical Industries. Manufacturing and Primary Industries*, Catalogue 46-216. Ministry of Supply and Services, Ottawa, Canada.

Thomas, J.F., M. Mukai, and B.D. Tebbens. 1968. Fate of airborne benzo(a)pyrene. *Environmental Science and Technology* **2**:33–39.

U.S. Environmental Protection Agency. 1980. *Ambient Water Quality Criteria Reports*. Office of Water Regulations and Standards, Washington, D.C.

U.S. International Trade Commission. 1960–1981. *Synthetic organic chemicals. U.S. production and sales*. U.S. Government Printing Office, Washington, D.C.

Varanasi, U., and D.C. Malins. 1977. Metabolism of petroleum hydrocarbons: accumulation and biotransformation in marine organisms. *In*: D.C. Malins (Ed.), *Effects of petroleum on arctic and subarctic marine environments and organisms. Volume II, Biological effects*. Academic Press, New York, pp. 175–270.

Versar. 1979. *Water-related environmental fate of 129 priority pollutants*. Vol. **II**. U.S. Environmental Protection Agency, Publication No. EPA-440/4-79-029b, Washington, D.C.

Walsh, G.E., K.A. Ainsworth, and L. Faas. 1977. Effects and uptake of chlorinated naphthalenes in marine unicellular algae. *Bulletin of Environmental Contamination and Toxicology* **18**:297–302.

Wiesenburg, D.A., G. Bodennec, and J.M. Brooks. 1981. Volatile liquid hydrocarbons around a production platform in the northwest Gulf of Mexico. *Bulletin of Environmental Contamination and Toxicology* **27**:167–174.

Woodward, D.F., P.M. Mehrle, Jr., and W.L. Mauck. 1981. Accumulation and sublethal effects of a Wyoming crude oil in cutthroat trout. *Transactions of the American Fisheries Society* **110**:437–445.

Zafiriou, O.C. 1977. Marine organic photochemistry previewed. *Marine Chemistry* **5**:497–522.

6
Chlorinated Pesticides

Chlorinated pesticides are a small but diverse group of artificially produced chemicals characterized by a cyclic structure and a variable number of chlorine atoms. Most members of the group are resistant to environmental degradation and relatively inert toward acids, bases, oxidation, reduction, and heat. The parent compounds often have a number of related analogs and isomers, which show significant variation in toxicity and persistence. In some instances, these isomers have been used to develop highly specific insecticides, such as γ-hexachlorocyclohexane, which shows low toxicity to plants and mammals.

DDT was first synthesized in 1874 and its insecticidal properties discovered in 1939. Technical DDT is a stable, white, amorphous powder composed of up to 14 analogs and isomers. In 1942, hexachlorocyclohexane (benzene hexachloride) was discovered to be an effective and simple insecticide. Of its isomers, γ-HCH has the greatest insecticidal activity and is marketed as lindane, whereas α-HCH and β-HCH are more toxic to mammals. Chlordane, a mixture of terpenoid compounds, was discovered in 1945 to be a highly effective residual insecticide. In 1948, the most active principle of chlordane, termed heptachlor, was developed, along with two other cyclodiene derivatives, aldrin and dieldrin. It has been subsequently shown that microbial conversion of heptachlor in the environment yields heptachlor epoxide, which exhibits toxicity equal to or greater than that of the parent compound whereas photochemical conversion yields the equally toxic photoheptachlor. Also during 1948, a product obtained by the chlorination of turpentine and containing a considerable number of chlorinated camphenes was sold as toxaphene.

Endrin was first marketed in 1951 as a highly effective insecticide, containing at least 95% ingredient in the active form. It has the advantage of not being as stable as many of the earlier insecticides and thus residues in the environment are relatively low. Three years later, a broad spectrum insecticide, named endosulfan, appeared on the market and, like its predecessor, was subject to environmental degradation. Technical endosulfan consists of about four parts of α-isomer and one part β-*trans* isomer. The α-isomer, which is somewhat more insecticidal, is slowly transformed to the more stable β form.

A number of other chlorinated pesticides have appeared on the market in recent years. Mirex was introduced commercially to North America in 1969 as an insecticide to control fire ants in the southern USA and as a flame retardant. It is moderately stable in the environment but does slowly break down to the highly toxic derivatives chlorodecane and photomirex. Methoxychlor, although known for some time, did not find widespread application until the North American ban on DDT in 1969. It is closely related to DDT, containing methoxy ($-OCH_3$) groups on the phenyl groups instead of chlorine. Recent years have also seen the development of a series of chlorinated herbicides, such as picloram, triallate, propachlor, chloranil, and dichlobenil. Most of these compounds, although toxic, break down rapidly in the environment, resulting in relatively low residue levels.

Production, Uses, and Discharges

Production and Uses

Production of several chlorinated pesticides such as DDT and dieldrin has decreased sharply during the last decade (Table 6.1). The decision to restrict use was based on the following factors: (i) persistence of the parent compound and metabolites, (ii) susceptibility to large-scale transport and volatilization, and (iii) extreme lipophilicity. In the USA, annual consumption of DDT decreased from a high of 70,000 metric tons in 1963 to 36,000 metric tons in 1969. Since 1972, the production of DDT has been effectively discontinued. Aldrin/dieldrin consumption also peaked during the mid-1960s, reaching 9000 metric tons, but decreased to 5000 metric tons in 1970 and is now negligible. At present, ~7300 metric tons of toxaphene are used annually in the USA, primarily for control of insects on nonfood crops. However, new U.S. Environmental Protection Agency restrictions that limit its use to scabies control in cattle and sheep will reduce the market to ~400 metric tons per year by 1986. Although the production/consumption of most other chlorinated insecticides has similarly decreased, there are only a few restrictions on the use of lindane (Table 6.1).

Table 6.1. Status and uses of some chlorinated pesticides in Canada and the USA.

Pesticide	Usage status	Uses
Aldrin/dieldrin	No registered use	Formerly used to control corn and citrus pests, termites, and in moth-proofing
Benzene hexachloride (BHC)	No registered use	Formerly used as a broad-spectrum insecticide
Chlordane	Major restrictions	Yermite control; formerly used in home and garden sprays and control of corn pests
Chlordecone	No registered use	Cockroach and ant control
DDT	Minor registered uses	Insect control on some fruits and vegetables; formerly used as a broad-spectrum insecticide
Endosulfan	Major restrictions	Insect control on fruit, vegetables, and tobacco
Endrin	Major restrictions	Lepidopteran control for cotton and other crops; formerly used to control tobacco worms
Heptachlor	Major restrictions	Control of termites and insect pests of nonfood plants, formerly used as a broad-spectrum insecticide
Lindane (γ-HCH)	Minor restrictions	Broad-spectrum insecticide
Methoxychlor	Major restrictions	Biting insect control in surface waters, garden sprays
Mirex	Major restrictions	Control of fire ants, flame retardant for plastics, generating smoke for military
Toxaphene	Increasing restrictions	Broad-spectrum insecticide for cotton and other nonfood crops; restricted to scabies control in cattle and sheep in 1986

Source: Metcalf (1981); Statistics Canada (1982).

Consumption of chlorinated insecticides in Canada has followed the same general trend as that in the USA. In several cases, data have been withheld and therefore true estimates of consumption cannot be made. Use of technical grade lindane has increased gradually over the last two decades, averaging only 16 metric tons in 1960 and >100 metric tons in 1976. There are no registered uses of aldrin/dieldrin in Canada and DDT consumption is low (<50 metric tons annually). Owing to the reluctance of suppliers to divulge recent sales data, it is not possible to estimate consumption of other chlorinated insecticides in Canada.

Most other western countries have either curtailed or limited the use of DDT. In Europe, endrin is limited to the control of rodents whereas lindane, toxaphene, and endosulfan find broader application (Heckman, 1981). Dieldrin is still used in many western nations as a moth-proofer for the textiles industry. In South Africa, there are no registered uses of DDT, endosulfan, and heptachlor (Van Dyk *et al.*, 1982). Aldrin, chlordane, dieldrin, and endrin find some application, whereas there are only a few restrictions on the use of lindane.

Discharges

Sprays are the principal means of insecticide application and use water as the main carrier, although volatile hydrocarbon sprays are occasionally employed. In past years, hydraulic nozzles with coarse atomization producing droplets 200–500 μm in diameter were used in application but have now been replaced by air blast and other atomizing nozzles producing droplets 30–80 μm in diameter. Such particles adhere to surfaces and thus have not been implicated in widespread drifting. By contrast, dusts with particles measuring 0.5–3.0 μm diameter are applied into a moving air stream and do not adhere well to substrates. Since this causes serious drift problems away from treatment areas, dusts no longer find widespread application.

In most instances, the actual spraying of insecticides over agricultural lands does not produce hazardous residues in surface waters. The bulk of insecticides in lakes and rivers generally originates from run-off from adjacent agricultural lands. The persistence of chlorinated insecticides in soils both extends and aggravates the problem of contaminated run-off. For example, dieldrin, chlordane, heptachlor epoxide, and DDT and its metabolites were found in two Ontario rivers several years after the use of these agents had either been terminated or limited (Frank, 1981). It was also shown that the total discharge of ΣDDT was relatively high (~10 kg yr^{-1}) compared with chlordane (~2 kg yr^{-1}), dieldrin (1–2 kg yr^{-1}), and other agents. Higher rates have been recorded for the Niagara River (ΣDDT 54 kg yr^{-1}), reflecting its water flow and the presence of pesticide manufacturing plants along the shore (Warry and Chan, 1981). As might be predicted, insecticide transport is generally greatest in the spring, corresponding to thaws, heavy rainfall, low ground cover, and high suspended solid loading in the water. Braun and Frank (1980) demonstrated that over 50% of the total organochlorine transport in 11 watersheds in Ontario occurred between January and April.

The use of chlorinated insecticides in controlling biting fly larvae has greatly decreased in many western nations in recent years. Temephos, abate, fenthion, and chlorpyrifos methyl are just a few of several newer agents that exhibit high specific toxicity and low persistence and therefore find relatively wide application. Methoxychlor is, however, still applied to several rivers in Canada for black-fly larvae control. During 1974, a 24% technical formulation was poured directly into the Athabasca River in northern Alberta for 15 min to give a final concentration of 0.3 mg L^{-1} in the river water (National Research Council of Canada, 1975). This method of application avoided spray drift, permitted accurate calculation of the total amount of methoxychlor entering the river, and resulted in 95%+ control of black flies. More recently (up to 1982), two 7.5-minute injections have been administered over specific larval rearing areas, thereby eliminating the need for whole-river application. In other provinces, methoxychlor may be applied by low dosage aerial spraying to achieve a theoretical concentration of 0.01 mg L^{-1} for 2 minutes.

Behavior in Natural Waters

Sorption

Sorption to bottom sediments is one of the most important factors determining the behavior of chlorinated pesticides in natural waters. Leshniowsky *et al.* (1970) was among the first to show that aldrin was quickly sorbed by bacterial floc and lake sediment, reaching an equilibrium in 20 and 10 minutes, respectively. Partition coefficients of 625 and 410 were calculated for the bacterial floc and sediment, suggesting that sorption was an important but not primary fate process in the aquatic environment (Kenaga and Goring, 1980). Other studies have shown that sorption of aldrin varied in the order: organic soil>clay>sand (Yaron *et al.*, 1967). Chlordane (both *cis*- and *trans*-isomers) at a concentration of 25 μg L^{-1} was effectively sorbed (>80%) by sediment (Oloffs *et al.*, 1973). In the absence of sediment, volatilization was the dominant process in the loss of chlordane from water. By analogy with DDT, the sediment partition coefficient for DDE was $\sim 10^5$ (Versar, 1979). The DDD content of some unperturbed sediments of Santa Barbara Bay increased by 50% from the level of 12 μg kg^{-1} in 1955 (Hom *et al.*, 1974). Similarly, sorption to suspended particulates with subsequent sedimentation was an important pathway in the fate

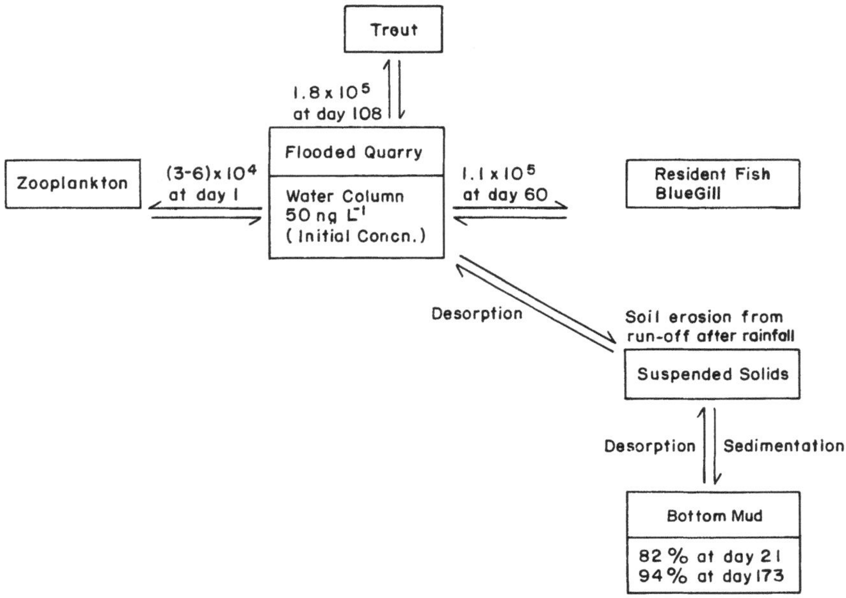

Figure 6.1. Dynamics of DDE in natural waters. (From Hamelink and Waybrant, 1976.)

of DDE in natural waters (Figure 6.1). From this study, it was concluded that (i) DDE was distributed in the water column by sorption-desorption processes of the sedimentable particulates, and (ii) decomposition of organic matter in the sediments facilitated the desorption of DDE.

The amount of DDT sorbed by clay was greater than that of heptachlor and dieldrin (Versar, 1979). Since the K_{oc} for DDT is 2.38×10^5 (Kenaga and Goring, 1980), sorption to particulate matter and eventual sedimentation seems to be an important pathway in the dynamics of DDT in the aquatic environment. The reported K values for the sorption of dieldrin to humic acid, soil, and silica sands are 228, 1.6×10^4, and 2–5 (Weil *et al.*, 1973). It seems that soil-derived suspended particles in water will have appreciable affinity for dieldrin. The low coefficients for sands are not unexpected considering their lack of organic content.

The α, β, and δ isomers of BHC have a relatively low affinity for surfaces, following the order $\delta \simeq \beta > \alpha$ (Tsukano, 1973). For soil with 1.9% organic carbon, $1/n$ was ~0.91 and $K = 10$–30; for soil with 5.2% organic carbon, $1/n = 0.71$–0.83 and $K = 30$–120 for all four isomers of BHC. These data suggest that BHC isomers other than the δ-isomer will not be sorbed significantly in natural waters. King *et al.* (1969) observed that an equilibrium in the sorption of lindane onto soils occurred in one hour, compared with 3 days for algae. In addition, sorption increased with the organic content of the soils. Sorption of lindane to sand reached a maximum within 4 hours and was largely independent of pH, time, or dissolved organic content.

Volatilization

The vapor pressures of chlorinated pesticides are classified into the following categories:

Low = ($0.1–0.9 \times 10^{-6}$ mm Hg)—DDT, endrin, and dieldrin
Medium = ($1.0–9.9 \times 10^{-6}$ mm Hg)—toxaphene and aldrin
High = ($10–99 \times 10^{-6}$ mm Hg)—chlordane and lindane
Very high = ($100–999 \times 10^{-6}$ mm Hg)—heptachlor

Pesticides with vapor pressures greater than 1×10^{-6} mm Hg at 20°C may volatilize readily from natural waters. Their rates of volatilization will depend upon the type of suspended solids present, the nature of sorptive bond, and temperature.

Half-lives for volatilization of aldrin from aquatic systems vary from a few hours to a few days: pure water (0.38 hours), water from San Francisco Bay (0.59 hours), the American River (0.60 hours), and the Sacramento River (0.60 hours), respectively (Table 6.2). Thus, volatilization could be the dominant fate process for aldrin in the environment. Similarly, in the absence of sediments or surfactants, chlordane is subject to significant volatilization. Oloffs and Albright (1974) recorded a 60% loss for both *cis*- and *trans*-isomers in 12 weeks with no metabolites being detected in the water. In

Table 6.2. Half-lives (hour) for volatilization of some typical chlorinated pesticides.

Medium	Aldrin	DDE	DDT	Dieldrin	Endosulfan	Lindane
Pure water[1]	0.38	0.67	3.9	7.7	—	—
Water from San Francisco Bay[1]	0.59	1.2	6.5	6.1	—	—
American River Water[1]	0.60	1.4	6.0	9.0	—	—
Sacramento River Water[1]	0.60	1.9	10.0	8.5	—	—
Calculated value[2]	7.7 days		3.1 days	1.5 yr	11 days	200 days

Source: [1]Singmaster (1975); [2]Versar (1979).
—, no data.

another study with *trans*-chlordane, a 70% reduction was reported in 20 hours (Bowman *et al.*, 1964).

DDD is less volatile than DDT and DDE; laboratory evaporation studies gave a relative ratio of volatilization of 10:3.3:1 for DDE:DDT:DDD, respectively (Singmaster, 1975). The calculated half-life of DDT ranges from a few hours to several weeks and that of DDD will be about one-third of DDT's value (Versar, 1979). Laboratory evaporation experiments using pure water, water from San Francisco Bay, the American River, and the Sacramento River at a water evaporation rate of 3.6 ± 0.2 ml h^{-1} yielded the following half-lives for the volatilization of DDE: 0.67, 1.2, 1.4, and 1.9 hours, respectively (Singmaster, 1975).

Based on evaporation studies, Tsukano (1973) claimed that there was a loss of 100%, 75%, 25%, and 25% of α-, γ-, β-, and δ-BHC isomers, respectively, in a 2-week period, where the water loss was about 80%. In contrast, Ernst (1977), from data on control aquaria in bioaccumulation studies, showed a quantitative recovery of α- and γ-isomers of BHC. Oloffs and Albright (1974) also claimed a slow volatilization for lindane.

Hydrolysis

Hydrolytic reactions are slow for most of the chlorinated pesticides. Wolfe *et al.* (1977) estimated the hydrolytic half-lives of DDD, DDE, and DDT from rate constants calculated from structure-activity relationship and literature data. The estimated values are in good agreement with the experimental findings with distilled water or in raw river water (Eichelberger and Lichtenberg, 1971). DDE is resistant to hydrolysis since it is the end product in the hydrolysis of DDT between pH 3–11. The second-order rate constant for the hydroxide ion-catalyzed hydrolysis of DDT at pH 9 is 9.9×10^{-3} M^{-1} sec^{-1}, which corresponds to a half-life of 81 days. Both isomers of endosulfan hydrolyze slowly at pH 5 with the rate increasing with pH. The data reported by Greve and Wit (1971) were from direct hydrolytic rate measurements, whereas that of Martens (1976) was from one set of data points on controls in biotransformation studies. After 10 days, the loss of endosulfan in controls as a function of pH was as follows:

pH	4.3	5.5	6.3	7	>8
Percent endosulfan lost	<1	2	8	28	>90

Endosulfan diol was the end product, and sorption to biomass reduced the degree of alkaline hydrolysis. From a microcosm study over a 33-day period,

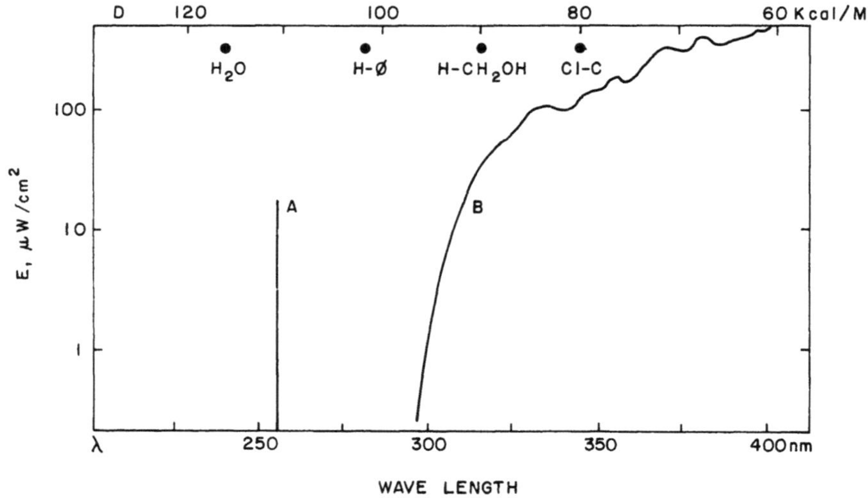

Figure 6.2. Energy distribution of the sunlight spectrum. A. Monochromatic mercury lamp at 254 nm. B. Black light lamp. (From Crosby, 1972.) © American Chemical Society. Reprinted with permission.

half-lives of 115 and 164 days were reported for α- and β-endosulfan isomers, respectively (Ali, 1978).

Photolysis and Oxidation

Chlorinated pesticides can undergo either direct or indirect photolysis. Most of the energy needed for environmental photolysis comes from solar radiation. Because of filtering by the ozone layer, energy of the sunlight reaching earth does not exceed 95 kcal/einstein ($\geq$300 nm). Approximate homolytic dissociation energies (kcal/mole) of some common bonds encountered in pesticides are given below:

C-H in benzene, 104; H-CH_2OH, 92; phenyl C-C1, ~80; H-OH, 118; and HO-OH, 48 Kcal/mole. Sunlight will deliver enough energy to cleave phenyl C-C1 and C-H bonds in aliphatic alcohol but not aromatic C-C and H-OH bonds (Figure 6.2). Most laboratory photochemical studies have been carried out at $\lambda = 254$ nm with energy = 112 kcal/einstein (Mercury lamp, line A, Figure 6.2). Consequently, the extent of photodecomposition taking place in the environment will depend on the light source, absorption maximum of pesticide, presence of photosensitizers, and altered bond strength resulting from substitution.

In general, chlorinated pesticides are resistant to chemical oxidation. However, aldrin may be oxidized to dieldrin by oxygen atoms or ozone under laboratory conditions.

Biotransformations

Chlorinated hydrocarbons are the most stable pesticides in the environment. Persistence of a compound is a measure of its resistance to degradation, but no pesticide is considered totally resistant to biological degradation. Table 6.3 lists the systems and metabolites in the biotransformation of chlorinated

Table 6.3. Biotransformation and metabolites of some chlorinated pesticides.

Compound	Organism(s)	Metabolites
DDD	Microbial systems	DDCO as the end product of the biological sequence
DDE	Microcosm study (snail, mosquito larvae, and mosquito fish)	3–14% polar metabolites
DDT	Anaerobic sewage sludge	TDE: $t_{1/2}$ = 4 days TDE, DBP, and DDE
	Sediment microbes	TDE, DDNS, and DDE
	Fresh water, sediment	TDE and DBP (major) and DDE, DDMU, DDMS, DDNU, DDM, and DBH (minor)
	Aquatic microorganisms	TDE, DDNS, and DDE
BHC	Soil-water mixture (incubated for 56 days)	% degradation of α-, β-, γ-, and δ-isomers = 90,70,95, and 50 δ-3,4,5,6 tetrachloro-1-cyclohexane was the product from α-BHC
	Washed cell suspension of *Clostridium sphenoides*	Total degradation of α- and γ-isomers in 4 and 2 h, respectively; δ-3,4,5,6-tetra chloro-1-cyclohexane the product from α-BHC
	Grass	Isomerization; γ→α→β-BHC
Lindane	Soil microorganisms	γ-PCCH, γ-TCCH, β-TCCH, PCB, 1,2,4,5-TCB, and 1,2,3,5-TCB
	Pseudomonas sp.	γ-TCCH, γ-PCCH, and δ-BHC
	Escherichia coli	γ-PCCH
	Clostridium sphenoides	γ-TCCH
	Clay loam soil and sewage sludge	benzene and 5% γ-TCCH

Source: Versar (1979).
γ-PCCH = 2,3,4,5,6-pentachloro-1-cyclohexane;
β,γ-TCCH = 3,4,5,6-tetrachloro-1-cyclohexane;
PCB = pentachlorobenzene; TCB = tetrachlorobenzene.

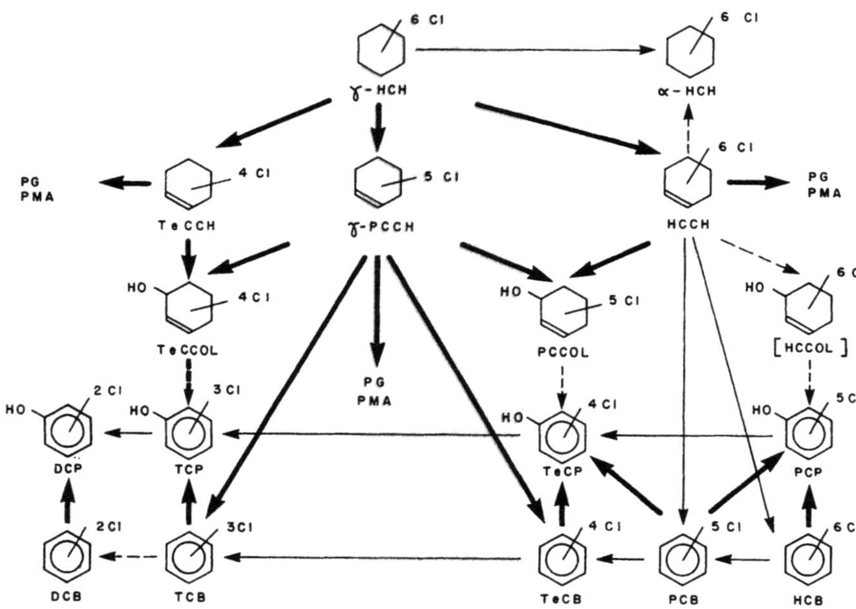

Figure 6.3. Degradation scheme of lindane in mammals. (From Engst *et al.*, 1978.) Abbreviations: DCB—dichlorobenzene; DCP—dichlorophenol; HCB—hexachlorobenzene; HCCH—hexachlorocyclohexene; HCCOL—hexachlorocyclohexenol; HCCH—1,2,3,4,5,6-hexachlorocyclohexane; PCB—pentachlorobenzene; PCCH—2,3,4,5,6-pentachlorocyclohexene; PCCOL—pentachlorocyclohexenol; PCP—pentachlorophenol; PG—(poly)chlorophenylglutathione; PMA—(poly)chlorophenylmercapturic acids; TCP—trichlorophenol; TeCB—tetrachlorobenzene; TeCCH—tetrachlorocyclohexene; TeCCOL—tetrachlorocyclohexenol; TeCP—tetrachlorophenol.

pesticides. Most of the metabolites of lindane can be produced by soil microbes and plants (Engst *et al.*, 1978). Some metabolites are more soluble in water than the incubated or applied lindane such as polychlorophenols. Although they are hydrophilic and easy to excrete, PCPs are potentially toxic. Several water-soluble lindane metabolites of unknown structure that are called conjugates have been identified. They are formed as end products of the possible detoxification mechanism where metabolites interact with mercapto compounds, glucuronic acid, and sulfuric acid, respectively. There are several degradation schemes proposed for lindane in mammals. The proposal by Engst *et al.* (1978) combines their own results with the literature data and has been continued on a wide experimental basis (Figure 6.3). Antidiuretics may interfere with the excretion of metabolites that might lead to toxic action (Smalley and Radeleff, 1971).

Residues

Air

Chlorinated pesticides enter the atmosphere primarily through spray drift during application, wind-blown dusts, and volatilization from treated surfaces. Heavy use of chlorinated pesticides generally means that spray drift is the most important means of entry to the atmosphere. In general, application by airplane results in more drift than application by ground equipment, and dusts drift more than sprays. For example, ΣDDT particles measuring 2 μm in diameter drifted about 35 km compared to 70 m for 50 μm droplets (Spencer, 1975). Wind-blown dusts are the most important source of entry to the atmosphere in areas where the use of chlorinated pesticides is either limited or curtailed. The dust source category will continue to contribute substantially to the world-wide redistribution of pesticides, particularly DDT and its derivatives, for many years to come. This partially accounts for the presence of detectable levels of pesticides in the sediments of remote Arctic lakes.

The decrease in use of chlorinated pesticides has resulted in a corresponding decline in airborne concentrations of most compounds. During the years of peak usage in the late 1960s, DDT residues generally ranged from 0.1 to 10 ng m^{-3} in rural areas of Canada, the USA, and Europe with some peak values exceeding 5000 ng m^{-3} (Pearce *et al.*, 1978). These values have now fallen to an average of 0.01–0.1 ng m^{-3}, and, consequently, the rate of deposition of DDT into natural waters is also low (Eisenreich *et al.*, 1981).

Comparable decreases have been widely reported for most other chlorinated insecticides. In Great Britain, total HCH levels in rain water averaged 0.084 ng m^{-3} during 1966–1967 and only 0.010 ng m^{-3} in 1975 (Wells and Johnstone, 1978). Although similarly low levels of γ-HCH are often reported for other areas of Europe, total annual deposition of airborne γ-HCH and α-HCH to the Great Lakes was relatively high, reaching 57.3 metric tons, compared with 47.4 for endosulfan and 1.7 for both DDT and dieldrin (Eisenreich *et al.*, 1981).

Water

Detectable levels of dissolved DDT, aldrin, and dieldrin are seldom found in lakes and rivers of Canada, the USA, and much of Europe, reflecting the low solubility and limited use of these compounds. However, occasionally high levels are still reported for the particulate fraction of samples, particularly if they are collected during periods of high flow and turbidity. For example, total DDT levels in the River Adige (Italy) reached 18 ng L^{-1} during the spring of 1978 (Galassi and Provini, 1981). A similar situation was reported for the Des Moines River (USA), during a period (1971–1973) when dieldrin use was decreasing (Figure 6.4). Detectable concentrations of dissolved lindane, α-HCH, and β-HCH are still reported for many waters in North America and Europe. Of 1400 samples collected from Western Canada, 96% contained α-HCH, averaging 5 ng L^{-1}, whereas the corresponding values for lindane were 58% and 1 ng L^{-1}, respectively (Gummer, 1979). Similarly, the average concentration of total-HCH in the Danube River (Czechoslovakia) was ~30 ng L^{-1} in 1972, falling to 20 ng L^{-1} by 1974 (Sackmauerová *et al.*, 1977). Comparable trends in residue levels are still reported for heptachlor, chlordane, toxaphene, and endosulfan in areas where these biocides are still used (Rihan *et al.*, 1978; Gummer, 1979).

Many of the chlorinated pesticides, which are restricted in the western world, still find general application in developing nations. Since the controls and safeguards, comparable to those in Europe and North America, are often not in place, high residues may be found in water. Osman *et al.* (1980), for example, reported an average endrin concentration of 1400 ng L^{-1} for River Nile (Egypt) water. Similarly, although now banned for general use, total-HCH in potable water exceeded 500 ng L^{-1} in at least two areas of Japan (Shinohara *et al.*, 1981).

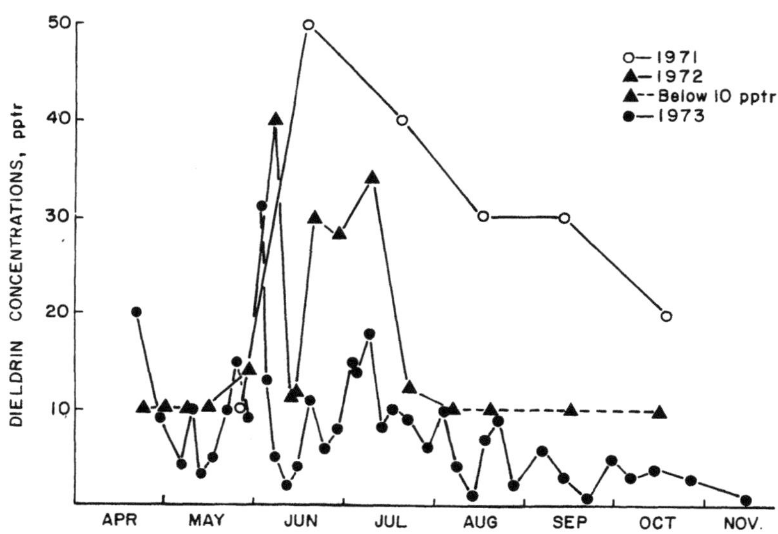

Figure 6.4. Total dieldrin concentrations in unfiltered Des Moines River water, Iowa, USA, between 1971 and 1973. (From Kellogg and Bulkley, 1976.)

Sediments

Sediments are the primary sink for organo-chlorine pesticides in both marine and fresh waters. Because the half-life of many of these agents is long, sediments will continue to be a significant source of contamination for many years to come. For example, total DDT residues in Lake Michigan averaged 11.9 $\mu g\ kg^{-1}$ in 1975, reaching a maximum of 125 $\mu g\ kg^{-1}$ (Frank *et al.*, 1981). These values were recorded 5 years after DDT use was restricted. Much higher concentrations averaging 94,000 $\mu g\ kg^{-1}$ were recorded in 1976 for sediments from the Southern California Bight near Los Angeles. This is one of the most seriously contaminated areas in the world, resulting in reproductive failure of local birds, mammals, and fish (Young and Heesen, 1978). Dieldrin is also commonly found in sediments, with some of the highest concentrations (20–430 $\mu g\ kg^{-1}$) on record occurring in two English Rivers in 1976 and 1977 (Brown *et al.*, 1979). Such levels were due to the use and discharge of dieldrin as a moth-proofer in the textile industry. Other organo-chlorine pesticides, although widely distributed, generally occur in low concentrations in sediments. This reflects the rate of usage and overall half-life of such agents.

Aquatic Plants and Invertebrates

Sorption of all chlorinated pesticides by aquatic plants is rapid and efficient. Under laboratory conditions, attainment of a steady state in tissue residues may occur within 30 min and result in sorption of >75% of the pesticide in culture water. The slope of the accumulation curve generally increases with the concentration of pesticide in water and with the rate of metabolism of the plant. Thus, sorption in nature should be greatest in the spring and early summer. CFs for plants generally follow the order DDE>DDT>dieldrin≥toxaphene≥aldrin≥endrin≥heptachlor>lindane. The half-life of these agents in tissues is highly variable and largely independent of species but generally follows the same order to that noted above. Canton *et al.* (1975) showed that depuration of α-HCH from *Chlorella pyrenoidosa* was essentially complete in 15 min, whereas more than 3 months was required for a comparable decline in DDT residues in several species of aquatic plants in an experimentally contaminated lake (Meeks, 1968). This is one example of why lindane continues to be used in most countries and DDT has been banned. A comparable situation can be found for other pesticides with a long half-life. For example, the application of toxaphene to a small mountain lake resulted in a decrease in residues in aquatic plants from only 0.39 to 0.21 $mg\ kg^{-1}$ over a year (Terriere *et al.*, 1966).

Residues in both freshwater and marine invertebrates have shown a marked decrease in recent years. In Mississippi (USA), ΣDDT levels in molluscs decreased by >50% within one year of the ban (Leard *et al.*,

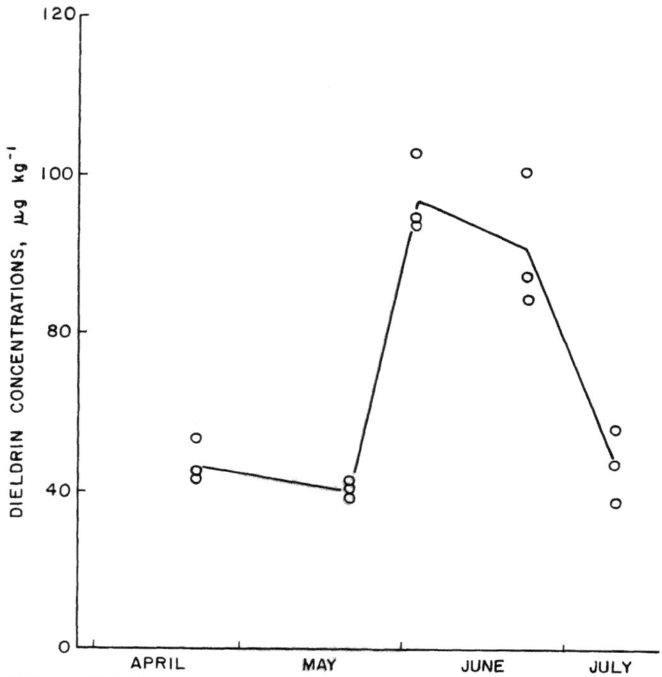

Figure 6.5. Mean dieldrin concentrations (wet weight) in the mayfly *Potamanthus* collected from the Des Moines River, Iowa, USA, during 1972. (From Kellogg and Bulkley, 1976.)

1980). Similarly, aldrin in zooplankton from Lake Paijanne (Finland) averaged 0.013 mg kg^{-1} dry weight in 1972 and was not detected in 1974 (Sarkka *et al.*, 1978). However, ΣDDT levels increased from 0.06 to 0.10 mg kg^{-1} dry weight during the same period, indicating that the lake sediments were continuing to release DDT to the food chain. During 1965–1972, several pesticides, including endrin, dieldrin, toxaphene, and DDT, were essentially ubiquitous in tissues of estuarine molluscs collected from the Atlantic and Pacific coasts of the USA (Butler *et al.*, 1978). By 1977, however, detectable pesticide residues were found in only 2 of 87 estuaries. Several hot-spots of DDT contamination still exist in various locations around the world (Young and Heesen, 1978). Because chlorinated pesticides are lipophilic, total residues in invertebrates vary with the amount of lipid present in tissues. This implies that factors which influence fat deposition in invertebrates, such as reproductive and nutritional condition, should have a significant albeit indirect effect on residues. Seasonal changes in pesticide levels have been noted for a wide range of fresh-water and marine invertebrates. In the Des Moines River (USA), dieldrin residues in the mayfly nymph *Potamanthus* increased from 40 $\mu g\ kg^{-1}$ wet weight in April to 90 $\mu g\ kg^{-1}$ in June (Figure 6.5). Similarly, dieldrin in the shrimp *Crangon vulgaris* collected from the Medway Estuary (UK) ranged from 4.9 to 7.4 μg kg^{-1} between November and March and 1.2–2.0 μg kg^{-1} between July and October (van den Broek, 1979). Such changes imply that biological monitoring programs involving invertebrates should consider seasonal variability in residues and lipid levels.

Fish

CFs for fish/water are highly variable and depend on test conditions, species of fish, and pesticide. In general, DDT and its metabolites have the greatest CFs, followed by toxaphene, aldrin, dieldrin, chlordane, heptachlor, heptachlor epoxide, endrin, and lindane (Table 6.4). Thus, under natural conditions, lindane seldom occurs at high concentrations in fish tissues, whereas significant adulteration of fish with DDT has been reported for numerous lakes, rivers, and estuaries. As with invertebrates, there has been a drop in DDT and dieldrin in tissues in recent years. Suns *et al.* (1981) reported a 64% decline in ΣDDT in spottail shiners collected in 1978 and

Table 6.4. Concentration factors reported for chlorinated pesticides in water and fish.

DDT	*p,p'-DDE*	*o,p'-DDT*	*p,p'-DDT*	*Dieldrin*
61,600[1]	51,000[2]	37,000[2]	29,400[2]	5800[1]
84,500[1]	181,000[6]			68,300[7]
260,000–4,400,000[4]				12,000[8]
1,100,000[5]				

Aldrin	*Endrin*	*Chlordane*	*Lindane*	*Endosulfan*
10,800[1]	4100[1]	11,400[1]	325[1]	330[10]
3100[1]	1400[1]	37,800[2]	180[2]	
	3700[3]	8300[1]	550[1]	
	6200[3]	16,000[9]		

Heptachlor	*Heptachlor epoxide*	*Toxaphene*
17,400[1]	14,400[1]	10,000[11]
9500[2]		52,000[12]
2200[1]		22,000[13]
4900[10]		

Sources: Kenaga (1980); Veith *et al.* (1979); Anderson and DeFoe (1980); U.S. Environmental Protection Agency (1980).

[1]Fathead minnow, 28-day exposure; [2]fathead minnow, 32-day exposure; [3]black bullhead, 32-day exposure; [4]lake whitefish, natural populations (Great Lakes); [5]yellow perch, natural populations (Great Lakes); [6]rainbow trout, 108-day exposure; [7]lake trout, 152-day exposure; [8]guppy, 32-day exposure; [9]sheepshead minnow, 189-day exposure; [10]sheepshead minnow, 28-day exposure; [11]brook trout, 140-day exposure; [12]fathead minnow, 98-day exposure; [13]channel catfish, 100-day exposure.

1979 from Lakes Erie and Ontario. By contrast, there has been a much less dramatic decline in residues in some areas despite the ban on many pesticides. For example, dieldrin levels in channel catfish from the Des Moines River (USA) averaged 61, 75, and 46 $\mu g\ kg^{-1}$ wet weight during 1971, 1973, and 1978, respectively (Leung *et al.*, 1981). The maintenance of these high levels was due to run-off from agricultural land that had been formerly treated with dieldrin.

The relatively high CFs and widespread use of chlordane, heptachlor, and heptachlor epoxide have led to elevated residues in many species. Veith *et al.* (1981), working on 22 watersheds near the Great Lakes (USA), reported average chlordane levels of up to 2680 $\mu g\ kg^{-1}$ and 240 $\mu g\ kg^{-1}$ in carp and channel catfish, respectively. Although this agent was not recorded in fish from the Des Moines River, heptachlor epoxide averaged 14 $\mu g\ kg^{-1}$in whole body and 14,000 $\mu g\ kg^{-1}$ in fat samples collected from omnivorous species. On the other hand, mean chlordane levels of 600 $\mu g\ kg^{-1}$ were found in the fat of herring collected from the Baltic Sea, despite the negligible use of the compound in Sweden and Finland (Jansson *et al.*, 1979, Moilanen *et al.*, 1982). This discrepancy was probably due to large-scale atmospheric transport from other parts of Europe. Given the preceding data, it is apparent that significant environmental contamination can result from the heavy use of chlordane, heptachlor, and related compounds.

Organochlorine pesticides accumulate in lipid deposits and other fatty tissues. Accordingly, muscle residues are often low compared with those in some organs and may not adequately reflect the extent of exposure to fish. Thus, monitoring programs involving fish should routinely analyze fat, muscle, and organs to obtain a more comprehensive view of the extent of

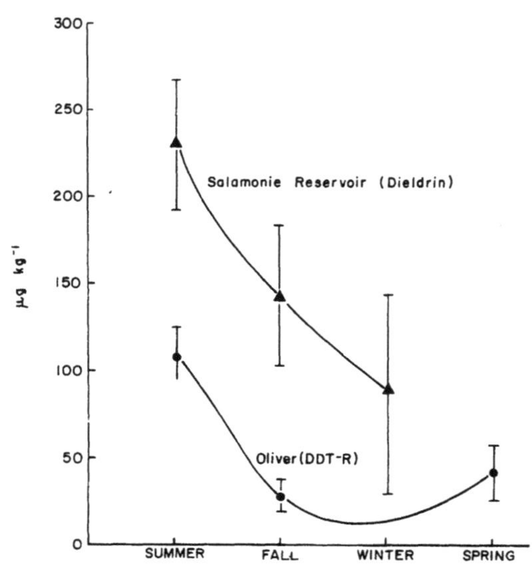

Figure 6.6. Seasonal changes in mean and range in pesticide levels (wet weight) in largemouth bass collected from Oliver Lake and the Salamonie Reservoir, USA, during 1972 and 1973. (From Vanderford and Hamelink, 1977.)

ambient pollution in receiving waters. Although chlorinated pesticides are lipophilic, it is often difficult to correlate whole body residues with lipid levels under natural conditions. This is due to the influence of other factors, such as age, size, metabolic rate, reproduction, and feeding conditions on residues.

Uptake occurs primarily through food in most species, but laboratory studies have also demonstrated that water is a significant route of contamination, at least under laboratory conditions. Rate of uptake depends on levels in the environment, duration of exposure, and temperature. This latter factor acts by altering the metabolic and hence uptake rate of fish. Under natural conditions, there is often a seasonal, temperature-induced cycle in pesticide levels (Figure 6.6). Residues generally increase with the age and size of fish for those pesticides with a long half-life. By contrast, lindane in tissues often decreases with the age/size of fish and does not show strong concentrations at higher trophic levels.

Toxicity

Aquatic Plants and Invertebrates

Chlorinated pesticides differ substantially in their toxicity to aquatic plants. Endrin is among the most toxic, inhibiting growth and carbon uptake at concentrations as low as 0.1–1.0 $\mu g\ L^{-1}$ (Menzel *et al.*, 1970). Equally low levels of dieldrin and aldrin may reduce growth, whereas DDT and DDE generally have little inhibitory effect below 1.0 $\mu g\ L^{-1}$ (Luard, 1973; Powers *et al.*, 1975). Development of tolerance to DDT and its derivatives is frequent, and there are several species that are apparently unaffected by DDT levels > 1000 $\mu g\ L^{-1}$ (Luard, 1973). Chlordane generally elicits toxic responses at concentrations < 50 $\mu g\ L^{-1}$ and is therefore more toxic than either heptachlor or toxaphene. However, this order is subject to considerable variability, due to the development of resistance, effects of other species-related variables, and differences in environmental conditions. Lindane and its isomers are generally the least toxic of the pesticides and probably pose little threat to aquatic plants in natural waters; the majority of species tolerate concentrations > 1000 $\mu g\ L^{-1}$ (Luard, 1973).

Aldrin, dieldrin, DDT, and endrin are highly toxic to marine and freshwater invertebrates, with LC_{50}s generally falling below 0.005 mg L^{-1} (Table 6.5). Endosulfan (α) may also be acutely toxic, depending on species, whereas chlordane, heptachlor, heptachlor epoxide, and toxaphene are intermediate in their effects on invertebrates. LC_{50}s for lindane are generally higher than any other agent, ranging from 0.005 to >1 mg L^{-1} (Table 6.5). Susceptibility to intoxication is species dependent. Although some organisms, such as ciliated protozoans, planarians, and oligochaetes may be

Table 6.5. Toxicity (mg L^{-1}) of chlorinated pesticides to invertebrates and fish

Species	Aldrin	Chlordane	Dieldrin	2,4'-DDD
Invertebrates				
Calpidium campylum[1,*] (protozoan)	>10	—	>10	>10
Daphnia magna[1] (cladoceran)	0.03	0.04–0.10	2.5	0.46
Daphnia pulex[2,†] (cladoceran)	0.028	0.029	0.250	0.003
Penaeus duorarum[3] (shrimp)	<0.001	<0.001	<0.001	—
Pteronarcys california[4] (stonefly nymph)	0.001	0.015	<0.001	0.380
Fish				
Fathead minnows[3,5]	0.03	0.05	0.02	—
Bluegill sunfish	0.01	0.02	0.01	—
Goldfish[5]	0.03	0.08	0.04	—
Guppies[5]	0.03	0.19	0.02	—
Mummichog[6]	0.01	—	0.02	—
Coho salmon[7]	0.05	0.06	0.01	—

Sources: [1]Dive *et al.* (1980); [2]Sanders and Cope (1966); [3]U.S. Environmental Protection Agency (1980); [4]Sanders and Cope (1968); [5]Henderson *et al.* (1959); [6]Livingston (1977); [7]Brown (1978).
*43 h.
†48 h.
—, no data.

essentially resistant ($LC_{50} > 10$ mg L^{-1}) to DDT, dieldrin, and other pesticides, LC_{50}s for other species fall below 0.0005 mg L^{-1} (Table 6.5). The extent and significance of such variability in nature is difficult to assess. Certainly, changes in the species composition of invertebrate communities following the accidental or intentional application of pesticides can often be related to this factor. It is also probable that differences in experimental conditions among studies account for some of the variability in toxicity. Apart from the more traditionally studied effects of temperature, pH, and chelators on toxicity, Bowman *et al.* (1981) reported that the type and quantity of solvent used to administer dieldrin to water produced a 10^2–10^3 difference in toxicity to the cladoceran *Daphnia pulex*. An additional factor that can influence toxicity is the development of resistance, particularly under natural conditions. For example, Naqvi and Ferguson (1970) collected the shrimp *Palaemonetes kadiakensis* from four lakes that had a different history of exposure to insecticides. The LC_{50}s of chlordane, endrin, (data are 96-h LC_{50}'s unless otherwise indicated)

2,4′-DDT	Endosulfan (α)	Endrin	Heptachlor	Heptachlor epoxide	Lindane	Toxaphene
>10	10	>10	10	3–6	10	—
0.004	0.24	0.9	0.05	0.12	1.25	0.01
0.004	—	0.020	0.042	—	0.460	0.015
<0.001	<0.001	0.04–0.6	<0.001	<0.001	—	0.001
0.007	0.002	<0.001	0.001	—	0.005	0.002
0.03	0.3–3.5	0.001	0.09	—	0.06	0.008
0.02	—	<0.001	0.02	—	0.08	0.004
0.03	—	0.002	0.23	—	0.15	0.006
0.04	—	0.002	0.11	—	0.14	0.02
0.01	—	<0.001	0.05	—	0.02	—
0.02	—	<0.001	0.06	—	0.05	0.009

and heptachlor showed a 7- to 25-fold variation among populations: 0.13–0.33, 0.001–0.01, and 0.04–0.27 mg L^{-1}, respectively.

Fish

The α- and β-isomers of endosulfan are particularly toxic with 24-h LC_{50}s ranging from 0.09 to 10 μg L^{-1} (Fox and Matthiessen, 1982). Although endosulfan is now restricted in Canada, the USA, and parts of Europe, it is still widely used as a replacement for DDT and dieldrin to control tsetse fly in parts of Africa. Extremely low rates of application (9.5 g ha^{-1}), resulting in lake-water concentrations of 0.2–4.2 μg L^{-1}, have produced significant fish kills (Fox and Matthiessen, 1982). Endosulfan sulfate and the primary metabolites of endosulfan are less toxic than the parent compound.

Endosulfan alcohol and endosulfan ether appear to be major detoxification products of endosulfan in fish.

Aldrin, dieldrin, endrin, and DDT are also highly toxic (Table 6.5) but no longer pose a widespread threat to adult fish in most Western nations. However, DDT/DDE residues, which remain in lake and river sediments, are apparently absorbed by the eggs and larvae of some species. Because these stages may be especially susceptible to pesticide intoxication, fish reproduction continues to be possibly hampered by persistent DDT/DDE residues in some natural waters. Toxaphene was formerly employed in the management of lakes as a coarse fish poison. It was cheap and effective but has been phased out in most Western nations because of its persistence. Lindane and its isomers are once again the least toxic of the pesticides (Table 6.5).

Symptoms of sublethal intoxication by pesticides include (i) inhibition of N-K-ATPase activity in fish gills and concomitant changes in blood plasma electrolyte levels and ability to osmoregulate, (ii) increase in leucocyte and erythrocyte counts and a decrease in plasma protein and hemoglobin levels, (iii) histological damage to gills, including separation of epithelium from the basement membrane, fusion of gill lamellae, and erosion of distal ends of gill filaments, and (iv) decline in the number of lymphoid cells in the spleen and decrease in the size of hepatocytes (Poels *et al.*, 1980). There may also be an increase in the blood pyruvate and lactic acid levels and inhibition of enzymes in liver and kidneys (Sastry and Sharma, 1979). These changes often have obvious external effects on fish, such as retardation of fin regeneration, development of goiters, and change in the discriminating ability and temperature selection. Such effects would clearly influence fish survival in nature and yet go largely undetected in routine monitoring programs.

Birds and Mammals

Thickness of the eggshells of several species of semiaquatic birds decreased between 1945 and the early 1970s. This trend occurred in many countries where DDT was used, resulting in a decline in populations. Successful nesting attempts by common egrets (*Casmerodius albus*) inhabiting a lagoon in California decreased by 46% between 1967 and 1970 (Faber *et al.*, 1972). Similarly, the number of young peregrine falcons (*Falco peregrinus*) produced annually in the Colville River area (Alaska) declined from 34 in 1967 to 9 in 1973 (Peakall *et al.*, 1975). The curtailment in use of DDT produced a moderately rapid increase in eggshell thickness to pre-1950 levels. Although this resulted in an increase in the size of many populations, some species such as peregrines are still rare throughout much of their former range. Thinning of eggshells can be related to the toxic effects of DDT and DDE on the shell gland. Exposure results in edema of villae, pyknosis of glandular epithelium, and cytoplasmic vacuolation of lining epithelium of the shell gland (Kolaja and Hinton, 1976). There may also be a decrease in Ca-ATPase activity resulting in decreased calcium transport across epithelia. Dieldrin and other pesticides are occasionally implicated in eggshell thinning, though they are a much less significant threat than DDT/DDE. Hoffman and Eastin (1982) reported that lindane and toxaphene were embryotoxic and teratogenic to mallard ducks and that the level of field application of toxaphene presented a potential hazard to reproduction.

Pesticides, particularly ΣDDT, are frequently found in high concentrations in aquatic mammals. O'Shea *et al.* (1980), for example, reported ΣDDT residues of up to 2700 mg kg^{-1} in the blubber of bottlenose dolphins from California. Although much lower levels (3.9 mg kg^{-1}) were reported for beluga from northern Canada, harbor seals inhabiting German coastal waters carried residues of up to 27 mg kg^{-1} (Addison and Brodie, 1973). Extremely high residues of ΣDDT have apparently been responsible for poor reproductive performance of marine mammals near California (Young and Heesen, 1978). There is also the possibility that biocides reduce the thickness of blubber, particularly in young animals, thereby decreasing survival. Overall, however, the maintenance of high pesticide levels seems to have substantially less toxic effect on mammals than birds.

Human Health

Aldrin, dieldrin, chlordane, DDT, DDE, heptachlor, lindane, and toxaphene are either known or suspected carcinogens (Sittig, 1980). Based on laboratory assays, the main sites of toxication in rats and mice are the liver, thyroid, and adrenal cortex. Occupational exposure to pesticides has also been implicated in the appearance of skin cancer and dermatitis. Similarly, the incidence of malignant tumors in terminally ill patients from the general population was significantly correlated with the concentration of DDE in the adipose tissue (Unger and Olsen, 1980). Apart from their carcinogenic properties, chlorinated pesticides have been implicated in the induction of arteriosclerotic cardiovascular disease, hypertension, and possibly diabetes. Lindane also causes hypoplastic anemia and bone marrow damage, resulting in decreased erythropoiesis. Several pesticides such as DDT, DDE, and dieldrin are also embryotoxic and teratogenic to several nonhuman species, though there are no well-documented cases of teratogenic effects of pesticides in humans. Pesticides enter the general population primarily through food. Potable water generally contains nondetectable levels of pesticides, reflecting their low solubility and the removal of suspended sediments during the treatment process. Consumption of adulterated fish and seafood products is an important source of contamination, though residues of DDT/DDE, dieldrin, and other chlorinated pesticides are also detected in many other foodstuffs (Sittig, 1980).

References

Addison, R.F., and P.F. Brodie. 1973. Occurrence of DDT residues in beluga whales (*Delphinapterus leucas*) from the Mackenzie Delta, N.W.T. *Journal of the Fisheries Research Board of Canada* **30**:1733–1736.

Ali, S. 1978. Degradation and environmental fate of endosulfan isomers and endosulfan sulfate in mouse, insect and laboratory model ecosystems. Ph.D. Thesis. University of Illinois, Chicago, 101 pp.

Anderson, R.L., and D.L. DeFoe. 1980. Toxicity and bioaccumulation of endrin and methoxychlor in aquatic invertebrates and fish. *Environmental Pollution (Series A)* **22**:111–121.

Bowman, M.C., F. Acree, Jr., C.S. Lofgren, and M. Beroza. 1964. Chlorinated insecticides: fate in aqueous suspensions containing mosquito larvae. *Science* **146**:1480–1481.

Bowman, M.C., W.L. Oller, T. Cairns, A.B. Gosnell, and K.H. Oliver. 1981. Stressed bioassay systems for rapid screening of pesticide residues. Part I. Evaluation of bioassay systems. *Archives of Environmental Contamination and Toxicology* **10**:9–24.

Braun, H.E., and R. Frank. 1980. Organochlorine and organophosphorus insecticides: their use in eleven agricultural watersheds and their loss to stream waters in southern Ontario, Canada, 1975–1977. *The Science of the Total Environment* **15**:169–192.

Brown, A.W.A. 1978. *Ecology of pesticides*. John Wiley, New York, 525 pp.

Brown, L., E.G. Bellinger, and J.P. Day. 1979. Dieldrin pollution in the River Holme catchment, Yorkshire. *Environmental Pollution* **18**:203–211.

Butler, P.A., C.D. Kennedy, and R.L. Schutzmann. 1978. Pesticide residues in estuarine mollusks, 1977 versus 1972—National Pesticide Monitoring Program. *Pesticides Monitoring Journal* **12**:99–101.

Canton, J.H., P.A. Greve, W. Slooff, and G.J. van Esch. 1975. Toxicity, accumulation, and elimination studies of α-hexachlorocyclohexane (α-HCH) with freshwater organisms of different trophic levels. *Water Research* **9**:1163–1169.

Crosby, D.C. 1972. The photodecomposition of pesticides in water. *In*: R.F. Gould (Ed.), *Fate of organic pesticides in the aquatic environment*. Advances in Chemistry Series No. 111, American Chemical Society, Washington, D.C., pp. 173–188.

Dive, D., H. Leclerc, and G. Persoone. 1980. Pesticide toxicity on the ciliate protozoan *Colpidium campylum*: possible consequences of the effect of pesticides in the aquatic environment. *Ecotoxicology and Environmental Safety* **4**:129–133.

Eichelberger, J.W., and J.J. Lichtenberg. 1971. Persistence of pesticides in river water. *Environmental Science and Technology* **5**:541–544.

Eisenreich, S.J., B.B. Looney, and J.D. Thornton. 1981. Airborne organic contaminants in the Greak Lakes ecosystem. *Environmental Science and Technology* **15**:30–38.

Engst, R., R.M. Macholz, and M. Kujawa. 1978. Confirmations of the degradation scheme of Gamma-hexachlorocyclohexane. *Die Nahrung*, **22**, 6, K29–K32.

Ernst, W. 1977. Determination of the bioconcentration potential of marine organisms. A steady state approach. *Chemosphere* **11**:731–740.

Faber, R.A., R.W. Risebrough, and H.M. Pratt. 1972. Organochlorines and mercury in common egrets and great blue herons. *Environmental Pollution* **3**:111–122.

Fox, P.J., and P. Matthiessen. 1982. Acute toxicity to fish of low-dose aerosol applications of endosulfan to control tsetse fly in the Odavango Delta, Botswana. *Environmental Research* **27**:129–142.

Frank, R. 1981. Pesticides and PCB in the Grand and Saugeen river basins. *Journal of Great Lakes Research* **7**:440–454.

Frank, R., R.L. Thomas, H.E. Braun, D.L. Gross, and T.T. Davies. 1981. Organochlorine insecticides and PCB in surficial sediments of Lake Michigan (1975). *Journal of Great Lakes Research* **7**:42–50.

Galassi, S., and A. Provini. 1981. Chlorinated pesticides and PCBs contents of the two main tributaries into the Adriatic Sea. *The Science of the Total Environment* **17**:51–57.

Greve, P.A., and S.L. Wit. 1971. Endosulfan in the Rhine River. *Water Pollution Control Federation Journal* **43**:2338–2348.

Gummer, W.D. 1979. *Pesticide monitoring in the prairies of Western Canada.* Water Quality Interpretive Report No. 4, Environment Canada, Inland Waters Directorate, Western and Northern Region, Water Quality Branch, Regina, Saskatchewan, 1979, 14 pp.

Hamelink, J.L., and R.C. Waybrant. 1976. DDE and lindane in a large-scale model lentic ecosystem. *American Fisheries Society Transactions* **105**:124–134.

Heckman, C.W. 1981. Long-term effects of intensive pesticide applications on the aquatic community in orchard drainage ditches near Hamburg, Germany. *Archives of Environmental Contamination and Toxicology* **10**:393–426.

Henderson, C., Q.H. Pickering, and C.M. Tarzwell. 1959. Relative toxicity of ten chlorinated hydrocarbon insecticides to four species of fish. *American Fisheries Society Transactions* **88**:23–32.

Hoffman, D.J., and W.C. Eastin, Jr. 1982. Effects of lindane, paraquat, toxaphene, and 2,4,5-trichlorophenoxyacetic acid on mallard embryo development. *Archives of Environmental Contamination and Toxicology* **11**:79–86.

Hom, W., R.W. Risebrough, A. Soutar, and D.R. Young. 1974. Deposition of DDE and polychlorinated biphenyls in dated sediments of the Santa Barbara basin. *Science* **184**:1199–1200.

Jansson, B., R. Vaz, G. Blomkvist, S. Jensen, and M. Olsson. 1979. Chlorinated terpenes and chlordane components found in fish, guillemot, and seal from Swedish waters. *Chemosphere* **4**:181–190.

Kellogg, R.L., and R.V. Bulkley. 1976. Seasonal concentrations of dieldrin in water, channel catfish, and catfish-food organisms, Des Moines River, Iowa. 1971–1973. *Pesticides Monitoring Journal* **9**:186–194.

Kenaga, E.E. 1980. Correlation of bioconcentration factors of chemicals in aquatic and terrestrial organisms with their physical and chemical properties. *Environmental Science and Technology* **14**:553–556.

Kenaga, E.E., and C.A.I. Goring. 1980. Relationship between water solubility, soil sorption, octanol-water partitioning, and concentration of chemicals in biota. *In*: J.G. Eaton, P.R. Parrish, and A.C. Hendricks (Eds.), *Proceedings of the 3rd Symposium on Aquatic Toxicology*, American Society for Testing and Materials, Philadelphia, pp. 78–115.

King, P.H., H.H. Yeh, P.S. Warren, and C.W. Randall. 1969. Distribution of pesticides in surface waters. *American Water Works Association Journal* **61**: 483–486.

Kolaja, G.J., and D.E. Hinton. 1976. Histopathologic alterations in shell gland accompanying DDT-induced thinning of eggshell. *Environmental Pollution* **10**:225–231.

Leard, R.L., B.J. Grantham, and G.F. Pessoney. 1980. Use of selected freshwater bivalves for monitoring organochlorine pesticide residues in major Mississippi stream systems, 1972–73. *Pesticides Monitoring Journal* **14**:47–52.

Leshniowsky, W.O., P.R. Dugan, R.M. Pfister, J.I. Frea, and C.I. Randles. 1970. Adsorption of chlorinated hydrocarbon pesticides by microbial floc and lake sediment and its ecological implications. *Proceedings of the 13th Conference on Great Lakes Research* **2**:611–618.

Leung, S.-Y., T.R.V. Bulkley, and J.J. Richard. 1981. Persistence of dieldrin in water and channel catfish from the Des Moines River, Iowa, 1971–73. *Pesticides Monitoring Journal* **15**:98–102.

Livingston, R.J. 1977. Review of current literature concerning the acute and chronic effects of pesticides on aquatic organisms. *CRC Critical Reviews in Environmental Control* **7**:325–351.

Luard, E.J. 1973. Sensitivity of *Dunaliella* and *Scenedesmus* (Chlorophyceae) to chlorinated hydrocarbons. *Phycologia* **12**:29–33.

Martens, R. 1976. Degradation of [8,9^{-14}C] endosulfan by soil microorganisms. *Applied and Environmental Microbiology* **31**:853–858.

Meeks, R.L. 1968. The accumulation of ^{36}Cl ring-labeled DDT in a freshwater marsh. *Journal of Wildlife Management* **32**:376–398.

Menzel, D.W., J. Anderson, and A. Randtke. 1970. Marine phytoplankton vary in their response to chlorinated hydrocarbons. *Science* **167**:1724–1726.

Metcalf, R.L. 1981. Insect control technology. *In*: M. Grayson and D. Eckroth (Eds.), *Kirk-Othmer encyclopedia of chemical technology*, Vol. **13**, Wiley, New York, pp. 413–485.

Moilanen, R., H. Pyysalo, K. Wickstrom, and R. Linko. 1982. Time trends of chlordane, DDT, and PCB concentrations in pike (*Esox lucius*) and Baltic herring (*Clupea harengus*) in the Turku archipelago, Northern Baltic Sea, for the period 1971–1982. *Bulletin of Environmental Contamination and Toxicology* **29**: 334–340.

Naqvi, S.M., and D.E. Ferguson. 1970. Levels of insecticide resistance in freshwater shrimp, *Palaemonetes kadiakensis*. *Transactions of the American Fisheries Society* **99**:696–699.

National Research Council of Canada. 1975. *Methoxychlor: its effects on environmental quality*. NRC Associate Committee on Scientific Criteria for Environmental Quality, Ottawa, Canada, 164 pp.

Oloffs, P.C., and L.J. Albright. 1974. Transport of some organochlorines in British Columbia waters. *Proceedings of the International Conference on Transportation of Persistent Chemicals in the Aquatic Ecosystem* **1**:89–92.

Oloffs, P.C., L.J. Albright, S.Y. Szeto, and J. Lau. 1973. Factors affecting the behavior of five chlorinated hydrocarbons in two natural waters and their sediments. *Journal of the Fisheries Research Board of Canada* **30**:1619–1623.

O'Shea, T.J., R.L. Brownell, Jr., D.R. Clark, Jr., W.A. Walker, M.L. Gay, and T.G. Lamont. 1980. Organochlorine pollutants in small cetaceans from the Pacific and south Atlantic oceans, November 1968–June 1976. *Pesticides Monitoring Journal* **14**:35–46.

Osman, M.A., B. Belal, A.M. Nomrossy, and A.M. Yousse. 1980. Organic contaminants in waters. *Journal of Environmental Science and Health* **B15**: 292–306.

Peakall, D.B., T.J. Cade, C.M. White, and J.R. Haugh. 1975. Organochlorine residues in Alaskan peregrines. *Pesticides Monitoring Journal* **8**:255–260.

Pearce, P.A., L.M. Reynolds, and D.B. Peakall. 1978. DDT residues in rainwater in New Brunswick and estimate of aerial transport of DDT into the Gulf of St. Lawrence, 1967–68. *Pesticides Monitoring Journal* **11**:199–204.

Poels, C.L.M., M.A. van der Gaag, and J.F.J. van de Kerkhoff. 1980. An investigation into the long-term effects of Rhine water on rainbow trout. *Water Research* **14**:1029–1035.

Powers, C.D., R.G. Rowland, R.R. Michaels, N.S. Fisher, and C.F. Wurster. 1975. The toxicity of DDE to a marine dinoflagellate. *Environmental Pollution* **9**: 253–262.

Rihan, T.I., H.T. Mustafa, and G. Caldwell, Jr. 1978. Chlorinated pesticides and heavy metals in streams and lakes of northern Mississippi water. *Bulletin of Environmental Contamination and Toxicology* **20**:568–572.

Sackmauerová, M., O. Pal'usová, and A. Szokolay. 1977. Contribution to the study of drinking water, Danube water and biocenose contamination with chlorinated insecticides. *Water Research* **11**:551–556.

Sanders, H.O., and O.B. Cope. 1966. Toxicities of several pesticides to two species of cladocerans. *Transactions of the American Fisheries Society* **95**:165–169.

Sanders, H.O., and O.B. Cope. 1968. The relative toxicities of several pesticides to naiads of three species of stoneflies. *Limnology and Oceanography* **13**: 112–117.

Sarkka, J., M.-L. Hattula, J. Janatuinen, and J. Paasivirta. 1978. Mercury and chlorinated hydrocarbons in plankton in Lake Paijanne, Finland. *Environmental Pollution* **16**:41–49.

Sastry, K.V., and S.K. Sharma. 1979. Endrin toxicosis on few enzymes in liver and kidney of *Channa punctatus* (Bloch). *Bulletin of Environmental Contamination and Toxicology* **22**:4–8.

Shinohara, R., A. Kido, and S. Eto, T. Hori, M. Koga, and T. Akiyama. 1981. Identification and determination of trace organic substances in tap water by computerized gas chromatography-mass spectrometry and mass fragmentography. *Water Research* **15**:535–542.

Singmaster, J.A. 1975. Environmental behavior of hydrophobic pollutants in aqueous solutions. Ph.D. Thesis. University of California at Davis, California, 143 pp.

Sittig, M. 1980. *Priority toxic pollutants. Health impacts and allowable limits.* Noyes Data Corporation, New Jersey, 370 pp.

Smalley, H.E., and R.D. Radeleff. 1971. Enhancement of insecticide toxicity by antidiuretic agent, diazoxide. *American Journal of Veterinary Research* **32**:345.

Spencer, W.F. 1975. Movement of DDT and its derivatives into the atmosphere. *Residue Reviews* **59**:91–117.

Statistics Canada. 1982. *Miscellaneous chemical industries. Manufacturing and primary industries.* Ministry of Supplies and Services, Ottawa, Canada.

Suns, K., C. Curry, G.A. Rees, and G. Crawford. 1981. *Organochlorine contaminant declines and their geographic distribution in Great Lakes spottail shiners* (Notropis hudsonius). Ontario Ministry of the Environment, Rexdale, Ontario, 18 pp.

Terriere, L.L., U. Kiigemagi, A.R. Gerlach, and R.L. Borovicka. 1966. The persistence of toxaphene in lake water and its uptake by aquatic plants and animals. *Journal of Agricultural and Food Chemistry* **14**:66–69.

Tsukano, Y. 1973. Factors affecting disappearance of BHC isomers from field rice soil. *Japan Agriculture Research Quarterly* **7**:93–97.

Unger, M., and J. Olsen. 1980. Organochlorine compounds in the adipose tissue of deceased people with and without cancer. *Environmental Research* **23**: 257–263.

U.S. Environmental Protection Agency. 1980. *Ambient water quality criteria reports*. Office of Water Regulations and Standards, Washington, D.C.

van den Broek, W.L.F. 1979. Seasonal levels of chlorinated hydrocarbons and heavy metals in fish and brown shrimps from the Medway estuary, Kent. *Environmental Pollution* **19**:21–38.

Vanderford, M.J., and J.L. Hamelink. 1977. Influence of environmental factors on pesticide levels in sportfish. *Pesticides Monitoring Journal* **11**:138–145.

Van Dyk, L.P., I.H. Wiese, and J.E.C. Muller. 1982. Management and determination of pesticide residues in South Africa. *Residue Reviews* **82**:37–124.

Veith, G.D., D.L. DeFoe, and B.V. Bergstedt. 1979. Measuring and estimating the bioconcentration factor of chemicals in fish. *Journal of the Fisheries Research Board of Canada* **36**:1040–1048.

Veith, G.D., D.W. Kuehl, E.N. Leonard, K. Welch, and G. Pratt. 1981. Polychlorinated biphenyls and other organic chemical residues in fish from major United States watersheds near the Great Lakes, 1978. *Pesticides Monitoring Journal* **15**:1–8.

Versar. 1979. *Water-related environmental fate of 129 priority pollutants*. Vol. **I**. U.S. Environmental Protection Agency, Publication No. EPA-440/4-79-029a, Washington, D.C.

Warry, N.D., and C.H. Chan. 1981. Organic contaminants in the suspended sediments of the Niagara River. *Journal of Great Lakes Research* **7**:394–403.

Weil, L., G. Dure, and K.E. Quentin. 1973. Adsorption of chlorinated hydrocarbons to organic particles and soils. *Zeitschrift fuer Wasser und Abwasser Forschung* **6**:107–112.

Wells, D.E., and S.J. Johnstone. 1978. The occurrence of organochlorine residues in rainwater. *Water, Air, and Soil Pollution* **9**:271–280.

Wolfe, N.L., R.G. Zepp, D.F. Paris, G.L. Baughman, and R.C. Hollis. 1977. Methoxychlor and DDT degradation in water: rates and products. *Environmental Science and Technology* **11**:1077–1081.

Yaron, B., A.R. Swoboda, and G.W. Thomas. 1967. Aldrin adsorption by soils and clays. *Journal of Agricultural and Food Chemistry* **15**:671–675.

Young, D.R., and T.C. Heesen. 1978. DDT, PCB, and chlorinated benzenes in the marine ecosystem off Southern California. *In*: R.L. Jolley (Ed.), *Water chlorination, environmental impact and health effects*, Vol. **2**, pp. 267–290.

7
Petroleum Hydrocarbons

Petroleum is a naturally occurring oily, flammable liquid composed principally of hydrocarbons (50–98%) with the remainder consisting of organic compounds, oxygen, nitrogen, sulfur, and traces of metal salts. Petroleum is usually found beneath the earth's surface and occasionally in pools above the surface. Unrefined petroleum is called crude oil. Petroleum is separated by distillation into the following four major fractions:

(i) straight-run gasoline, boiling point (b.p.) <200°C;
(ii) middle distillate, b.p. ~185–345°C, which is further fractionated into kerosene, heating oils, and a variety of fuels for diesel, jet, rocket, and gas turbine engines;
(iii) wide-cut gas oil, b.p. ~345–540°C, which is further separated into waxes, lubricating oils, and feed stock for catalytic cracking to gasoline; and
(iv) residual oil, which may be asphaltic.

The physical properties and chemical composition of petroleum vary widely depending on its origin. It could range from a colorless liquid (mainly gasoline) to a heavy, black, tarry material (high in asphalt). Although most crudes are black, some are amber, red, or brown with greenish fluorescence. Their specific gravity is in the range of 0.82–0.95.

The hydrocarbon types found in petroleum are (i) paraffins, (ii) cycloparaffin (naphthenes or cyclohexanes), and (iii) aromatics. Paraffins range from methane to *n*-hexacontane ($C_{60}H_{122}$, a microcrystalline wax) of both straight- and branched-chain alkanes. Higher boiling fractions are high in saturated alkanes while commercial paraffin mainly consists of straight-chain alkanes of C_{22}–C_{30} atoms. Cycloparaffins are mainly monocyclic and polycyclic compounds with 5 or 6 carbon atoms in a ring. Examples are cyclopentane, cyclohexane, alkylcyclopentane, alkyl cyclohexanes, *trans*-decahydronaphthalene, *cis*-bicyclo [3,3,0] octane, and other higher cyclic hydrocarbons. Smaller amounts of aromatic compounds consist chiefly of alkylbenzenes such as toluene, the xylenes, and *p*-cymene. Higher boiling fractions contain predominantly fused-ring polynuclear aromatics (alkyl naphthalenes) and some linked-ring aromatics (biphenyls).

Heavy oil and bitumen generally contain a large proportion of high molecular weight (max. 10,000) asphaltenes. These compounds consist of a core of five stacked sheets, with each sheet composed of an average of 16 condensed aromatic rings. Contained within or attached to this core are aliphatic hydrocarbons, oxygen, sulfur, nitrogen, and trace metals such as nickel and vanadium. In part because of their extremely low solubility, asphaltenes are essentially nontoxic to aquatic life.

Sulfur is found in all crude oil in concentrations ranging from trace to 8%, with most averaging 0.5–1.5%. It occurs as elemental sulfur, hydrogen sulfide, and carbonyl sulfide (COS), as well as in association with various aliphatic and aromatic compounds. Nitrogen is generally less common than sulfur, averaging <0.1–1.6%. At least 35 different compounds have been identified and are concentrated in the higher boiling fractions. Although low in oxygen, most crude oils contain numerous heavy metals within a concentration range of 0.001–1000 mg L^{-1}. Oil deposits also often occur along with salt water, and, consequently, metal salts are present in emulsified water.

Production, Uses, and Discharges

Production

Petroleum displaced coal as the main source of energy in the world in 1965. Since that date, world petroleum consumption has doubled to the point where it now serves ~50% of the world's energy demand. This rapid growth has been due to the ease with which petroleum can be discovered, produced, transported, and refined. However, it has been estimated that production of conventional petroleum will peak in a few decades, possibly by the year 2000. Although production of nonconventional sources of petroleum, such as tar sands, oil shales, and synthetic oil will increase, these sources will not make up the short-fall in supply.

Beginning in 1980, world demand for oil began to decline as a result of recession and conservation (Figure 7.1). Production by the 13 members of OPEC fell from a high of 31 million barrels a day in 1977 to just below 20 millions barrels in 1982. Kuwait alone cut production by 70% to 650,000

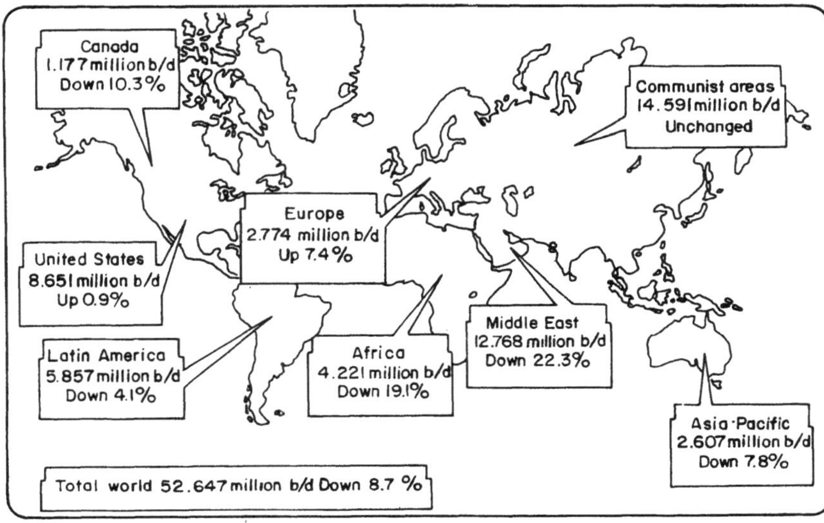

Figure 7.1. World oil production by region, first half 1982. (From *Oil and Gas Journal*, March 8, 1982: 99–103.)

bbl (barrels) daily while Nigerian production fell by 30% to 1,600,000 barrels per day. In fact, overall world production in 1982 was ~6 million barrels a day below the level of 1979. Such decreases, if prolonged, will obviously extend the life of conventional petroleum reserves. The corresponding decline in the world price for oil may also make the development of some nonconventional sources uneconomical in the short-term.

Proven world reserves of oil increased steadily from 1950 to 1970 but in more recent times have remained relatively constant at $\sim 5.7 \times 10^{11}$ bbl (Figure 7.2). Middle Eastern countries account for some 54% of this total, followed by the USSR at 10.1% and the USA at 5.9%. Based on historical trends in discovery, the ultimate world reserve of oil is probably near 2×10^{12} bbl (Drew, 1982). Few if any large new fields will be discovered, although the potential for Arctic and deep-water offshore regions is not fully known. Therefore, expansion of the world's proven reserves will depend largely on the discovery of additional reserves in known fields and use of secondary and tertiary recovery methods.

Alberta is the main producer of Canadian petroleum, accounting for $\sim 1077 \times 10^3$ bbl a day (Table 7.1). This is considerably in excess of the output from Saskatchewan ($\sim 127 \times 10^3$ bbl daily), the second largest producer. In the USA, the largest proven reserves (9.25×10^9 bbl) are in Alaska, followed by Texas (7.69×10^9 bbl), California (3.47×10^9 bbl), and Louisiana (2.89×10^9 bbl) (Drew, 1982). Historically, Texas has been the most important producer and will continue to be so for some time.

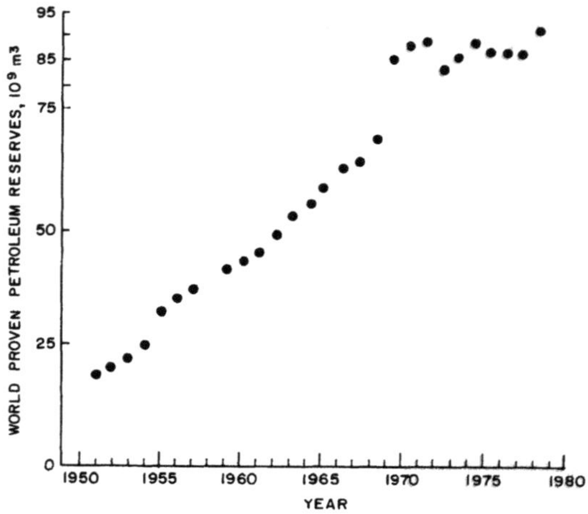

Figure 7.2. Yearly levels of world proven reserves of petroleum. To convert m^3 to bbl, multiply by 6.20. (From Drew, 1982.)

However, advances in technology, together with decreasing world reserves, will probably increase the utilization of Alaskan reserves. European production is restricted primarily to the North Sea. Output in 1980 was 2.2×10^6 bbl daily and, unlike many areas, increased to 2.7×10^6 bbl a day in 1982 (*Oil and Gas Journal*, 1982).

Petroleum is the world's primary source of energy and petrochemical feedstock. Although crude oil has few uses, it is refined to produce gasoline, jet fuel, industrial and domestic fuel oil, kerosene, solvents, and diesel fuel, as well as feedstocks such as ethylene, propylene, butenes, isoprene, and butadiene. These feedstocks form the basis of production of elastomers, plastics, and artificial fibers. In the USA, ~41% of refinery yield is automotive and aviation gasoline (Hoffman, 1982). Fuel oil is the next

Table 7.1. Total Canadian petroleum production (bbl $\times 10^3$ per day).

	1981	1982
Alberta (conventional)	1094	1077
Alberta (synthetic)	118	111
Saskatchewan	136	127
British Columbia	39.9	37.7
Manitoba	9.8	9.4
E. Canada and NWT	4.5	4.5
Total Crude	1402	1367
LPGs	229	228
Total production	1631	1595

Source: Oilweek (October 1982).

largest product at ~32%, followed by jet fuel (6.7%), petrochemical feed (4.6%), still gas (3.7%), and asphalt (3.0%). Many petroleum products are blended or otherwise modified from primary stocks as they come from refineries. For example, naphtha may be sold as a solvent to a dry-cleaner, as a mineral spirit to paint makers or as a petrochemical feedstock.

Refined products of crude petroleum are not pure chemical compounds but mixtures of hydrocarbons. The three major refining processes are crude distillation, cracking (catalytic, thermal, or hydro cracking) of the larger compounds into smaller ones, and chemical synthesis of the desired product properties by reforming, alkylation, and isomerization.

Discharges

The total input of petroleum hydrocarbons into the world's oceans during the early 1980s was ~4.6×10^6 metric tons per year (Table 7.2). Of this, river run-off is most significant, accounting for 1.6×10^6 tons yr^{-1}, followed by the transportation category (0.8×10^6 tons yr^{-1}), which includes loading and unloading of tankers and ballast discharge. Although rivers also constitute the main input of oil into coastal waters of the UK and other European countries (Table 7.2), the discharge of ballast water appears to be the main source of oil slicks observed in the Arabian Sea and approaches to the Persian Gulf (Oostdam, 1980). In fact, the number of oil slicks observed in this area is immense, exceeding 6500 (Figure 7.3). Oostdam (1980) reported that the volume of the largest slick observed in the Arabian Sea exceeded 54,000 m^3. For comparison, the spill of the *Torrey Canyon* in 1967

Table 7.2. Estimated inputs of petroleum hydrocarbons in the world's oceans and coastal water of the UK during the early 1980s.

	World's oceans (10^6 tons yr^{-1})	UK waters (10^4 tons yr^{-1})
Rivers	1.6	5.0
Transportation	0.8	1.5
Natural seepage	0.6	ND
Municipal/industrial wastewater	0.45	1.3
Refineries	0.02	1.1
Offshore production	0.2	0.2
Atmosphere	0.2	ND
Others	0.7	0.3
Total	4.6	9.4

Sources: Hileman (1981); Read and Blackman (1980).
ND, no data.

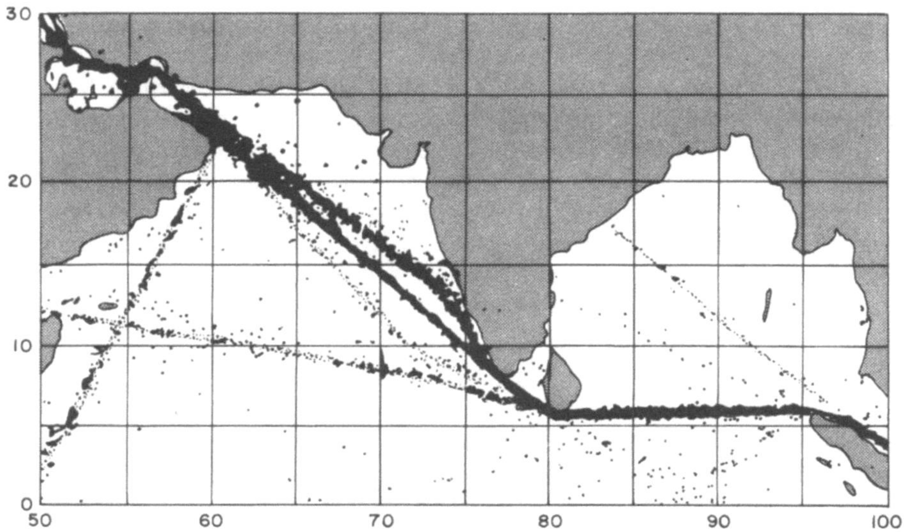

Figure 7.3. Points of observations of oil slicks and other floating pollutants in the Indian Ocean. Reprinted with permission from *Marine Pollution Bulletin*, vol. 12, R.S. Gupta and T.W. Kureishy, Present state of oil pollution in the northern Indian Ocean. © 1981, Pergamon Press, Ltd.

contained 117,000 m^3, whereas the *Amoco Cadiz* spill in 1978 amounted to 213,000 m^3.The largest discharge in world history, the Ixtoc blowout in the Gulf of Mexico, released ~530,000 m^3 to the environment.

Urban run-off, together with municipal/industrial discharges, are the main sources of petroleum hydrocarbons in most rivers. Eganhouse and Kaplan (1981) estimated that the world emission rate from these sources was ~0.4 kg yr^{-1} $person^{-1}$ and that run-off from urban rivers contributes over 2.5 times as much petroleum to the oceans as nonurban rivers. It was also shown that, in the Los Angeles basin, the annual emission rate of hydrocarbons from paved roads was 2.9 metric tons km^{-2} yr^{-1} compared with a basin-wide average of 1.3 metric tons km^{-2} yr^{-1} for all roads. Whipple and Hunter (1979) showed that a major source of hydrocarbons from Philadelphia was crankcase oil from garages and oil from car wash businesses and parking lots.

Substantial quantities of hydrocarbons may be released to the environment from natural deposits of oil, heavy oil, shale oil, and bitumen. Although such discharges generally have no detectable effect on residue levels in the oceans, there may be significant accumulation of hydrocarbons in certain fresh waters. One example is the Athabasca River in northern Canada, where oil

Figure 7.4. Oil sands and heavy oil deposits in Alberta, Canada. (From McRory, 1982.)

and grease levels average 850 $\mu g\ L^{-1}$ and have reached >3000 $\mu g\ L^{-1}$ on occasion (Figures 7.4, 7.5). The natural deposits of bitumen in this area are of a magnitude (812 billion barrels) that permits the operation of major extraction plants (up to 190,000 barrels per day). Since the bitumen is located at and near the surface, natural river flow and periodic flooding cause significant erosion, particularly during the summer. Using the average concentration of oil and grease in the Athabasca and its mean annual discharge rate of 702 $m^3\ sec^{-1}$, the total amount of oil and grease transported downstream is ~1.7×10^7 kg yr^{-1}. For comparison, the tar sands plants are licensed to discharge 165,000 kg annually, or about 1% of the total river burden.

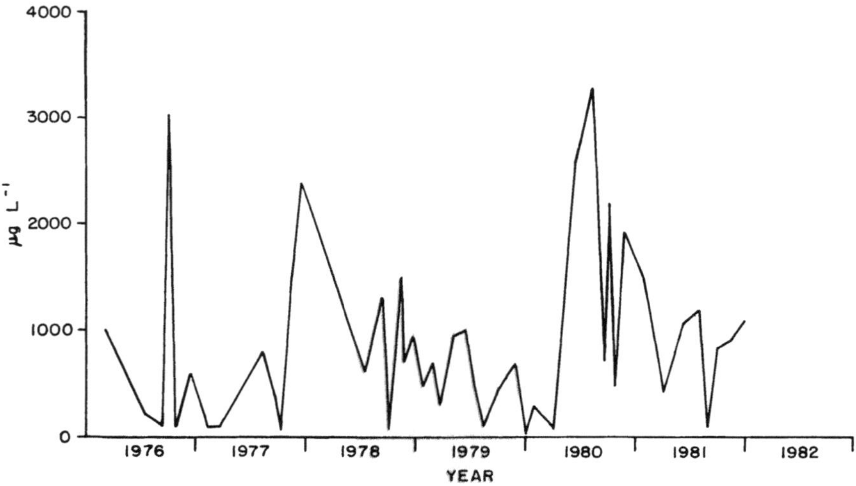

Figure 7.5. Concentration of oil and grease in the Athabasca River, Alberta, Canada. (From Naquadat, 1976–81.)

Behavior in Natural Waters

Sorption

The rate of sorption of petroleum hydrocarbons to sediments is dependent upon temperature (Meyers and Oas, 1978) and organic content of the sediment (Meyers and Quinn, 1973). Laboratory experiments showed that dried sediment of size fraction 50–500 μm rapidly sorbed petroleum hydrocarbons at concentrations of 1–5 mg L^{-1}: 94% loss in 1 hour and 99% loss after 48 hours from the water column (Knap and Williams, 1982). Although approximately 70% of this reduction was due to sorption to sediment, volatilization was responsible for the residual loss. Since the sediment was stirred continuously except prior to sampling, the sorption rates should be considered as maximum rates. The aliphatic fraction sorbed to the sediment faster than the aromatic fraction, reflecting differences in their aqueous solubilities. Desorption of hydrocarbons from sediments is generally slow (Meyers and Quinn, 1973), and this explains the considerable buildup of petroleum hydrocarbons around outfall areas. However, resuspension of sediments through dredging and other rehabilitation processes might release some of the sorbed material back into the aquatic environment.

Sorption to suspended particulate matter and eventual deposition on the shores of rivers and estuaries, along with other natural physico-chemical processes, tend to remove the oil introduced into surface waters (Levy,

1971). Sorption to weeds such as *Sargassum* affects the distribution of pelagic tarballs within the Georgia and Florida gulf streams (Cordes *et al.*, 1980). The tarballs range in size from tiny flecks up to a maximum diameter of ~2 cm and have 30% content of polycyclic aromatic hydrocarbons. The principal compound is perylene with small amounts of fluoranthene and pyrene. The presence of abundant amounts of perylene reflects the high resistance to natural weathering processes of tarballs. Tarballs have a residence time of ~1 year in North Atlantic ocean and represent ~10–30% of the original petroleum crude (Butler *et al.*, 1973; Levy and Walton, 1976).

Reduced filtration of sea water through sandy beaches containing oil may be caused by retardation of interstitial water flow owing to decrease in pore space (McLachlan and Harty, 1981). The weathered oil forms mixtures of sand in oil with no pore space available for water flow. In most cases (excepting heavy spills), oil reduces water filtration only to a small extent and is unlikely to remain long enough on the beach to be weathered. Thus the magnitude of the effects depends on the volume of oil spill, state of weathering, location on the beach, and degree of admixture with sand.

Volatilization

The rate of volatilization of petroleum hydrocarbons depends to a large degree on the rate of aeration of the ambient water. Volatilization is likely to remove *n*-alkanes $<C_{18}$, but there is also some loss of hydrocarbons up to the 4-ring systems. Studies have been conducted to characterize the crude oil weathered under different simulated environmental conditions such as water agitation and sun exposure (Riley *et al.*, 1980). A combination of light and water sprayed upon the surface of the oil produced the largest relative decreases in volatile saturated hydrocarbons (C_{12} to C_{26}) and most aromatic (naphthalene to 2,3,6-trimethyl naphthalene) hydrocarbons. After 24 days, monoaromatic hydrocarbons and saturated hydrocarbons from C_8 to C_{10} were not detectable in all three weathered oils. Similar results have been reported for No. 2 fuel oil in laboratory studies (Zurcher and Thuer, 1978). The decreases in the content of most aromatic compounds were less for oil protected from light, water agitation, and air circulation. Retention of aromatic hydrocarbons appeared to be related to their molecular weight as shown by the enrichment of 3-ring aromatics such as 3,6-dimethyl phenanthrene in weathered oil relative to the original crude oil (Riley *et al.*, 1980). These data indicate that 4- and 5-ring polynuclear aromatic hydrocarbons (carcinogenic and mutagenic) will also be enriched in weathered oil. Since these compounds are sparingly soluble in water, they are not likely to affect the water column species. Thus, weathered oil mixing with sediments may be more hazardous to benthic species than contamination from fresh oil.

Chemical dispersion of oil spills is used to accelerate natural weathering processes by spreading higher concentrations to greater depths and widths. This process greatly enhances the volatilization of petroleum hydrocarbons to the atmosphere and is due to the mixing of dispersed droplets having high specific surface areas in near-surface water. An untreated slick may be less susceptible to volatilization because its lower surface/volume ratio tends to suppress transport by diffusion of volatile hydrocarbons. McAuliffe *et al.* (1980) found that slicks treated with a delayed dispersion (2 hours) were only slightly higher (≤ 1.1 mg L^{-1}) in oil concentrations than the untreated oil sampled immediately after discharge, whereas slicks sprayed immediately with a dispersant showed highest oil concentrations of 3 mg L^{-1} (La Rosa crude oil) and 18 mg L^{-1} (Murban crude oil), respectively. The less effective dispersion in the delayed treatment was due to increased viscosities resulting from rapid weathering of volatile hydrocarbons. The dispersed oil weathered rapidly, with volatilization of C_1–C_{10} hydrocarbons greatly exceeding their dissolution in water.

Biotransformation

Microbial Metabolism. Although crude petroleum contains a variety of hydrocarbons varying in volatility and molecular weight, long-term degradative processes such as microbial degradation involve only the nonvolatile residue. Virtually all crude oils and hydrocarbons are vulnerable to microbial degradation under favorable conditions. Bhosle and Mavinkurve (1980–1981) isolated 23 hydrocarbon-utilizing bacteria (*Vibrio* and *Pseudomonas*) and one yeast from water and sediment samples using enrichment techniques. *Bacillus, Candida*, and *Arthrobacter* species exhibited the widest range of hydrocarbon-utilizing profiles. In marine environments, microorganisms are the major biotic factors in cycling petroleum hydrocarbons.

Degradation studies of crude oil in Arctic tundra ponds showed that (i) oil alone did not increase the number of heterotrophic or oil-degrading indigenous microflora over a short period of 28 days, (ii) weekly addition of oleophilic phosphate at 0.1 mM significantly stimulated the microflora in the presence or absence of oil, (iii) addition of inorganic phosphate failed to induce this effect, and (iv) microflora mineralized the hydrocarbon part of the phosphate before the polyaromatic fraction (Bergstein and Vestal, 1978). Similar results with oleophilic nitrogen and phosphorus-containing fertilizers were reported earlier (Atlas and Busdosh, 1976).

Production of surface-active compounds by many microorganisms is believed to be concomitant with their growth on oil and hydrocarbons (Cooper and Zajic, 1980). Although some chemical dispersants used in treating oil spills enhance bacterial metabolism of crude oils (Traxler and Bhattacharya, 1978; Marty *et al.*, 1979), others interfere with microbial activity and cause population shifts (Griffiths *et al.*, 1981). Bacteria isolated from Atlantic Ocean sediments utilized a mixed hydrocarbon substrate (Walker *et al.*, 1976a,b). The artificial mixture contained *n*-alkanes, cyclohexanes, and aromatic hydrocarbons. All hydrocarbons, including naphthalene, phenanthrene, benz[a]anthracene, perylene, and pyrene showed some degradation. The bacteria were mainly various species of *Pseudomonas* and *Actinobacter*.

Bitumen found in the Athabasca oil sands area (Canada) resembles the residue left after the microbial degradation of conventional crude oils (Rubinstein *et al.*, 1977). The bulk of the material is (i) complex cyclic and branched aliphatic, (ii) aromatic hydrocarbons and heteroatomic, and (iii) high molecular weight compounds (Selucky *et al.*, 1977). These factors suggest that Athabasca bitumen should be fairly resistant to microbial degradation. Studies on the hydrocarbon-degrading potentials of indigenous micro-organisms in that area showed that (i) the plate count was sensitive to demonstrate the increase in adapted mixed microbial populations, (ii) biodegradation potentials for ^{14}C-hexadecane and ^{14}C-naphthalene substrates increase within the oil sands formation indicating a significant increase in microbial activity, (iii) the biodegradation potentials showed a marked seasonal variation, and (iv) the rivers of the Athabasca oil sands areas can, under aerobic conditions and suitable nutritional regime, support an active hydrocarbon-oxidizing microbial community (Wyndham and Costerton, 1981a). *In situ* colonization of bitumen surfaces and *in vitro* degradation studies showed that bituminous hydrocarbons were readily colonized *in situ* in both summer and winter by microbial populations native to Athabasca oil sands. In addition, isolates from the sediments of the Athabasca River and tributaries can utilize all fractions of Athabasca bitumen except the asphaltene fraction, and saturated, aromatic, and first polar fractions of the bitumen were preferentially degraded (Wyndham and Costerton, 1981b).

Animal Metabolism

In general, petroleum hydrocarbons are transformed by vertebrates in two steps (Figure 7.6). The oxidation metabolism of aromatic and aliphatic compounds is carried out by a cytochrome-P_{450}-dependent mixed function oxidase (MFO) system located in the endoplasmic reticulum of the cell. The MFO system primarily renders the insoluble or sparingly water-soluble compounds into water-soluble and excretable metabolites. Although this system effectively detoxifies certain xenobiotics, compounds such as PAH and alkenes are metabolized to intermediates that are highly toxic, mutagenic, or carcinogenic to the host. These electrophilic epoxides chemically interact with DNA and RNA or with genetic substrates. Microsomal epoxide hydrases and glutathione-S-transferases detoxify the active epoxides into diols and catalyze the formation of glutathione (GSH)

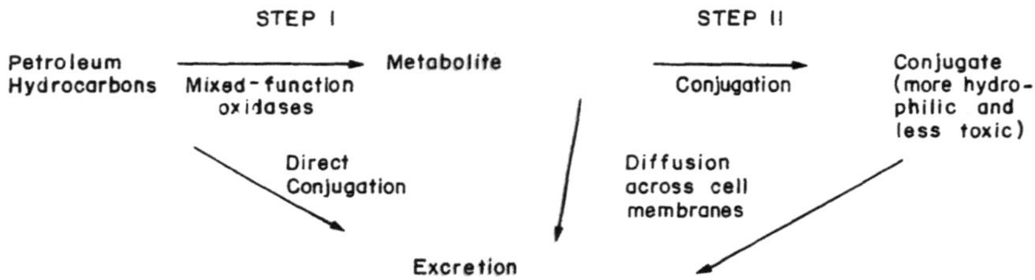

Figure 7.6. Transformations of petroleum hydrocarbons in vertebrates. (From Connell and Miller, 1981.)

conjugates, respectively. Nonenzymatic pathways include reactions with GSH, solvolysis to yield diols, and isomerization of arene oxides to phenols (Corner, 1975; Bend *et al.*, 1977). Conjugating enzyme systems mediate the reaction of the oxidation products (epoxides and phenols) with hydrophilic compounds such as glucoronic acid to yield highly polar compounds. These compounds include glucoronides, glycosides and mercapturates that are readily excreted in bile and urine (Roubal *et al.*, 1977; Varanasi *et al.*, 1979). The MFO system is located in the liver and sometimes in other organs of many vertebrates and invertebrates. Methods have been developed to identify the MFO system in nonmammalian species including many aquatic organisms (Neff, 1979). Many aquatic species were found to metabolize PAHs into polar metabolites. Production of dieldrin was also a good indicator of MFO activity in fresh-water invertebrates (Khan *et al.*, 1974). Exogenous factors such as temperature, season, and salinity and endogenous factors such as age, sex, nutritional state, and physiological state can influence the levels of MFO system and the associated enzymatic activities (Singer and Lee, 1977).

The MFO system is also known as the aryl hydrocarbon hydroxylase (AHH) or drug-metabolizing system in mammals. In fish, as in mammals, most MFO activity is localized in the liver (2.404 μmoles of B[a]P hydroxylase = AHH) and in minor amounts in kidney (0.026) and heart (0.006) (Pederson *et al.*, 1974). Many studies have shown the presence of various oxygenases in fish (Bend *et al.*, 1977; Stegeman, 1978). AHH is present in many marine fish species from different habitats and life stories (Payne, 1977). Several fish species including rainbow trout can hydroxylate benzo[a]pyrene and naphthalene. Quantitative data on AHH activity based on B[a]P hydroxylase activity, benzphetamine demethylase activity, 7-ethoxycoumarin deethylase activity, and cytochrome P_{-450} content in vertebrates, crustaceans, and bivalves are available in the literature (Vandermeulen and Penrose, 1978; Philpot *et al.*, 1976). Specific enzyme activities derived from single substrate measurements are limited in their application to complex mixtures of petroleum hydrocarbons (Malins, 1977a, b). MFO absence or activity could determine hydrocarbon retention in organisms and food chain accumulation (Connell and Miller, 1981). Epoxide hydrase activity has been detected in teleosts, elasmobranchs and invertebrates (Bend *et al.*, 1977; James *et al.*, 1974). Generally vertebrates demonstrated significant *in vitro* enzyme activity with all the substrates tested with the exception of the little skate. Invertebrates also demonstrated significant activity.

Excretion of the electrophilic intermediates as conjugated derivatives has been well identified in terrestial mammals as well as in marine species. All marine species investigated are able to conjugate the oxide intermediates to GSH-derivatives (Bend *et al.*, 1977). Generally, invertebrates show lower activity than vertebrates. In addition, wide variation in glutathione-S-transferase activity has been observed beween species and substrates used. Reasonable experimental evidence now exists for the enzyme-mediated biotransformation of petroleum hydrocarbons in several marine fish and invertebrate species.

Residues

Water and Sediments

Extraordinarily high levels of petroleum hydrocarbons in water have been reported following oil spills. Boehm and Fiest (1982), for example, found that the concentration of high molecular weight ($>C_{10}$) hydrocarbons averaged 10,600 $\mu g\ L^{-1}$ within several hundred meters of the Ixtoc I blowout in the Gulf of Mexico (probably the largest spill in history). After the wreck of the *Amoco Cadiz* in 1978, residues of up to 240 $\mu g\ L^{-1}$ were found in sea water from Northwest Brittany (Law, 1981). This can be compared to an average concentration of $\sim$3 $\mu g\ L^{-1}$ reported for coastal waters off Guernsey, some 200 km from the spill site. Similarly, levels of $<$1–395 and 4–229 $\mu g\ L^{-1}$ have been found in the Arabian Sea and Bay of Bengal, respectively (Gupta and Kureishy, 1981). Since both bodies of water carry heavy tanker traffic, occasionally high levels will appear as a result of controlled/uncontrolled discharges.

Regulated waste discharge from drilling platforms, oil terminals, and refineries result in moderately high residues in localized areas. For example, the average concentration of volatile liquid hydrocarbons (C_6–C_{14}) immediately below the discharge of a production platform in the Gulf of Mexico was 130 $\mu g\ L^{-1}$ decreasing to $\sim$ 4$\mu g\ L^{-1}$ at a distance of 50 m from the site (Wiesenburg *et al.*, 1981). Similarly, in the Mersey Estuary (UK), total hydrocarbon concentrations ranged up to 74 $\mu g\ L^{-1}$, compared with residues of 24–60 $\mu g\ L^{-1}$ for other British estuaries (Law, 1981). Burns and Smith (1981) reported that maximum hydrocarbon residues in coastal bays in southern Australia ranged up to 23 $\mu g\ L^{-1}$ whereas levels in the Baltic Sea and Gulf of St. Lawrence are generally <5 μg L^{-1} (Ehrhardt, 1981; Levy and Walton, 1976)

The natural background level of hydrocarbons in sediments is extremely variable, ranging from ~1 to >500 mg kg^{-1}. Low levels are generally found in offshore marine areas while localized hydrocarbon deposits can produce large residues in some fresh waters. For example, Strosher and Peake (1979) reported a maximum concentration of 943 mg kg^{-1} in the sediments of the Athabasca River upstream of the mining operations. This level is comparable to that found in various estuaries and bays receiving waste from refineries and heavy industry. However, in the aftermath of oil spills, residues have exceeded 5000 mg kg^{-1}.

Since hydrocarbons are subjected to weathering and microbial breakdown, most low molecular compounds disappear from spilled oil, leaving tar-like substances. This material is washed up on beaches and therefore provides an opportunity to monitor the total amount of tar in the oceans. Using this method, Knap *et al.* (1980) concluded that, during the 1970s, there was a 15% increase in the amount of tar on Bermuda beaches despite reported improvements in tanker design and operation. This possibly indicates that the amount of oil discharged to the Atlantic has increased in recent years. Cordes *et al.* (1980), using the same method, concluded that oil from the Ixtoc blowout in the Gulf of Mexico had not been deposited on the beaches and Florida via the Gulf Stream.

Aquatic Invertebrates and Fish

Typical of many organic pollutants, petroleum hydrocarbons are sorbed rapidly by aquatic invertebrates until a steady state or equilibrium in concentrations is achieved. The rate of uptake depends primarily on the exposure concentration, but temperature and other environmental factors may alter the metabolic rate of the animal and hence rate of uptake. Most petroleum hydrocarbons are lipophilic and thus the maximum level achieved during the steady-state phase depends on the body lipid content, as well as exposure concentration (Figure 7.7). Depuration from tissues is generally rapid but once again depends on temperature. Burns and Smith (1981), working in relatively warm Australian waters, found that ~90% of hydrocarbons in the mussel *Mytilus* was eliminated within 3 weeks. By contrast, the same species required 14 weeks for 90% clearance of fuel oil under European winter conditions (Blackman and Law, 1980).

Following oil spills, hydrocarbon levels in the muscle tissue of invertebrates may increase to 100–500 mg kg^{-1} wet weight (Clark and MacLeod, 1977). In areas of chronic pollution, residues are substantially lower, falling within the range of 1 to ~150 mg kg^{-1}, whereas in contaminated waters, concentrations are generally <5 mg kg^{-1} (Clark and MacLeod, 1977). However, much higher levels, exceeding 1000 mg kg^{-1}, have been reported

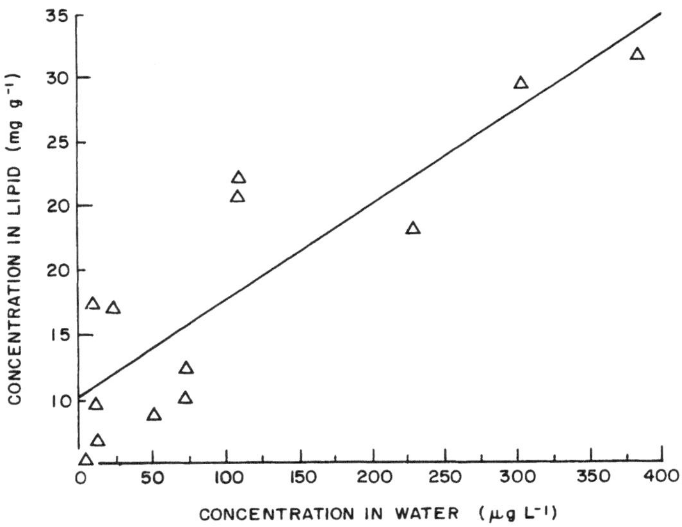

Figure 7.7. Concentration of petroleum hydrocarbons in mussels and water (from K.A. Burns and J.L. Smith, Biological monitoring of ambient water quality . . . , in *Estuarine and Coastal Shelf Science*, vol. 13, 1981:433–433. © Academic Press Inc. (London) Ltd. Used with permission

for the fat of various species, reflecting the lipophilic nature of these substances.

Fish absorb hydrocarbons primarily from the water, although ingestion of tainted food also leads to tissue contamination. Rate of uptake depends directly upon exposure concentration and lipid content of the fish. Apparently, the various components of petroleum hydrocarbons are sorbed at different rates and are also selectively deposited in specific tissues. For example, Whittle *et al.* (1977) treated juvenile herring with ^{14}C-hexadecane and ^{14}C-benzo[a]pyrene to determine the sites of deposition of these compounds. It was then shown that 59% of the hexadecane was found in the muscle, whereas 8% and 3% occurred in the mesenteric fat and stomach fat, respectively. By contrast, the corresponding values for benzo[a]pyrene were 0.1, 0.02, and 87.1%, respectively. In most cases, the rate of depuration is rapid but once again depends on the chemical composition of the oil, lipid content of the fish, and environmental factors such as temperature. Although the maximum half-life of *n*-alkanes in kerosene-tainted mullet was ~18 days, the concentration of naphthalenes in gulf killifish exposed to No. 2 fuel oil declined to nondetectable levels within 15 days (Connell and Miller, 1981).

The concentration of petroleum hydrocarbons in muscle tissue of fish collected from the open ocean generally ranges from <0.5 to 5 mg kg^{-1} wet weight (Clark and MacLeod, 1977). Substantially larger residues, exceeding 100 mg kg^{-1}, are occasionally reported for off-shore collections and, although the exact source of contamination is not known, high lipid levels and movement into coastal waters probably aggravate the hydrocarbon burden. Inshore populations naturally contain relatively high levels in the muscle. Fish from Lake Maracaibo (Venezuela), for example, carried burdens ranging up to 40 mg kg^{-1} (saturates and aromatics), compared with 375 mg kg^{-1} for herring collected from the Firth of Clyde (Clark and MacLeod, 1977). Comparable, and even higher levels, are often reported for liver tissue, reflecting its relatively high lipid content. For example, Atlantic cod from the North Atlantic contained residues of up to 345 mg kg^{-1}, whereas Greenland halibut and ocean perch inhabiting the same area had burdens of 230 and 110 mg kg^{-1}, respectively (Clark and MacLeod, 1977).

Toxicity

Aquatic Plants and Invertebrates

Petroleum hydrocarbons have a highly variable influence on aquatic plants. At low concentrations, these substances often have a stimulatory effect on growth of many species, whereas at higher concentrations, the effect may be greatly reduced. Federle *et al.* (1979) reported that treatment of phytoplankton with Prudhoe Bay (Alaska) crude oil at concentrations of 30–200 $\mu g\ L^{-1}$ resulted in a 90–100% decrease in primary production within 5 days. Thereafter, there was a shift in the species composition of the plankton, followed by a 50% recovery in the primary production rate. Such changes in species composition have also been noted under natural conditions and probably have some effect on the rest of the food chain, though this is a difficult point to confirm outside the laboratory.

As might be expected, plant responses to chronic/acute exposure depend on species, environmental conditions, and the type of oil. Hsiao *et al.* (1978) found that the rate of primary production of Beaufort Sea phytoplankton varied by more than 100% following exposure to four different types of oil (Figure 7.8). Similarly, the flagellate *Euglena gracilis* grew in culture water containing up to 10% oil whereas the growth of the green algae *Scenedesmus quadricauda* was halted at the same concentration (Dennington *et al.*, 1975). Such variability implies that *in situ* impacts on algal populations will be difficult to predict without site-specific data on ameliorating factors.

In general, aquatic invertebrates are substantially more sensitive to petroleum hydrocarbons than algae (Mahoney and Haskin, 1980). LC_{50}s of 1–5 mg L^{-1} have been reported for the water-soluble fraction of many types of crude oil, but this range can be extended to <0.1–100 mg L^{-1} depending

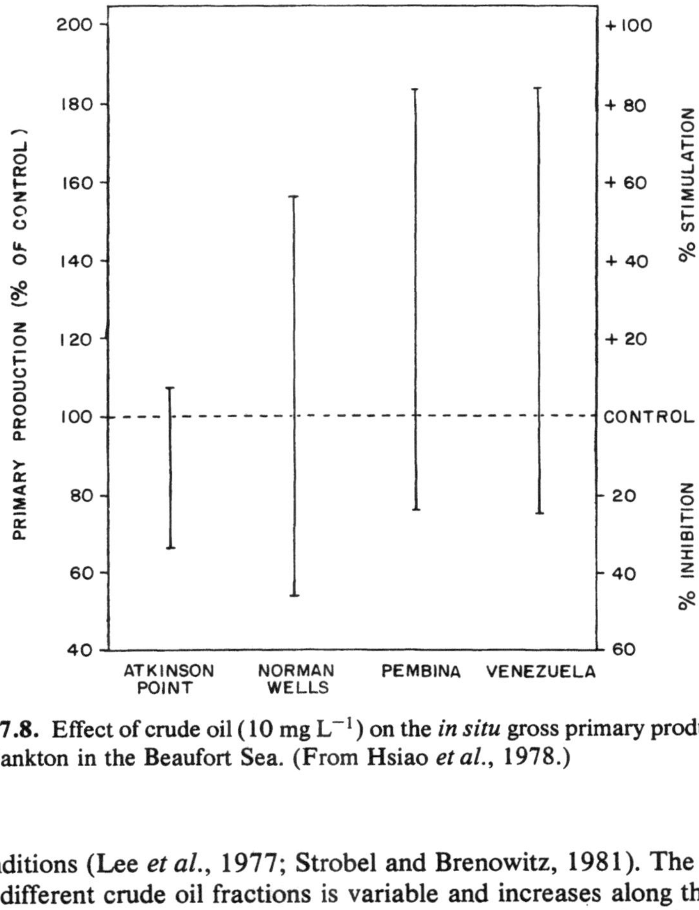

Figure 7.8. Effect of crude oil (10 mg L^{-1}) on the *in situ* gross primary production of phytoplankton in the Beaufort Sea. (From Hsiao *et al.*, 1978.)

on conditions (Lee *et al.*, 1977; Strobel and Brenowitz, 1981). The toxicity of the different crude oil fractions is variable and increases along the series paraffins, naphthenes, olefins, and aromatics. Within each series, the smaller molecules are more toxic than the larger; thus octane and decane are acutely toxic while dodecane and the higher paraffins are almost nontoxic. As a reflection of these differences, Donahue *et al.* (1977) reported that the toxicity of four crude oils to barnacle larvae increased in the order Venezuela>Kuwait>Alaska>South Louisiana.

Weathered crude oil is not acutely toxic to invertebrates. Essentially all of the toxic/volatile components are rapidly dissipated to the atmosphere upon weathering, whereas the water-soluble fraction is transported to the water column where it is susceptible to various ameliorating processes. Lee *et al.* (1980) exposed zooplankton from the Gulf of Mexico to crude oil from the 1979 Ixtoc blowout. The oil was collected 2 months after the blowout from the Gulf of Mexico and was not toxic at concentrations of up to 14 mg L^{-1}, the highest concentration tested. Similarly Wong *et al.* (1981) showed that survival of the cladoceran *Daphnia pulex* exposed to 24-hour weathered crude oil at 10 mg L^{-1} was similar to that of controls and significantly higher than that recorded for unweathered oil.

Exposure to crude oil induces numerous effects on the reproductive system of a host of invertebrate species. Renzoni (1975) showed that the spermatozoa of molluscs were highly sensitive to various crude oils, thereby depressing the rate of successful fertilization and increasing the frequency of developmental abnormalities. Similarly, Barry and Yevich (1975) reported that exposure of the mollusc *Mya arenaria* to crude oil resulted in the development of neoplastic lesions in gonadal tissue, as well as in the kidney, urinary pores, heart, and epibranchial chamber. Comparable lesions have been found in the reproductive tissue and mucous secretory cells of the coral *Manicina areolata* (Peters *et al.*, 1981).

Fish

Susceptibility of fish to crude oils is highly variable, depending on the source of oil and its chemical composition, test conditions, and species under investigation. Craddock (1977), in his review of the effects of petroleum on fish, reported that the lethal concentration of fresh crude oil to adults varies from 90 to 18,000 mg L^{-1}, whereas the corresponding range for eggs and larva is 0.1–100 mg L^{-1}. However, mortality can be induced in adults at concentrations as low as 0.5 mg L^{-1} following long-term (90-day) exposure (Woodward *et al.*, 1981). As with invertebrates, the low boiling point and more soluble aromatic compounds are most toxic. Accordingly, the lethal concentration of the water-soluble fraction of crude oil in short-term tests varies from 5–50 to 0.1–1.0 mg L^{-1} for adults and egg larvae, respectively (Craddock, 1977). It is known that the volatile aromatic fraction dissipates rapidly to the atmosphere, usually within 24 hours, thereby reducing potential long-term impacts of oil spills on aquatic ecosystems. However, under ice conditions, there may be little volatilization until the spring breakup. Thus, the consequences of an oil spill during the winter, particularly in Arctic/subarctic areas, are substantially greater than those for temperate-tropical areas.

Treatment of fish with low levels of crude oil generally induce the activity of aryl hydrocarbon hydroxylases in the liver, gills, and other tissues (Payne and Fancey, 1982). These enzymes are indicative of metabolism of aromatic compounds through hydroxylation to less acutely toxic forms such as arene and alkene oxides. Thus the level of enzyme activity in fish livers generally increases with the duration and level of exposure to hydrocarbons (Nava and Engelhardt, 1982). However, this response is species specific and, in some instances, there may be a depression or no change in enzyme activity (Ridlington *et al.*, 1982). The activity of various plasma enzymes, such as alkaline phosphatase, also increases following exposure to hydrocarbons, whereas there is often a decrease in the hemoglobin content of blood (Lebsack *et al.*, 1980).

Exposure to low levels of oil also results in numerous changes in the structure of tissues. Hawkes and Stehr (1982) reported that treatment of surf smelt embryos with Cook Inlet crude oil at 54 $\mu g\ L^{-1}$ produced necrotic neurons in the forebrain and neuronal layer of the retina. In addition, the ellipsoid and myoid regions of the receptor cells were also necrotic, whereas chinook salmon that were fed pellets containing 5 mg kg^{-1} petroleum hydrocarbons showed only minor changes to the structure of the intestine (Hawkes *et al.*, 1980). However, addition of chlorinated biphenyls to this diet produced sloughing of the epithelia, indicative of an interactive effect. There was also distinct subcellular inclusions of the columnar cells of the mucosa. Such changes in tissue structure, although not acutely lethal, contribute to the onset of behavioral symptoms such as reduced feeding activity and predator avoidance that in turn lead to a decline in fish populations under natural conditions.

Semiaquatic Birds and Mammals

The unintentional consumption of oil during feeding and grooming is one of the main causes of mortality of semiaquatic birds and mammals following spills. Baker *et al.* (1981) reported that the death of 13 otters *Lutra lutra* near on oil terminal in the Shetlands was due to hemorrhagic gastroenteropathy associated with the ingestion of oil. Although similar results have been reported for various fur seals, mammals that rely for thermal insulation on subcutaneous fat have a relatively low risk of consuming oil while grooming. In fact, there appears to be only one reported instance of death of naked seals under these circumstances (Baker *et al.*, 1981).

Birds regularly pluck and preen oiled feathers. This inevitably leads to the ingestion of oil, but may also accelerate heat loss and reduce buoyancy. Thus, species that inhabit the colder limits of their range are particularly susceptible to hypothermia and starvation induced by oil contamination (Levy, 1980). Apparently, the same effects can be induced by vegetable oil spills (McKelvey *et al.*, 1980). Randall *et al.* (1980) reported that the cleaning of penguins *Spheniscus demersus* resulted in a 32% mortality rate but, of those released, 87% molted and reproduced normally.

Ingested oil produces a number of subacute effects such as reduced fecundity and changes in enzyme activity, tissue structure, and organ/body weight. For example, mallard hens fed a diet containing 2% South Louisiana crude oil showed a delay in egg laying and a decline in egg production, shell thickness and hatchability (Vangilder and Peterle, 1980). The same diet also led to a reduction in the weight and yolk protein and energy content of the eggs. Exposure of mallard ducklings to food containing 5% No. 2 fuel oil reduced weight gains and feather development in early life whereas treatment at 0.5% had no appreciable effect on development (Figure 7.9). Gorsline and Holmes (1982) reported that female mallards consuming petroleum-contaminated food showed a significant increase in the activity of hepatic

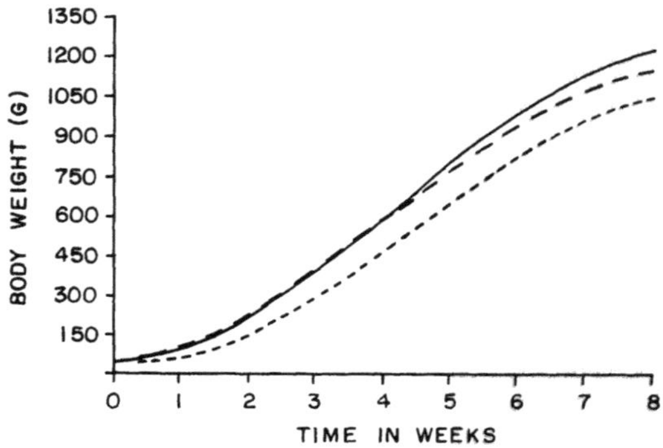

Figure 7.9. Weekly body weight gain in mallard ducklings fed No. 2 fuel oil from hatching to 8 weeks of age. Solid line represents control ducklings, long dash line represents ducklings fed a 0.15% oil diet, and short dashed line represents ducklings fed a 5.0% diet. (From Szaro *et al.*, 1981.)

aryl carbon hydroxylases, used to detoxify hydrocarbons. The progeny of these females similarly exhibited elevated enzyme activity in proportion to the level of exposure for the mothers. Although chronically treated birds may also produce high levels of plasma corticosterone, acute exposure generally leads to a decline in hormone levels (Gorsline and Holmes, 1981). This reflects dysfunction of the adrenocortical cells in the brain and is often correlated with an increase in the weight of the adrenal gland (Peakall *et al.*, 1982). Dysfunction of the kidney and liver in acutely exposed specimens generally elevates the activity of plasma alanine aminotransferase and carbamoyltransferase (Szaro *et al.*, 1981). Both of these enzymes are probably involved in the metabolism and detoxification of high molecular weight organic compounds.

Human Health

Petroleum hydrocarbons contain a wide range of substances that are potential health hazards. These include aliphatic compounds, monocyclic aromatics, polycyclic aromatics, and heavy metals. Most petroleum products have an objectionable taste and odor well below the level that might induce chronic toxicity in humans. Thus, the most frequently reported symptoms of exposure are relatively mild and include nausea, vomiting, and diarrhea.

Since petroleum products are ubiquitous, low-level exposure through food and water is inevitable for a major portion of the population. Although this is potentially significant considering that some substituents of petroleum are carcinogenic, the overall level of exposure does not appear to constitute a significant threat to human health. For example, the known carcinogen benzo[a]pyrene generally occurs at concentrations of 0–20 $\mu g\ kg^{-1}$ in uncooked seafoods (Neff, 1979). For comparison, charcoal-broiled steak may contain up to 50 μg B[a]P kg^{-1}, smoked meat up to 5 $\mu g\ kg^{-1}$, and vegetable oils up to 35 $\mu g\ kg^{-1}$. Similarly, treated drinking water constitutes $<0.5\%$ of the total B[a]P food intake throughout most of Canada and the

USA (Pucknat, 1981). Although definitive epidemiological data are not available, one might anticipate that fishery products and drinking water obtained from nonindustrial areas do not constitute a major source of carcinogens.

References

Atlas, R.M., and M. Busdosh. 1976. Microbial degradation of petroleum in the arctic. *In*: J.M. Sharpley and A.N. Kaplan (Eds.), *Proceedings of the 3rd International Biodegradation Symposium*. Applied Science Publishers, London, pp. 79–85.

Baker, J.R., A.M. Jones, T.P. Jones, and H.C. Watson. 1981. Otter *Lutra lutra* L. mortality and marine oil pollution. *Biological Conservation* **20**:311–321.

Barry, M., and P.P. Yevich. 1975. The ecological, chemical and histopathological evolution of an oil spill site. Part III. Histopathological studies. *Marine Pollution Bulletin* **6**:171–173.

Bend, J.R., M.O. James, and P.M. Dansette. 1977. *In vitro* metabolism of xenobiotics in some marine animals. *In*: H.F. Kraybill, C.J. Dawe, J.C. Harsharger, and R.G. Tardiff (Eds.), *Aquatic pollutants and biologic effects with emphasis on neoplasia. Annals of New York Academy of Science* **298**: 505–521.

Bergstein, P.E., and J.R. Vestal. 1978. Crude oil biodegradation in arctic tundra ponds. *Arctic* **31**:158–169.

Bhosle, N.B., and S. Mavinkurve. 1980–1981. Hydrocarbon-utilising microorganisms from Dona Paula Bay, Goa. *Marine Environmental Research* **4**:53–58.

Blackman, R.A.A., and R.J. Law. 1980. The Eleni V oil spill: fate and effects of the oil over the first twelve months. *Marine Pollution Bulletin* **11**:217–220.

Boehm, P.D., and D.L. Fiest. 1982. Subsurface distributions of petroleum from an offshore well blowout. The Ixtoc I blowout, Bay of Campeche. *Environmental Science and Technology* **16**:67–74.

Burns, K.A., and J.L. Smith. 1981. Biological monitoring of ambient water quality: the case for using bivalves as sentinel organisms for monitoring petroleum pollution in coastal waters. *Estuarine and Coastal Shelf Science* **13**:433–443.

Butler, J.N, B.F. Morris, and J. Sass. 1973. *Pelagic tar from Bermuda and the Sargasso Sea*. Special Publication No. 10., Bermuda Biological Station, 346 pp.

Clark, Jr., R.C., and W.D. Macleod, Jr. 1977. Inputs, transport mechanisms, and observed concentrations of petroleum in the marine environment. *In*: D.C. Malins (Ed.), *Effects of petroleum on arctic and subarctic marine environments and organisms. Volume I. Nature and fate of petroleum*. Academic Press, New York, pp. 91–223.

Connell, D.W., and G.J. Miller. 1981. Petroleum hydrocarbons in aquatic ecosystems—behaviour and effects of sublethal concentrations: Part 1. *CRC Critical Reviews in Environmental Control* **11**:37–104.

Cooper, D.G., and J.E. Zajic. 1980. Surface-active compounds from microorganisms. *Advances in Applied Microbiology* **26**:229–253.

Cordes, C., L. Atkinson, R. Lee, and J. Blanton. 1980. Pelagic tar off Georgia and Florida in relation to physical processes. *Marine Pollution Bulletin* **11**: 315–317.

Corner, E.D.S. 1975. The fate of fossil fuel hydrocarbons in marine animals. *Proceedings of the Royal Society of London Series B* **189**:391–413.

Craddock, D.R. 1977. Acute toxic effects of petroleum on arctic and subarctic marine organisms. *In*: D.C. Malins (Ed.), *Effects of petroleum on arctic and subarctic marine environments and organisms. Volume II. Biological effects.* Academic Press, New York, pp. 1–93.

Dennington, V.N., J.J. George, and C.H.E. Wyborn. 1975. The effects of oils on growth of freshwater phytoplankton. *Environmental Pollution* **8**:233–237.

Donahue, W.H., R.T. Wang, M. Welch, and J.A.C. Nicol. 1977. Effects of water-soluble components of petroleum oils and aromatic hydrocarbons on barnacle larvae. *Environmental Pollution* **13**:187–202.

Drew, L.J. 1982. Petroleum (resources). *In*: M. Grayson and D. Eckroth (Eds.), *Kirk-Othmer encyclopedia of chemical technology*, Vol. **17**. Wiley, New York, pp. 132–142.

Eganhouse, R.P., and I.R. Kaplan. 1981. Extractable organic matter in urban stormwater runoff. 1. Transport dynamics and mass emission rates. *Environmental Science and Technology* **15**:310–315.

Ehrhardt, M. 1981. Organic substances in the Baltic Sea. *Marine Pollution Bulletin* **12**:210–213.

Federle, T.W., J.R. Vestal, G.R. Hater, and J.C. Miller. 1979. Effects of Prudhoe Bay crude oil on primary production and zooplankton in arctic tundra thaw ponds. *Marine Environmental Research* **2**:3–18.

Gorsline, J., and W.N. Holmes. 1981. Effects of petroleum on adrenocortical activity and on hepatic naphthalene-metabolizing activity in mallard ducks. *Archives of Environmental Contamination and Toxicology* **10**:765–777.

Gorsline, J., and W.N. Holmes. 1982. Ingestion of petroleum by breeding mallard ducks: some effects on neonatal progeny. *Archives of Environmental Contamination and Toxicology* **11**:147–153.

Griffiths, R.P., T.M. McNamara, B.A. Caldwell, and R.Y. Morita. 1981. A field study of the acute effects of the dispersant Corexit 9527 on glucose uptake by marine microorganisms. *Marine Environmental Research* **5**:83–91.

Gupta, R.S., and T.W. Kureishy. 1981. Present state of oil pollution in the northern Indian Ocean. *Marine Pollution Bulletin* **12**:295–301.

Hawkes, J.W., E.H. Gruger, Jr., and O.P. Olson. 1980. Effects of petroleum hydrocarbons and chlorinated biphenyls on the morphology of the intestine of chinook salmon (*Oncorhynchus tshawytscha*). *Environmental Research* **23**: 149–161.

Hawkes, J.W., and C.M. Stehr. 1982. Cytopathology of the brain and retina of embryonic surf smelt (*Hypomesus pretiosus*) exposed to crude oil. *Environmental Research* **27**:164–178.

Hileman, B. 1981. Offshore oil drilling. *Environmental Science and Technology* **15**:1259–1263.

Hoffman, H.L. 1982. Petroleum products. *In*: M. Grayson and D. Eckroth (Eds.), *Kirk-Othmer encyclopedia of chemical technology*, Vol. **17**. Wiley, New York, pp. 257–271.

Hsiao, S.I.C., D.W. Kittle, and M.G. Foy. 1978. Effects of crude oils and the oil dispersant Corexit on primary production of arctic marine phytoplankton and seaweed. *Environmental Pollution* **15**:209–221.

James, M.O., J.R. Fouts, and J.R. Bend. 1974. *In vitro* epoxide metabolism in some marine species. *Bulletin Mt. Desert Island Biological Laboratory* **14**:41–46.

Khan, M.A.Q., R.H. Stanton, and G. Reddy. 1974. Detoxication of foreign chemicals by invertebrates. *In*: M.A.Q. Khan and J.P. Bederka (Eds.), *Survival in toxic environments*. Academic Press, New York, pp. 177–201.

Knap, A.H., and P.J.L. Williams. 1982. Experimental studies to determine the fate of petroleum hydrocarbons from refinery effluent on an estuarine system. *Environmental Science and Technology* **16**:1–4.

Knap, A.H., T.M. Iliffe, and J.N. Butler. 1980. Has the amount of tar on the open ocean changed in the past decade? *Marine Pollution Bulletin* **11**:161–164.

Law, R.J. 1981. Hydrocarbon concentrations in water and sediments from UK marine waters, determined by fluorescence spectroscopy. *Marine Pollution Bulletin* **12**:153–157.

Lebsack, M.E., A.D. Anderson, K.F. Nelson, and D.S. Farrier. 1980. Sublethal effects of an *in situ* oil shale retort water on rainbow trout. *Toxicology and Applied Pharmacology* **54**:462–468.

Lee, G.F., R.A. Hughes, and G.D. Veith. 1977. Evidence for partial degradation of toxaphene in the aquatic environment. *Water, Air, and Soil Pollution* **8**: 479–484.

Lee, W.Y., A. Morris, and D. Boatwright. 1980. Mexican oil spill: a toxicity study of oil accommodated in seawater on marine invertebrates. *Marine Pollution Bulletin* **11**:231–234.

Levy, E.M. 1971. The presence of petroleum residues off the east coast of Nova Scotia, in the Gulf of St. Lawrence, and the St. Lawrence river. *Water Research* **5**:723–733.

Levy, E.M. 1980. Oil pollution and seabirds: Atlantic Canada 1976–77 and some implications for northern environments. *Marine Pollution Bulletin* **11**:51–56.

Levy, E.M., and A. Walton. 1976. High seas oil pollution: particulate petroleum residues in the North Atlantic. *Journal of the Fisheries Research Board of Canada* **33**:2781–2791.

Mahoney, B.M., and H. H. Haskin. 1980. The effects of petroleum hydrocarbons on the growth of phytoplankton recognized as food forms for the eastern oyster, *Crassostrea virginica* Gmelin. *Environmental Pollution (Series A)* **22**: 123–132.

Malins, D.C., Ed. 1977a. *Effects of petroleum on arctic and subarctic marine environments and organisms. Volume I. Nature and fate of petroleum*, 321 pp. *Volume II, Biological effects*. Academic Press, New York, 500 pp.

Malins, D.C. 1977b. Metabolism of aromatic hydrocarbons in marine organisms. *In*: H.F. Kraybill, C.J. Dawe, J.C. Harshbarger, and R.G. Tardiff (Eds.), *Aquatic pollutants and biological effects with emphasis on neoplasia. Annals of the New York Academy of Science* **298**:482–496.

Marty, D., A. Bianchi, and C. Gatellier. 1979. Effects of three oil spill dispersants on marine bacterial populations. I. Preliminary study. Quantitative evolution of aerobes. *Marine Pollution Bulletin* **10**:285–287.

McAuliffe, C.D., J.C. Johnson, S.H. Greene, G.P. Canevari, T.D. Searl. 1980.

Dispersion and weathering on chemically treated crude oils on the ocean. *Environmental Science and Technology* **14**:1509–1518.

McKelvey, R.W., I. Robertson, and P.E. Whitehead. 1980. Effect of non-petroleum oil spills on wintering birds near Vancouver. *Marine Pollution Bulletin* **11**: 169–171.

McLachlan, A., and B. Harty. 1981. Effects of oil on water filtration by exposed sandy beaches. *Marine Pollution Bulletin* **12**:374–378.

McRory, R.E. 1982. *Oil sands and heavy oils of Alberta*. Department of Energy and Natural Resources, Edmonton, Alberta, 94 pp.

Meyers, P.A., and T.G. Oas. 1978. Comparison of associations of different hydrocarbons with clay particles in simulated seawater. *Environmental Science and Technology* **12**:934–937.

Meyers, P.A., and J.G. Quinn. 1973. Association of hydrocarbons and mineral particles in saline solution. *Nature* **244**:23–24.

Naquadat. 1976–1981. *Surface water quality data*. Canada Inland Waters Directorate, Environment Canada, Ottawa.

Nava, M.E., and F.R. Engelhardt. 1980. Compartmentalization of ingested labelled petroleum in tissues and bile of the American eel (*Anguilla rostrata*). *Bulletin of Environmental Contamination and Toxicology* **24**:879–885.

Neff, J.M. 1979. *Polycyclic aromatic hydrocarbons in the aquatic environment. Sources, fates and biological effects*. Applied Science Publishers, England, 262 pp.

Oil and Gas Journal. 1982. Penn Well Publishing Co., Tulsa, Oklahoma.

Oilweek. 1982. October 25. Maclean Hunter Publication, Calgary, Alberta.

Oostdam, B.L. 1980. Oil pollution in the Persian Gulf and approaches. 1978. *Marine Pollution Bulletin* **11**:138–144.

Payne, J.F. 1977. Mixed function oxidases in marine organisms in relation to petroleum hydrocarbon metabolism and detection. *Marine Pollution Bulletin* **8**:112–116.

Payne, J.F., and L.L. Fancey. 1982. Effect of long term exposure to petroleum on mixed function oxygenases in fish: further support for use of the enzyme system in biological monitoring. *Chemosphere* **11**:207–213.

Peakall, D.B., D.J. Hallett, J.R. Bend, G.L. Foureman, and D.S. Miller. 1982. Toxicity of Prudhoe Bay crude oil and its aromatic fractions to nestling herring gulls. *Environmental Research* **27**:206–215.

Pedersen, M.G., W.K. Hershberger, and M.R. Juchau. 1974. Metabolism of 3,4-benzopyrene in rainbow trout (*Salmo gairdneri*). *Bulletin of Environmental Contamination and Toxicology* **12**:481–486.

Peters, E.C., P.A. Meyers, P.P. Yevich, and N.J. Blake. 1981. Bioaccumulation and histopathological effects of oil on a stony coral. *Marine Pollution Bulletin* **12**: 333–339.

Philpot, R.M., M.O. James, and J.R. Bend. 1976. Metabolism of benzo(a)pyrene and other xenobiotics by microsomal mixed function oxidases in marine species. In: *Sources, effects and sinks of hydrocarbons in the aquatic environment*. American Institute of Biological Sciences, Washington, D.C., pp. 184–199.

Pucknat, A.W. 1981. *Health impacts of polynuclear aromatic hydrocarbons*. Environmental Health Review No. 5., Noyes Data Corporations, Park Ridge, New Jersey, 271 pp.

Randall, R.M., B.M. Randall, and J. Bevan. 1980. Oil pollution and penguins—is cleaning justified? *Marine Pollution Bulletin* **11**:234–237.

Read, A.D., and R.A.A. Blackman. 1980. Oily water discharges from offshore North Sea installations: a perspective. *Marine Pollution Bulletin* **11**:44–47.

Renzoni, A. 1975. Toxicity of three oils to bivalve gametes and larvae. *Marine Pollution Bulletin* **6**:125–128.

Ridlington, J.W., D.E. Chapman, B.L. Boese, and V.G. Johnson. 1982. Petroleum refinery wastewater induction of the hepatic mixed-function oxidase system in Pacific staghorn sculpin. *Archives of Environmental Contamination and Toxicology* **11**:123–127.

Riley, R.G., B.L. Thomas, J.W. Anderson, and R.M. Bean. 1980. Changes in the volatile hydrocarbon content of Prudhoe Bay crude oil treated under different simulated weathering conditions. *Marine Environmental Research* **4**:109–119.

Roubal, W.T., T.K. Collier, and D.C. Malins. 1977. Accumulation and metabolism of carbon-14 labelled benzene, naphthalene and anthracene by young coho salmon. *Archives of Environmental Contamination and Toxicology* **5**:513–529.

Rubinstein, I., O.P. Strausz, C. Spyckerelle, R.J. Crawford, and D.W.S. Westlake. 1977. The origin of the oil sand bitumens of Alberta: a chemical and microbiological simulation study. *Geochimica et Cosmochimica Acta* **41**:1341–1353.

Selucky, M.L., Y. Chu, T. Ruo, and O.P. Strausz. 1977. Chemical composition of Athabasca bitumen. *Fuel* **56**:369–381.

Singer, S.C., and R.F. Lee. 1977. Mixed function oxygenase activity in blue crab, *Callinectes sapidus*: tissue distribution and correlation with changes during molting and development. *Biological Bulletin* **153**:377–386.

Stegeman, J.J. 1978. Influence of environmental contamination on cytochrome P-450 mixed-function oxygenases in fish: implications for recovery in the Wild Harbor Marsh. *Journal of the Fisheries Research Board of Canada* **35**: 668–674.

Strobel, C.J., and A.H. Brenowitz. 1981. Effects of bunker C oil on juvenile horseshoe crabs (*Limulus polyphemus*). *Estuaries* **4**:157–159.

Strosher, M.T., and E. Peake. 1979. *Baseline states of organic constituents in the Athabasca River system upstream of Fort McMurray*. Prepared for the Alberta Oil Sands Environmental Research Program by Environmental Sciences Centre (Kananaskis), The University of Calgary. AOSERP Report 53, 71 pp.

Szaro, R.C., G. Hensler, and G.H. Heinz. 1981. Effects of chronic ingestion of No. 2 fuel oil on mallard ducklings. *Journal of Toxicology and Environmental Health* **7**:789–799, ©Hemisphere Publishing Corporation.

Traxler, R.W., and L.S. Bhattacharya. 1978. Effects of a chemical dispersant on microbial utilization of petroleum hydrocarbons. *In*: L.T. McCarthy, Jr., G.P. Lindblom, and H.F. Walters (Eds.), *Chemical dispersants for the control of oil spills*. American Society for Testing and Materials, ASTM STP 659, pp. 181–187.

Vandermeulen, J.H., and W.R. Penrose. 1978. Absence of aryl hydrocarbon hydroxylase (AHH) activity in three marine bivalves. *Journal of the Fisheries Research Board of Canada* **35**:643–647.

Vangilder, L.D., and T.J. Peterle. 1980. South Louisiana crude oil and DDE in the diet of mallard hens: effects on reproduction and duckling survival. *Bulletin of Environmental Contamination and Toxicology* **25**:23–28.

Varanasi, U., D.J. Amur, and P.A. Treseler. 1979. Influence of time and mode of exposure on biotransformation of naphthalene by juvenile starry flounder (*Platichthys stellatus*) and rock sole (*Lepidopsetta bilineata*). *Archives of Environmental Contamination and Toxicology* **8**:673–692.

Walker, J.D., J.J. Calomiris, T.L. Herbert, and R.R. Colwell. 1976a. Petroleum hydrocarbons: degradation and growth potential for Atlantic Ocean sediment bacteria. *Marine Biology* **34**:1–9.

Walker, J.D., P.A. Seesman, T.L. Herbert, and R.R. Colwell. 1976b. Petroleum hydrocarbons: degradation and growth potential of deepsea sediment bacteria. *Environmental Pollution* **10**:89–99.

Whipple, Jr., W., and J.V. Hunter. 1979. Petroleum hydrocarbons in urban runoff. *Water Resources Bulletin* **15**:1096–1105.

Whittle, K.J., J. Murray, P.R. Mackie, R. Hardy, and J. Farmer. 1977. Fate of hydrocarbons in fish. *Rapports et Proces Verbaux des Reunions Conseil International pour Exploration de la Mer* **171**:139–142.

Wiesenburg, D.A., G. Bodennec, and J.M. Brooks. 1981. Volatile liquid hydrocarbons around a production platform in the northwest Gulf of Mexico. *Bulletin of Environmental Contamination and Toxicology* **27**:167–174.

Wong, C.K., F.R. Engelhardt, and J.R. Strickler. 1981. Survival and fecundity of *Daphnia pulex* on exposure to particulate oil. *Bulletin of Environmental Contamination and Toxicology* **26**:606–612.

Woodward, D.F., P.M. Mehrle, Jr., and W.L. Mauck. 1981. Accumulation and sublethal effects of a Wyoming crude oil in cutthroat trout. *Transactions of the American Fisheries Society* **110**:437–445.

Wyndham, R.C., and J.W. Costerton. 1981a. In vitro microbial degradation of bituminous hydrocarbons and in situ colonization of bitumen surfaces within the Athabasca oil sands deposit. *Applied and Environmental Microbiology* **41**: 791–800.

Wyndham, R.C., and J.W. Costerton. 1981b. Heterotrophic potentials and hydrocarbon biodegradation potentials of sediment microorganisms within the Athabasca oil sands deposit. *Applied and Environmental Microbiology* **41**:783–790.

Zurcher, F., and M. Thuer. 1978. Rapid weathering processes of fuel oil in natural waters: analyses and interpretations. *Environmental Science and Technology* **12**:838–843.

8
Phenols

Phenols are a diverse group of organic chemicals consisting of a basic benzene ring and one or more hydroxyl groups. Simple phenol (hydroxybenzene, C_6H_5OH) was isolated in 1834 from coal tar. Although this was the only source of phenol until World War I, synthetic production gradually grew in importance and, by 1930, exceeded natural production. Today almost all phenol is manufactured by sulfonation of benzene and hydrolysis of the sulfonate. The more complex phenols are obtained by replacing one or more of the hydrogen atoms attached to the benzene ring with various atoms (such as chlorine) or more complex substituents (such as methyl or nitro molecules).

The carbon atom to which the hydroxyl group is attached is numbered 1 and the numbering proceeds clockwise around the ring. For one substitution (in addition to the hydroxyl group) there are three isomers. *Ortho* (*o*) means that the second group is attached to carbon atom 6 or 2; *meta* (*m*) means that the second group is attached to C3 or C5; *para* (*p*) means that the second group is on C4. If two or more substituent groups are present, then they are identified by the appropriate carbon numbers. Substitution of the OH group into a naphthalene ring yields naphthol. Because of their frequent use in the early years, several phenols were known by trivial names. For example, dihydric phenols (dihydroxy benzenes) exist in three isomers. The trivial names for the *ortho, meta*, and *para* isomers are catechol, hydroquinone, and resorcinol, respectively.

Production, Uses, and Discharges

Production and Uses

Phenol is one of the most important and versatile industrial organic chemicals. It is used in many commercial products including resins, nylons, plasticizers, antioxidants, oil additives, polyurethanes, drugs, pesticides, explosives, dyes, and gasoline additives. Consequently, phenol production is high and increasing in many industrialized countries. In the USA, total production exceeded 1150×10^3 metric tons in 1981 compared with 968×10^3 metric tons in 1982 (U.S. International Trade Commission, 1975–1982). Production is slightly lower in Western Europe whereas in Japan, $\sim 300 \times 10^3$ metric tons are synthesized annually (Table 8.1). Canadian consumption is relatively low, increasing from 1100 metric tons in 1975 to 72,100 metric tons in 1980 (Statistics Canada, 1981). Although no data are available for the USSR, production in Eastern Europe currently exceeds 360×10^3 metric tons per year (Table 8.1.).

All 19 possible chlorinated phenols are commercially available. Monochlorophenol, consisting of two isomers, is used mainly in the production of higher chlorinated phenols. Combined annual production in the USA, 18,800 metric tons, was considerably less than the 39,000 tons reported for 2,4-dichlorophenol. This latter compound is used primarily in the manufacture of 2,4-D. Production of the two main isomers of trichlorophenol and

Table 8.1. World production capacity of synthetic phenol in 1978.

Country	Production (10^3 metric tons)	Country	Production (10^3 metric tons)
North America		Western Europe	
Canada	50	Belgium/Luxembourg	140
Mexico	25	France	175
USA	1589	FRG	420
Total	1664	Italy	280
South America		UK	267
Argentina	12	Total	1282
Brazil	65		
Total	77	Eastern Europe	>360
Far East		Australia	17
Japan	289		
Other	22		
Total	311	World Total	>3711

Source: Thurman (1982).

of 2,3,4,6-tetrachlorophenol is relatively low, <7,000 metric tons. All three compounds find use as biocides and preservatives. In addition, 2,4,5-trichlorophenol is used in the production of 2,4,5-T and related products, whereas 2,4,6-trichlorophenol is used to manufacture tetrachlorophenol. Since the chemical processes involved in the synthesis of trichlorophenol may lead to the formation of TCDD (Chapter 10), 2,4,5-T and the higher chlorinated phenols may be contaminated with this compound. Similarly, pentachlorophenol production in the USA is relatively high, 22,000 metric tons, and is applied on land and water, making it an important source of TCDD in the environment.

Only one plant (Edmonton) manufactures chlorinated phenols in Canada (Jones, 1981). The installation can produce up to 1800 and 450 metric tons of pentachlorophenol and 2,3,4,6-tetrachlorophenol per year, respectively. Although 2,4-dichlorophenol was formerly produced at the site, no other chlorinated phenol has been manufactured in Canada. Imports of chlorinated phenols into Canada have increased from 558 to 625 metric tons between 1976 and 1980.

Production and capacity data for alklyl phenols are generally not available (Reed, 1980). However, it is known that US production of 4-methyl-2,6-dibutyl phenol is ~16,000 metric tons per year compared with 19,000 and 10,000 metric tons for Europe and Japan, respectively. Similarly, since 4,4-isopropylidenediphenol is the major outlet for phenol, output is also relatively high, amounting to 160,000 metric tons annually in the USA. Nonylphenol, used primarily as an intermediate in the manufacture of nonionic surfactants, was produced at a rate of 69,000 metric tons in the USA during 1981, whereas the corresponding figures for *o*-cresol were 15,000 and 14,000 metric tons.

Nitrophenol isomers are used primarily as intermediates for the production of dyes, pigments, preservatives, pesticides, pharmaceuticals, and rubber chemicals. Nitrophenol may also be inadvertently produced by microbial or photodegradation of pesticides that contain the nitrophenol moiety. Approximately 6800 metric tons of 2-nitrophenol are produced annually in the USA compared with ~19,000 metric tons for 4-nitrophenol. Although production data for 3-nitrophenol are not available, the amount is probably <500 metric tons. Of the six possible isomers of dinitrophenol, 2,4-dinitrophenol is the most important commercially. It is used as an intermediate in the production of dyes, explosives, wood preservatives, and photochemicals. Current output now exceeds 500 metric tons per year. Production data for the remaining isomers of dinitrophenol are not available.

There are six possible isomers of trinitrophenol, but commercial use is restricted to 2,4,6-trinitrophenol, also known as picric acid. Annual production, although not known, is presumed to be small. Similarly, there is either little or no production and use of the dinitrocresols in Canada and the USA.

Discharges

Since phenol is a high-volume industrial chemical with multiple uses, it is commonly found in a variety of municipal and industrial discharges (Table 8.2). Some of the highest reported concentrations are from petroleum refineries and other hydrocarbon processing industries. Although residues are much lower in municipal effluents, the total amount of material discharged from these sources constitutes a considerable input of phenol to the environment. Chlorinated phenols are commonly discharged from industries using chlorination as part of their process chemistry. Consequently, relatively high residues have been reported for effluents from pulp and paper mills, wood preservation plants, and certain chemical industries (Table 8.2). Chlorination of sewage apparently leads to the production of simple chlorophenols at concentrations of up to at least 1.7 $\mu g\ L^{-1}$ (Jolley *et al.*, 1978). The more complex compounds have also been reported from municipal wastes but at low levels (Table 8.2).

Relatively little information is available on the concentrations of other phenols in industrial and municipal discharges. Jungclaus *et al.* (1978) reported the presence of at least nine alkylphenols in the wastewater of a specialities chemical plant. Of these 2,6-di-*t*-butylphenol was most common, reaching a concentration of 800 $\mu g\ L^{-1}$, followed by 2,4-di-*t*-butylphenol (max. 600 $\mu g\ L^{-1}$) and 2,4-di-*t*-amylphenol (400 $\mu g\ L^{-1}$). Webb *et al.* (1973) reported that 4,6-dinitro-*o*-cresol reached a maximum level of 18,000 $\mu g\ L^{-1}$ in the effluent of another specialty chemical plant, whereas the corresponding residues for nitrophenol and 2-nitro-*p*-cresol were 9300 and 1400 $\mu g\ L^{-1}$, respectively. Substantially lower levels were reported for seven nitro- and alkylphenols originating from the effluent of an unspecified chemical plant (Lopez-Avila and Hites, 1980). Although quantitative data are not available, it is known that combustion and pyrolysis of building materials containing phenolic resin produce methlyphenols, dimethylphenols, and trimethylphenols. Such materials include foam insulation and adhesives in laminates.

As might be expected, the type of feedstock chemical and industrial process play a key role in determining the presence of phenols in wastewater. For example, the nitration of benzene and toluene to produce nitrobenzene and dinitrotoluene also leads to the incidental formation of nitrophenol, dinitrophenol, and dinitro-*o*-cresol. Similarly, alkylphenols and methylphenols may be produced during alkylation and solvent extraction of toluene, xylene, and C_8–C_9 alkylphenols. Wise and Fahrenthold (1981) suggested that most industrial processes were not sources of priority pollutants because the processes do not involve critical precursor/process combinations. In addition, synthetic production methods generally lead to an increase in complexity of priority pollutant molecules. These in turn exhibit variable toxicity and persistance, which may be comparable to related priority pollutants.

Table 8.2. Concentration ($\mu g\ L^{-1}$) of various phenols in municipal and industrial wastewater.

Compound	Concentration	Source
Phenol	0.03–20	Treated sewage, European cities
	200–3,016,000	3 petroleum refineries (USA)
	3200	2 coal gasification plants (USA)
	38,000–1,240,000	Coke plant (USA)
	10–300	Specialty chemical plant (USA)
2,4-Dichlorophenol	51–330	Chemical plant (Vancouver)
	<0.1	Wood preservation plant (British Columbia)
	<0.1	Landfill leachate (Vancouver)
2,6-Dichlorophenol	220	Chemical plant (Vancouver)
	2.4	Wood preservation plant (British Columbia)
	1.2–5.6	Landfill leachate (Vancouver)
2,4,5-Trichlorophenol	<0.05	Treated sewage, 4 plants (Vancouver)
	0.5–2400	2 chemical plants (Vancouver)
	<0.05	Wood preservation plant (British Columbia)
	0.05–2	Landfill leachate (Vancouver)
2,4,6-Trichlorophenol	<0.05–1	Treated sewage, 4 plants (Vancouver)
	<0.05–3120	3 chemical plants (Vancouver)
	0.5–1	Wood preservation plant (British Columbia)
	0.4–1.0	Landfill leachate (Vancouver)
	25–115	Pulpmill effluent (Vancouver)
2,3,4,6-Tetrachlorophenol	0.6–28	Treated sewage, 4 plants (Vancouver)
	1.2–8270	4 wood preservation plants (British Columbia)
	0.3–166	3 chemical plants (Vancouver)
	0.2–0.8	Landfill leachate (Vancouver)
Pentachlorophenol	0.5–4.7	Treated sewage, 4 plants (Vancouver)
	0.25–1.3	Treated sewage, 6 cities (Ontario)
	0.05–2760	Wood preservation plants (British Columbia)
	5,400,000	Chemical plant (Philadelphia)
	0.6–42	Landfill leachate (Vancouver)

Sources: Jones (1981); Buikema *et al.* (1979); Waggot and Wheatland (1978).

Behavior in Natural Waters

Phenol is fairly soluble in water and nonpolar solvents such as benzene and oils. Because of the electronegative character of the phenyl group, phenol is weakly acidic ($pK_a = 10.02$). Alkaline salts of phenol are also readily soluble in water. In general, the melting and boiling points of chlorophenols increase and the volatility decreases with increasing number of chlorine atoms in the ring. In addition, the aqueous solubility of these compounds decreases and solubility in nonpolar solvents such as benzene and petroleum ether increases (moderate to highly soluble) with increasing substitution of chlorine atoms in the ring. The organoleptic properties of the chlorophenols are manifested by imparting odor to water and tainting to fish. Chlorophenols are stronger acids than phenol as seen by the decreasing pK_a (−ve logrithm of the acid dissociation constant) with increasing chlorine content of the benzene ring. Nitrophenols are also more acidic than phenol and acidity increases with the number of nitro groups. The physical properties of the isomeric nitrophenols vary among themselves.

Sorption

The log P (octanol/water) of phenol indicates only a weak affinity for organic detritus and organic-rich sediments. Laboratory experiments showed an almost 100% desorption of phenol from a thin-layer of montmorillonite clay exposed to 40% humidity for one week (Saltzman and Yariv, 1975). This observation rules out any sorption to microcrystalline clay particles in natural waters. Phenol was an ineffective flocculant for soils and clays, implying its inability to aggregate the organic-inorganic components in water (Chang and Anderson, 1968). The increasing partition coefficient with chlorine atom substitution suggests a relatively greater affinity for the organic part of the sediments and suspended solids.

Characteristics of clay minerals are determined by the evolutionary age of the clays which, in turn, are determined by weathering. The surface area is largest for clay minerals of intermediate age whose interlamellar spaces are rigidly bonded together (kaolinite) (Table 8.3.) For montmorillonite, the lattice expansion with concomitant exposure of interlamellar surfaces accounts for the large surface area. The variation in CEC values is due to differences in accessibility to basal planes with particle size variation. On the other hand, kaolinite, with a highly compacted structure of 1:1–silica tetrahedral–alumina octahedral sheets, has no interlamellar sorbing area. The sorption is entirely spherical and increases with decreasing particle size. Illite falls in between montmorillonite and kaolinite in geological formations (Table 8.3).

Table 8.3. Characteristics of pertinent clay fractions found in the environment.

Name of the clay	Occurrence	CEC in meq 100 g^{-1}	Location of CEC	Surface area in m^2 g^{-1} by water vapor method	Area in nm^2 per exchange site
Kaolinite	podzols and tropical soils	3–15	External	16.6	0.387
Illite	many soil types	10–40	External	92.0	0.566
Montmorillonite	many soil types	95–150	20% external 80% in Interlamellar Spaces	750.0	1.545

Kaolinite has a 1:1 layer structure, whereas illite and montmorillonite have a 2:1 layer structure with interlamellar spaces. In montmorillonite, the spaces are expandable, whereas in illite they are not.
CEC = cation exchange capacity
Source: With permission from *Journal of Theoretical Biology*, vol. 66, 527–540, 1977, S. Ramamoorthy and G.G. Leppard, Fibrillar pectin and contact cation exchange at the root surface. © Academic Press Inc. (London) Ltd.

Sorption of 2,4-dichlorophenol followed the order bentonite, illite, and kaolinite, correlating with decreasing specific surface area (Aly and Faust, 1966). Although it was concluded that sorption to sedimentary clays in surface waters is not a significant pathway in the fate of chlorophenols, sorption to organic detritus may be important. Pentachlorophenol (PCP) has high log P (octanol/water) of 5.01 and low aqueous solubility of 0.014–0.08 g L^{-1} indicating favorable sorption onto organic-rich sediment and suspended solids. In fact, PCP sorbs to soils only in acidic conditions, and the amount sorbed correlates directly with the organic content of the soil. Choi and Aomine (1974a) concluded that (i) sorption of PCP to soils involves neutral molecules or anion and precipitation of molecules in the micelle and external liquid phase, and (ii) pH is the governing factor for PCP sorption to soil. Further study by the same authors (1974b) showed that the soil type determined the mode of PCP sorption, either as molecules or as anions. It must be noted here that the study was conducted in a hexane medium where PCP solubility was much higher and amenable to the analytical detection of changes in concentration. Hence, apparently inactive sorbents such as sediments in natural waters could sorb PCP substantially in a localized acidic regime created by anaerobic microbial processes. From the data available on PCP sorption to 14 different soils and 4 different clays, a positive correlation was shown between the PCP sorption coefficient and soil factors such as cation exchange capacity, heat of wetting by water, and heat of wetting by toluene (Kaufman, 1976). It was suggested that PCP sorption was based on the charge attraction between the OH^- group induced by polarization and surface charge of the soil particle. Also, PCP was more mobile in high pH soils and least mobile in acidic soils. Nitrophenols have relatively low log P values, suggesting only a weak affinity for organics or organic-coated material in water. However, 4-nitrophenol was irreversibly sorbed to montmorillonite clay (Saltzman and Yariv, 1975). The reported effective flocculating capacity of 2-nitrophenol for clays and solids in water rules out any uncertainty in similarity to the 4-isomer because of steric hindrance in the former. Other substituted phenols such as methyl phenols, because of their moderate partition coefficients, might be sorbed by organic detritus in natural waters. In general, sorption of organic compounds involves weak bonding (only a few Kcal/mole) or physical adsorption of monolayers leading to multiple layers. Rarely is a chemical bond involved in their interaction.

Volatilization

The low vapor pressure and high aqueous solubility of phenol indicate little or no volatilization from water. This is reflected in the phase distribution ratio of phenol between vapor and liquid at atmospheric pressure, 1.8 (Hakuta, 1975). Gaseous phenol sorbed onto montmorillonite clay was totally desorbed in a week (Saltzman and Yariv, 1975). Phenols such as 2-chlorophenol with a vapor pressure of 2.2 torr at 20°C will also have a tendency to volatilize from the water column, but their high aqueous solubility and high solvation resulting from their acidic nature will retard such loss. Hence, volatilization will not be a significant fate process for most chlorophenols in the aquatic environment.

Oxidation and Hydrolysis

Phenol undergoes oxidation in aqueous solution by molecular oxygen. This suggests a significant possibility of nonphotolytic oxidation in highly aerated waters. Higher chlorinated phenols will be resistant to oxidation under natural environmental conditions. Oxidation of 4-nitrophenol has been reported under laboratory conditions yielding a complex mixture, the principal products being hydroquinone and 1,4-benzoquinone (Gunther *et al.*, 1971). However, it is difficult to relate this reaction to natural surface waters.

Because of the high negative charge-density of the aromatic ring, covalent bond units are fairly resistant to hydrolysis. Extreme conditions such as the presence of concentrated alkali and temperatures in the range 130–200°C are needed to substitute a Cl atom with a OH group as in the formation of hexachlorobenzene to pentachlorophenol. Therefore, hydrolysis in natural waters will not be a significant fate process for phenols. However, nitrophenols sorbed to montmorillonite clay may undergo hydrolysis within the clay structure to yield benzoquinones (Saltzman and Yariv, 1975).

Photolysis

Although the λ_{max} of undissociated phenol is 270 nm, metal-coordinated phenol is anionic and has λ_{max} extending to 310 nm. Thus, phenol could undergo environmental photolysis either involving the phenolate anion or energy transfer through photosensitization of the undissociated molecule. Phenol forms a reddish, high molecular weight material on exposure to sunlight and air, reflecting the formation and photolysis of an oxygen-phenol charge-transfer complex. Experimental irradiation of phenol at 254 nm in the presence of oxygen yielded a phenoxy radical intermediate that subsequently gave substituted biphenyls, hydroquinone, and catechol (Joschek and Miller, 1966). Environmental photooxidative degradation of phenol into hydroquinone occurs under natural sunlight and commercial sun lamps. The possible loss due to direct oxidation or volatilization was not considered in these studies. Substitution, such as with methyl groups, increases the benzene rings' susceptibility to electrophilic attack and, in turn, photodegradation. Smog chamber studies have shown that *m*-xylene and toluene have photolytic half-lives of ~4 hours and ~12 hours, respectively, in metropolitan airsheds (Laity *et al.*, 1973). The effect of the hydroxyl group is greater than that of the methyl group. Hence phenol, on entry into the troposphere, will be degraded within a few hours. With increasing acidity of the compound (decreasing pK_a), the proportion of anion compared to the unionized molecule increases. This is environmentally significant since the anion absorbs well beyond 310 nm (sunlight spectrum) leading to photolytic reactions. For many substituted phenols with pK_a values in the range of 4–8, the absorption maximum will primarily be that of the corresponding anion. This includes di-, tri-, tetra- and pentachlorophenols, nitrophenol, and dinitrophenols. The photodegradation products, if not available, should be similar to those identified for analogous compounds. The conditions under which phenols photolyze and their breakdown products are listed in Table 8.4. Figure 8.1 illustrates the proposed photolytic pathways for PCP degradation in dilute aqueous solution of 100 mg L^{-1}, resulting in the formation of chlorinated phenols, tetrachlorodihydroxyl benzenes, and nonaromatic fragments like dichloromaleic acid.

Table 8.4. Photolysis of phenols and their breakdown products.

Compound	Light source	Experimental conditions	Products
2-Chlorophenol	313 nm	Aqueous alkali	2-hydroxy phenol
2-Chlorophenol	—	Aqueous solution (1000 mg L^{-1}) in presence of H_2O_2	Photodegraded to below odor threshold level (2 mg L^{-1})
4-Chlorophenol	254 and >290 nm	Aqueous alkali	Tarry material
	254 and >290 nm	Cyanide in aqueous solution	4-cyano phenol
2,4-Dichlorophenol	Solar radiation	10 days, aerated	Totally degraded to a black material, a polymer of 2-hydroxy benzoquinone
2,4-Dichlorophenol	>290 nm	In the presence of riboflavin (photosensitizer)	A mixture of tetra-chloro-phenoxy phenols and tetra-chlorodihydroxy biphenyls. no polymeric products
2,4,6-Trichlorophenol	—	In the presence of an electron acceptor	detected 2,6-dichlorophenoxyl semiquinone radical anion; semiquinones rapidly disproportionate to 2,6-dichlorobenzoquinone and 2,6-dichlorohydroquinone
Pentachlorophenol	Sunlight	2% solution of sodium salt of PCP (NaPCP)	50% degraded in 10 days; major products: chloranilic acid, yellow compound ($C_{12}HO_4Cl_7$) identified as 3,4,5-trichloro-6-(2′-hydroxy-3′,4′, 5′,6′-tetra chloro-

			phenoxy-)-O-benzoquinone, tetrachlororesorcinol, ($C_{12}HO_4Cl_7$) 2,5-dichloro-3 hydroxy-6-pentachloro-phenoxy-*p*-benzoquinone, an orange-red compound ($C_{12}H_2O_5Cl_6$)2,6-dichloro-3-hydroxy-5-(2′,4′,5′,6′-tetrachloro-3′-hydroxyphenoxy)-*p*-benzoquinone and a yellow compound ($C_{18}H_2O_6Cl_{10}$) 3,5-dichloro-4-(2,3,5,6-tetrachloro-4 hydroxy phenoxy)-6-(2,3,4,5-tetra chloro-6-hydroxyphenoxy)-O-benzoquinone
	Summer sunlight or UV light, $\lambda = 300$–450 nm	Aqueous solution (100 mg L^{-1}) irradiated for 7, 20, 30 days	After 7 days, chlorophenols, tetrachlorodihydroxy benzenes, and nonaromatic fragments such as dichloromaleic acid, after 20 days, hydroxylated trichlorobenzoquinones, trichlorodiols, dichloromaleic acid, Cl^- ions and CO_2
	Sunlight	Accidental release into a fresh-water lake	2,3,5,6- and 2,3,4,5-tetrachlorophenol
4-Nitrophenol	Sunlight	Aqueous solution of 200 mg L^{-1} for a period of 2 months	Totally degraded in 1–2 months; principal products hydroquinone, 4-nitrocatechol, and a dark, acidic polymer nitro group reduced to amino and azo group
Nitrofen, nitroaromatic pesticide	Sunlight	In aqueous solution containing 10% methanol	

Source: Versar (1979).

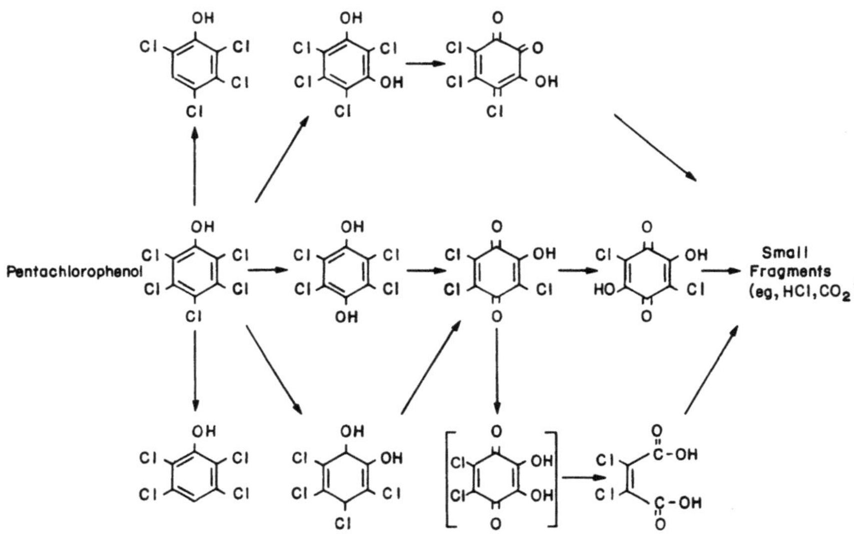

Figure 8.1. Proposed pathways for photolytic degradation of pentachlorophenol in aqueous solution (From Wong and Crosby, 1978.)

Biotransformations

Microbial Metabolism. Although chlorophenols have been used as antifungal or antimicrobial agents, they are known to be detoxified and degraded by microorganisms such as fungi. Several factors control the metabolism of chlorophenols.

(i) An aromatic ring with the halogen atom in the *meta* position is resistant to breakdown.

(ii) Fungi, including some woodrot fungi, may produce the enzymes tyrosinase and peroxidases that can degrade PCP. The stability of the benzene ring increases with increasing number of chlorine atoms. Inactivation of the hydroxyl group is the primary detoxifying mechanism.

(iii) The rate of breakdown of chlorophenols by microorganism depends on "adaptation time." *Ortho*- and *para*-chlorophenols at 1 mg L^{-1}, when added to domestic sewage, did not degrade up to 30 days at 20°C (Versar, 1979), but a similar concentration of the same CPs was degraded at the same temperature in polluted surface waters. It was concluded that only specialized microflora adapted to chlorophenols will degrade them. In a review of microbial metabolism of PCP, Reiner *et al.*, (1978) proposed a hypothetical pathway for the biodegradation of PCP by a bacterial culture (KC-3) (Figure 8.2).

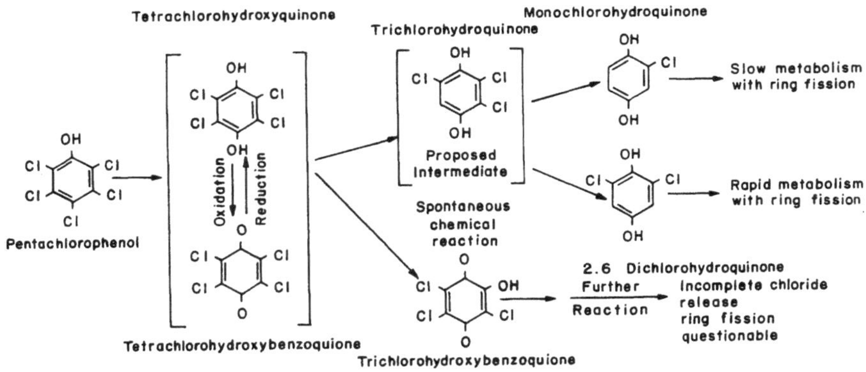

Figure 8.2. Proposed pathway for the microbial degradation of pentachlorophenol in a bacterial culture. (From Reiner *et al.*, 1978.)

Metabolism of Pentachlorophenol. PCP is susceptible to accumulation and metabolism by plants. Hague *et al.* (1978) reported a 3% accumulation of applied ^{14}C–PCP in rice plants. Half of that amount was removed by continuous extraction in 3 days whereas the remainder was tightly bound to plant material, not extractable by methanol. About 90% of the extractable part was the parent compound (PCP), ~9% unidentified conjugated products, and ~1% dechlorinated product. Degradation of PCP in soils is characterized by dechlorination and ring cleavage, yielding a variety of products (Figure 8.3).

Although several early studies on fish reported a lack of oxidation or conjugation systems for excretion of PCP metabolites, more recent investigations have revealed that several species have the capacity to form ether glucoronides and sulfates. Kobayashi and Akitake (1975) demonstrated that sulfate conjugation is one of the most general detoxification mechanisms for phenolic compounds in goldfish. Other studies have revealed the presence of both sulfate and glucoronide conjugation systems in fish, as in mammals, for the clearance of phenolic compounds. Although enzyme activity in the liver generally decreases with increasing Cl atoms, the activity being lowest for PCP, some marine organisms bioaccumulate lower chlorinated phenols. Figure 8.4 presents the schematic pathways for chlorophenols in fish. Both branchial and biliary excretion seem to be major routes compared with renal excretion.

PCP is rapidly metabolized in rats, resulting in rapid elimination from the liver, kidney, and blood. The dechlorination mediated by liver chromosal enzymes is enhanced by pretreatment with inducing agents such as phenobarbitol. Initially, hydrolytic dechlorination takes place, yielding tetrachloro-*p*-hydroquinone, which reductively dechlorinates further to

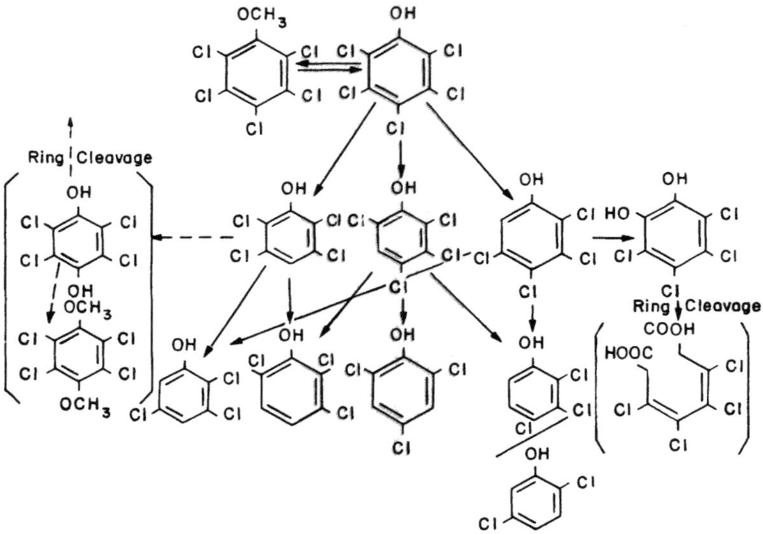

Figure 8.3. Proposed pathway for the degradation of pentachlorophenol in soil. (From Kaufmann, 1978.)

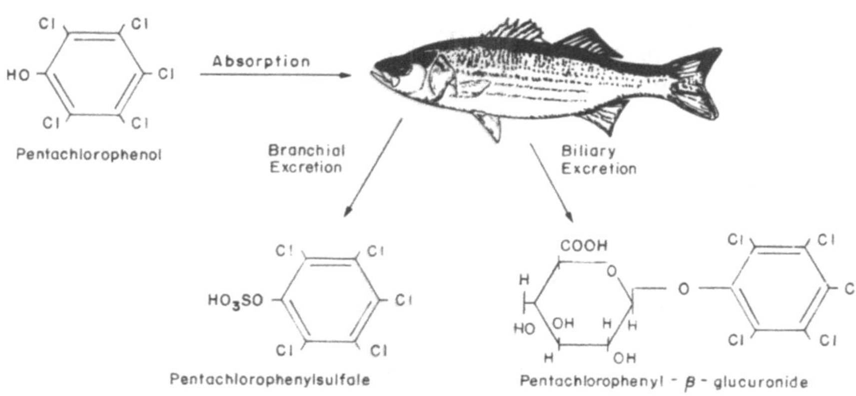

Figure 8.4. Metabolic pathways for pentachlorophenol in fish. (From Kobayashi, 1978.)

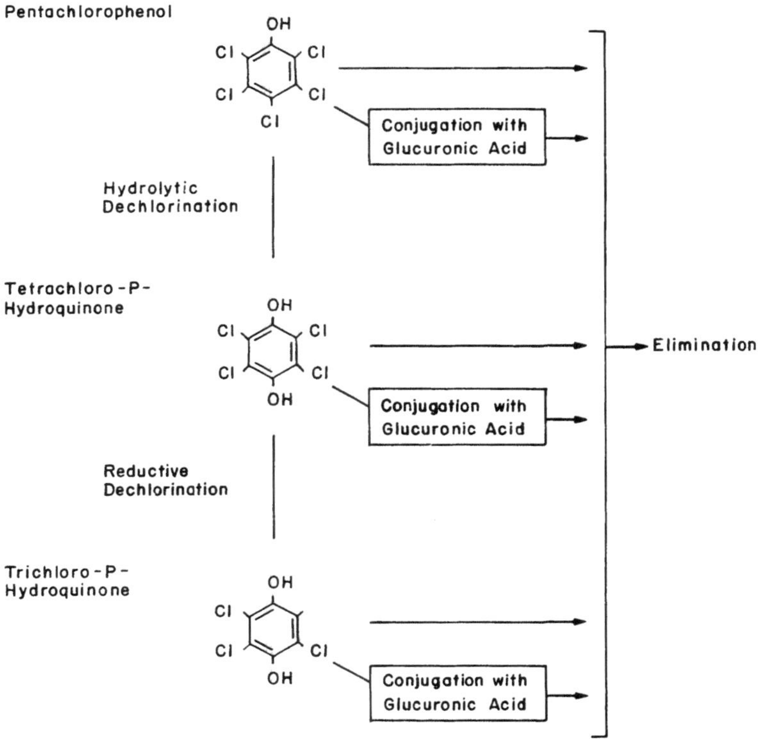

Figure 8.5. Pathways for metabolic degradation of pentachlorophenol in rats. (From Ahlborg, 1977.)

trichloro-*p*-hydroquinone. Figure 8.5 shows the metabolic pathways of PCP in rats.

Residues

Air and Water

Although relatively little is known about the concentration of phenols in the atmosphere, many of these compounds probably occur over cities and industrial areas and are subsequently precipitated into lakes and rivers. During 1979, a chemical spill occurred in Missouri as a result of a train derailment (U.S. Environmental Protection Agency, 1980). The concentration of 2-chlorophenol at the site reached a maximum of 1 mg m^{-3} on the day of the wreck, decreasing to <0.0015 mg m^{-3} within 2 months. Similarly,

Bevenue *et al.* (1972) reported pentachlorophenol residues of 0.002 to 0.27 $\mu g\ L^{-1}$ in the snow of Maunakea Summit. This resulted in pentachlorophenol levels of ~0.010 $\mu g\ L^{-1}$ in Lake Waiau, fed almost exclusively by the summit snows. Arsenault (1976) found that pentachlorophenol occurred at relatively high levels (max 0.3 mg m^{-3}) in the air of wood-preservation plants. Although such industries likely emit pentachlorophenol into the atmosphere, the overall significance of such discharges to aquatic systems is not known.

Some of the highest values of PCP in water (306–895 $\mu g\ L^{-1}$) have been reported for the Hayashida River (Japan), which receives uncontrolled input of leather tannery wastes (Yasuhara *et al.*, 1981). Elevated concentrations (10–100 $\mu g\ L^{-1}$) have also occurred in the Pawtuxet River (USA) owing to discharge from a chemical plan (Jungclaus *et al.*, 1978), whereas residues of 1–6 $\mu g\ L^{-1}$ were reported for the Delaware and Mississippi Rivers (Sheldon and Hites, 1978). A survey of raw water supplies in the USA showed that phenol was present in only 2 of 110 samples (U.S. Environmental Protection Agency, 1980).

Chlorinated phenols are widely distributed in surface waters and occur at highly variable concentrations, depending on waste source. For example, Piet and De Grunt (1975) reported that total monochlorophenol levels in rivers and coastal waters of the Netherlands ranged from 3 to 20 $\mu g\ L^{-1}$ compared with 0.01–1.5 and 0.003–0.1 $\mu g\ L^{-1}$ for dichlorophenols and trichlorophenols, respectively. Marked weekly and monthly variations in these levels were largely related to rainfall and water flow (Wegman and Hofstee, 1979). Jones (1981), having reviewed much of the Canadian literature, reported that the release of effluent from a pulp and paper mill resulted in average dichlorophenol and trichlorophenol levels of 4 and 13 $\mu g\ L^{-1}$, respectively, in the coastal waters of Lake Superior. In addition, water collected from stream mouths and inshore areas of Lake Ontario contained pentachlorophenol ranging from <0.005 to 1.40 $\mu g\ L^{-1}$ with transient levels of up to 23.0 $\mu g\ L^{-1}$ being recorded after periods of heavy rainfall.

Relatively little information is available on the presence or absence of nitrophenols, methylphenols, and alkyl phenols in water. One of the few available reports (Jungclaus *et al.*, 1978) indicated that the concentrations of 11 alkyl phenols in the Pawtuxet River (USA) ranged from <1 to 6 $\mu g\ L^{-1}$. A similar range in values was reported for 4 alkyl phenols in the Delaware River receiving both municipal and industrial wastes (Sheldon and Hites, 1978). Although nitrophenols were not found in either river, such compounds, especially trinitro-phenol, may be detected in rivers with input from munitions factories (U.S. Environmental Protection Agency, 1980). Methylphenols probably occur at detectable levels in systems receiving runoff from asphalt roadways and effluents from municipal/industrial outfalls. However, these compounds are rapidly biodegraded and therefore do not appear to be persistent or widespread contaminants of water.

Sediments

Sediments are a major sink for many phenols, and, in general, increasing substitution leads to an increase in their persistence in sediments. Eder and Weber (1980) reported that pentachlorophenol residues in the Weser Estuary (FRG) averaged 13 μg kg^{-1} dry weight compared with 0.3–1.5 μg kg^{-1} for dichloro- trichloro-, and tetrachlorophenol. Similarly, discharges from a wood preservation plant into the coastal waters of British Columbia (Canada) resulted in an average pentachlorophenol concentration of 65 μg kg^{-1} (Jones, 1981). These can be compared with averages of 96 and 26 μg kg^{-1} found for tetrachlorophenol and trichlorophenol, respectively, at the same location. The accidental release of pentachlorophenol into a small stream produced peak sediment levels of 1100–1300 μg kg^{-1} (Pierce *et al.*, 1977). This was followed by a decline to ~900 and 100 μg kg^{-1} within 2 and 4 months of the incident, respectively. Although sediment enrichment factors, relating concentrations in sediments to those in the overlying water, are variable, values generally range from 10 to 500 (Wegman and van den Broek, 1983).

Phenol, because of its biodegradability, is seldom detected in sediments. The same may be generally said about a number of other compounds such as the cresols, phenylphenol, and nonylphenol. By contrast, some of the alkylphenols occur at remarkably high concentrations, particularly near waste discharge sites. For example, Jungclaus *et al.* (1978) reported that two isomers of dibutylphenol reached 100–150 mg kg^{-1} in the Pawtuxet River, whereas the corresponding maximum values for tributylphenol and dibutylmethyl phenol were 25 and 60 mg kg^{-1}, respectively. Although comparable residues probably occur in the sediments of other industrial zone rivers, the environmental implications of such levels are not known at this time.

Aquatic Plants and Invertebrates

Uptake of most phenols by aquatic plants is initially rapid, followed by attainment of a steady state in residues. This may occur within 10–20 minutes of initial treatment, depending on compound, species, and exposure concentration (Figure 8.6). Uptake rate probably increases with temperature in most species but is inversely related to pH. Although data are not available for most compounds, Virtanen and Hattula (1982) reported concentration factors of 200–4500 for three species of algae exposed to 2,4,6-trichlorophenol for 21–36 days.

Invertebrates may sorb phenols directly from the water or through their food. Uptake from water is initially rapid and is again followed by a steady state in residues (Figure 8.7). The resulting concentration factors are highly variable, depending on species, compound, and test conditions. Schimmel

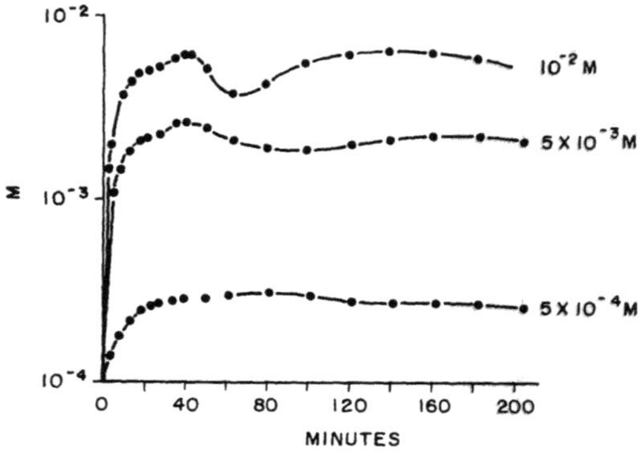

Figure 8.6. Resorcinol accumulation in *Nitella* cells exposed to different media concentrations. (From Stom *et al.*, 1980.)

et al. (1978) exposed oysters *Crassostrea virginica* to pentachlorophenol and found CFs of 41–78. Although these are comparable to the values reported for other species such as shrimp *Palaemonetes pugio*, substantially higher CFs (740–3020) were found when the snail *Lymnaea stagnalis* was exposed to 2,4,6-trichlorophenol (Virtanen and Hattula, 1982). Depuration appears to be moderately rapid in most species, regardless of compound. For example the half-lives of 3,5-diethylphenol, 3-nitrophenol, phenol, and 4-aminophenol in the crayfish *Astacus leptodactylus* were approximately 8, 20, 45, and <45 hours, respectively (Nagel and Urich, 1981). Similarly, the half-life of pentachlorophenol in oysters is <24 hours (Figure 8.7).

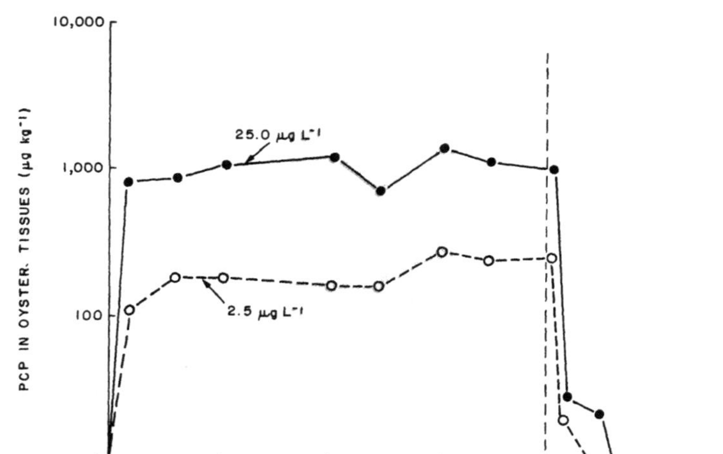

Figure 8.7. Uptake of pentachlorophenol by oysters *Crassostrea virginica* exposed for 28 days, then allowed to depurate in pentachlorophenol-free water. (From Schimmel *et al.*, 1978.)

Relatively little is known about the significance of food as a source of phenols. Some compounds, such as 2,4,6-trichlorophenol and tetrachloroguaiacol, can be concentrated through the food chain, whereas others, including pentachlorophenol and tetrachlorocatechol, decreased at the higher trophic levels. Although it might be expected that highly substituted compounds would concentrate through the food chain, no laboratory data are available to confirm this point.

Fish

Fish may contain a wide range of phenols in their tissues. Some of the more commonly detected compounds include chlorinated phenols, their methylated derivatives, and alkylphenols. Paasivirta *et al.* (1980) measured the concentrations of 12 polyhalogenated compounds in fish from Lake Paijanne (Finland), which has a number of pulp mills situated on its shore. Only six compounds were regularly detected, the most common of which were 2,4,6-trichlorophenol, 4,5,6-trichloroguaiacol, and tetrachloroguaiacol (Table 8.6). Similarly, discharge of pulpmill wastes into Atlantic coastal waters (Canada) resulted in the adulteration of fish tissues with highly chlorinated phenols, whereas fish inhabiting the Fraser River (Canada) were contaminated with phenols from wood preservation plants (Table 8.5). The methylated derivative of pentachlorophenol (pentachloroanisole) has been found in fish from the Detroit River (USA) and the fish from 15 of 26 rivers sampled near the Great Lakes (Kuehl, 1981). Lower chlorinated anisoles and pentachloroanisole have also been found in fish from the Arkansas River and in fish exposed to municipal wastes (Veith *et al.*, 1979).

Initial uptake of phenols from water is rapid and followed by a steady state in concentrations, usually within one day of exposure. The resulting concentration factors are highly variable and may exceed 15,000 in the case of phenol. This may be compared with the CF of 770 reported for pentachlorophenol, 1850 for 2,4,5-trichlorophenol, and 180 for 3-nitrophenol (Veith *et al.*, 1979). Although data are limited, a number of factors may influence the uptake and concentration of phenols in tissues, one of the most important being pH. Kobayshi and Kishino (1980), for example, found that exposure of goldfish to pentachlorophenol resulted in a decrease in CFs from 125 to 10 as pH increased from 5.5 to 9.0. This can be attributed to an increase in the concentration of undissociated pentachlorophenol at lower pHs. It is also probable that chelators reduce the rate of uptake of phenols to fish, though there are few data to confirm this point.

Rate of depuration from fish tissues is extremely variable. Phenol has a particularly short half-life (<1 hour), as does 3-nitrophenol (~1 hour). Several other compounds, such as 3,5-diethylphenol and 4-aminophenol, also have a short half-life (5–7 hours), but a number of the alkylphenols persist in tissues for extended periods of time. For example, McLeese *et al.*

Table 8.5. Average concentration ($\mu g\ kg^{-1}$ wet weight) of phenols in the muscle of fish collected from a lake, river and coastal water.

Lake Paijanne (Finland)	Pike	Roach
2,4,6-Trichlorophenol	15.5	30.3
2,3,4,6-Tetrachlorophenol	15.0	9.0
Pentachlorophenol	6.9	8.8
4,5,6-Trichloroguaiacol	12.4	27.6
Tetrachloroguaiacol	116.2	13.5
Tetrachlorocatechol	12.6	17.9

Fraser River (British Columbia)	Staghorn sculpin	Northern squawfish	Largescale sucker	Dolly varden
Tetrachlorophenol	30.9	3.5	<3	<3
Pentachlorophenol	59.9	16.8	15.6	34.9

Atlantic Coast (Canada)*	Tomcod	Smelt	Winter flounder	Alewife
2,4-Dichlorophenol	2900	5050	2070	ND
2,4,6-Trichlorophenol	1500	340	750	ND
Pentachlorophenol	3100	4800	5240	820

*$\mu g\ kg^{-1}$ lipid.
ND, not detected.
Sources: Paasivirta *et al.* (1980); Jones (1981).

(1981) reported that the half-life of *p*-dodecylphenol in Atlantic salmon was 690 days, compared with 4, 10, and 4 days for *p*-butylphenol and *p*-nonyphenol, respectively.

Toxicity

Aquatic Plants and Invertebrates

Phenols have a highly variable effect on aquatic plants (Table 8.6). Some compounds may actually stimulate growth at low concentrations, whereas others are consistently toxic. The LC_{50} for simple phenol, 2,4-dimethyl phenol, and 2-chlorophenol generally range from 10 to 500 mg L^{-1}. The 4-chlorophenol isomer is substantially more toxic, as are the more highly chlorinated derivatives of phenol. This trend of increasing toxicity with degree of substitution is also reflected in the effects of catechols (dihydroxybenzenes) on algae. Dence *et al.* (1980) reported that the growth of

Table 8.6. Acute toxicity (mg L^{-1}, 96-h LC_{50}) of phenols to aquatic algae, invertebrates, and fish.

Compound	Algae Fresh-water chlorophytes*	Invertebrates: Cladoceran *Daphnia magna*	Invertebrates: Shrimp *Mysidopsis bahia*	Fish: Bluegill sunfish	Fish: Fathead minnow	Fish: Mixed species
Phenol	10–30	36.4	—	16.4	36	4.2–44.5
2-Chlorophenol	500	4.4	—	8.2	12.3	8.1–58.0
4-Chlorophenol	4.8	4.4	29.7	3.8	—	3.8–14.0
2,4-Dichlorophenol	—	2.6	—	2.0	8.2	2.0–13.7
2,4,5-Trichlorophenol	1.2	2.7	3.8	0.4	—	0.4–0.9
2,4,6-Trichlorophenol	5.9	6.0	—	0.3	9.0	0.3–9.0
2,3,5,6-Tetrachlorophenol	2.7	0.6	21.9	0.2	—	—
2,3,4,6-Tetrachlorophenol	0.6	0.3	—	0.1	—	0.1–0.5
Pentachlorophenol	1.0–2.7	0.7–0.8	0.1–5.1	0.1	0.2	0.06–1.7
o-Cresol	—	16.0	—	21	13	12.6–23.2
p-Cresol	—	13.5	—	12	19	12–19
4-Chloro-6-methylphenol	92.6	0.3	0.1	2.3	0.03	—
2,4-Dichloro-6-methylphenol	—	0.4	—	1.6	—	—
2,4-Dimethylphenol	500	2.1	—	7.8	16.8	5–17
4-Nitrophenol	4.2	22	7.2	8.3	60.5	7.8–17
2,4-Dinitrophenol	9.2	4.1	4.9	0.6	16.7	0.3–17
2,4,6-Trinitrophenol	41.7	85	19.7	170	—	—
2,4-Dinitro-6-methylphenol	50	3.1	—	0.2	2.0	—

Sources: LeBlanc (1980); Buccafusco *et al.* (1981); U.S. Environmental Protection Agency (1980); Buikema *et al.* (1979); Ademan and Vink (1981).
**Selenastrum capricornutum* and *Chlorella pyrenoidsa*.
—, no data.

Chlorella pyrenoidosa was inhibited by simple catechol at a concentration of 20–40 mg L^{-1}, compared with 2–4 mg L^{-1} and 0.1 mg L^{-1} for 4,5-dichlorocatechol and trichlorocatechol, respectively.

Phenols probably inhibit algae through interference of the flow of electrons in oxidative phosphorylation by accepting electrons from reduced FAD. Hydroxylated phenols are also cofactors of peroxidases, whereas in *Nitella* spp, inhibition of cyclosis is probably due to phenolic binding of the sulfide and thiol groups (Buikema *et al.*, 1979). Huber *et al.* (1982) showed that exposure of *Lemna minor* to pentachlorophenol inhibited glutamate dehydrogenase activity but had no effect on alanine aminotransferase activity. Growth stimulation of algae may be due to bacterial decomposition of phenols followed by utilization of the degradation products by algae.

In general, phenols are moderately to slightly toxic to invertebrates (Table 8.6). As might be expected, simple phenol is among the least hazardous, but increasing substitution, particularly with chlorine, leads to greater toxicity. Other compounds that are moderately toxic to invertebrates include the chloro-methyl phenols and some alkyl phenols. McLeese *et al.* (1981) reported 96-h LC_{50}'s of 0.1–1.9 mg L^{-1} for shrimp *Crangon septimspinosa* exposed to hexylphenol, heptylphenol, nonylphenol and dodecylphenol, whereas the corresponding LC_{50}'s for butylphenol and pentylphenol ranged from 1.3 to 5.2 mg L^{-1}. The various guaiacol derivatives are generally slightly toxic, as are the veratroles (Dence *et al.*, 1980).

Symptoms of sublethal intoxication by phenols are numerous and may include a reduction in fecundity and survival of progeny. Under natural conditions, this can result in a change in the development and structure of invertebrate communities owing to the differential response of various species to phenols. Since algal metabolism is also influenced by phenols, there may be a reduction in oxygen levels in surface waters, leading to further changes in the density of sensitive species. Exposure of invertebrates to simple phenol generally causes an increase in oxygen consumption with phenol concentration. This is correlated with a corresponding increase in energy consumption that, under natural conditions, would lead to a decrease in growth. Although the toxic mechanism is not well understood, it is known that exposure to phenols produces hyperplasia of gill filaments that then interferes with oxygen uptake. There may also be a decrease in the number of red blood cells and increase in the number of cells with poikilocytosis, anisocytosis, and schizocytosis. Phenols have been implicated in neural dysfunction and decrease in the rate of synthesis of ATP in invertebrates.

Fish

Toxicity of phenols to fish generally increases with the degree of substitution on the carbon ring (Table 8.6). Thus, the LC_{50} of phenol is ~25 mg L^{-1} compared with ~1 mg L^{-1} for pentachlorophenol. Similarly, the corresponding levels for 3,5-dichlorocatechol and tetrachlorocatechol are 2.9

and 1.1 mg L^{-1}, respectively. In general, the parasubstituted compounds are more toxic than *ortho* and *meta* compounds. For example, the LC_{50}s of 4 (*para*)-chlorophenol and 2 (*ortho*)-chlorophenol are 3.8–14 and 8.1–58 mg L^{-1}, respectively. This reflects the relatively greater ease with which *para*-compounds can be transported across the gill membrane. Alkylphenols, such as *p*-hexylphenol, *p*-nonylphenol, and *p*-dodecylphenol, are also moderately toxic to fish, with LC_{50}s ranging from 0.13 to 0.19 mg L^{-1}.

Biochemical changes in fish exposed to phenols include (i) increased energy utilization and catabolism of fatty acids resulting in increased oxygen consumption, (ii) denaturation of enzymes such as succinic dehydrogenase, acid and alkaline phosphatases, fumarase, and cytochrome oxidase, and (iii) changes in blood chemistry. Acute exposure to phenols leads to structural damage to tissues such as necrosis of the pharynx, liver, and gills, hypertrophy of the spleen, and internal hemorrhages. Gill damage may consist of initial stripping of the epithelium from the secondary lamellae, leaving the naked pilla cells. Liver damage is characterized by formation of vacuoles, enlargement of nucleii of some cells, and deformation of the nuclear membrane of some cells. Webb and Brett (1973) reported that the growth and energy conversion efficiency of sockeye salmon exposed to sodium pentachlorophenol decreased at concentrations as low as 1.7 $\mu g\ L^{-1}$. The 96-h LC_{50} in these studies was 63 $\mu g\ L^{-1}$ and the swimming performance of salmon remained unaffected up to a concentration fo 47 $\mu g\ L^{-1}$.

The toxicity of polar compounds is inversely related to the pH of the media. This is due to the fact that, at low pH, the percentage of relatively toxic undissociated phenol increases. Kobayashi and Kishino (1980) demonstrated that the LC_{50} for goldfish exposed to pentachlorophenol ranged from 0.06 to 2.2 mg L^{-1} as pH increased from 6 to 9. Similarly, the toxicity of dinitrophenol to three fish species varied by a factor of 2–4 within a pH range of 6.0–8.8 (Dalela *et al.*, 1980). In most instances, the effect of pH on toxicity depends on the level of ionization of individual phenolic compounds. This is why pentachlorophenol (a highly polar compound) is strongly influenced by pH compared with weakly polar/neutral compounds such as dinitrophenol and chlorophenol. Relatively little is known about the effects of water hardness, salinity, and their interactions with pH on the toxicity of phenols. The data available to date are often contradictory, and, consequently, it is not possible to draw conclusions about the effect of hardness/salinity on phenol toxicity. By contrast, it is known that combinations of phenol, dinitrophenol, and pentachlorophenol generally interact synergistically in their effect on fish, whereas phenol, copper, zinc, and nickel show an additive effect.

Human Health

Instances of phenolic poisoning following the consumption of contaminated water or fish are extremely rare. In one case, a train derailment in Wisconsin resulted in the adulteration of well water with phenol (U.S. Environmental Protection Agency, 1980). Maximum residues of ~1100 mg L^{-1} were recorded, resulting in a daily exposure of 10–240 mg/person in those who used the well water for drinking. Although this resulted in an increase in the incidence of diarrhea and mouth sores, there was no significant effect on serum glutamic oxalacetic transaminase, bilirubin, creatine, uric acid, glucose, or cholesterol levels. There do not appear to be any reported cases of significant intoxication of humans following ingestion of fish or water contaminated with highly substituted phenols.

Simple phenol is probably not carcinogenic or teratogenic but apparently induces mutations in various bacteria and yeast (Sittig, 1980). Although the same may be said about most chlorinated phenols, some of these substances, notably trichlorophenol, may be contaminated with TCDD, a known teratogen. Methylated derivatives of phenol are both carcinogenic and mutagenic, whereas the majority of nitrophenols are mutagenic but not carcinogenic.

References

Ademan, D.M.M., and I.G.J. Vink. 1981. A comparative study of the toxicity of 1,1,2-trichloroethane, dieldrin, pentachlorophenol and 3,4 dichloroaniline for marine and fresh water organisms. *Chemosphere* **10**:533–554.

Ahlborg, U.G. 1977. *Metabolism of chlorophenols: studies on dechlorination in mammals*. Swedish Environment Protection Board PM 895, Stockholm.

Aly, O.M., and S.D. Faust. 1966. Studies on the fate of 2,4-D and ester derivatives in natural surface waters. *Journal of Agricultural Food Chemistry* **12**:541–546.

Arsenault, R.D. 1976. *Pentachlorophenol and contained chlorinated dibenzodioxins in the environment: a study of environmental fate, stability, and significance when used in wood preservation*. Presented at the American Wood-Preservers' Association Annual Meeting, Atlanta, Georgia, April 25–28. pp. 122–148.

Bevenue, A., J.N. Ogata, and J.W. Hylin. 1972. Organochlorine pesticides in rainwater, Oahu, Hawaii, 1971–1972. *Bulletin of Environmental Contamination and Toxicology* **8**:238–241.

Buccasfusco, R.J., S.J. Ells, and G.A. LeBlanc. 1981. Acute toxicity of priority pollutants to bluegill (*Lepomis marcochirus*). *Bulletin of Environmental Contamination and Toxicology* **26**:446–452.

Buikema, Jr., A.L., M.J. McGuinniss, and J. Cairns. 1979. Phenolics in aquatic ecosystems: a selected review of recent literature. *Marine Environmental Research* **2**:87–181.

Chang, C.W., and J.U. Anderson. 1968. Flocculation of clays and soils by organic compounds. *Soil Science Society of America Proceedings* **32**:23–27.

Choi, J., and S. Aomine. 1974a. Adsorption of pentachlorophenol by soils. *Soil Science and Plant Nutrition* **20**:135–144.

Choi, J., and S. Aomine. 1974b. Mechanisms of pentachlorophenol adsorption by soils. *Soil Science and Plant Nutrition* **20**:371–379.

Dalela, R.C., S. Rani, and S.R. Verma. 1980. Influence of pH on the toxicity of phenol and its two derivatives pentachlorophenol and dinitrophenol to some fresh water teleosts. *Acta Hydrochimica et Hydrobiologia* **8**:623–629.

Dence, C., C.J. Wang, and P. Durkin. 1980. *Toxicity reduction through chemical and biological modification of spent pulp bleaching liquors*. U.S. Environmental Protection Agency, Publication No. EPA-600/2-80-039 Cincinnati, Ohio.

Eder, G., and K. Weber. 1980. Chlorinated phenols in sediments and suspended matter of the Weser estuary. *Chemosphere* **9**:111–118.

Gunther, R., W.G. Filby, and K. Eiben. 1971. Hydroxylation of substituted phenols: an ESR study in the $Ti^3 +/H_2O_2$ system. *Teterahedron Lett.* 251–254.

Hague, A., I. Schenunert, and F. Korte. 1978. Isolation and identification of a metabolite of pentachlorophenol ^{-14}C in rice plants. *Chemosphere* **7**:65.

Hakuta, T. 1975. Vapor-liquid equilibriums of pollutants in sea water. II. Vapor-liquid equilibriums of phenolic substance water systems. *Nippon Kaisui Gakkai-Shi* **28**:379–385.

Huber, W., V. Schubert, and C. Sautter. 1982. Effects of pentachlorophenol on the metabolism of the aquatic macrophyte *Lemna minor L. Environmental Pollution (Series A)* **29**:215–223.

Jolley, R.L., G. Jones, W.W. Pitt, and J.W. Thompson. 1978. Chlorination of organics in cooling waters and process effluents. *In*: R.L. Jolley (Ed.), *Water chlorination: environmental impact and health effects*. Vol. **I**. Ann Arbor Science Publishers, Ann Arbor, Michigan, pp. 105–138.

Jones, P.A. 1981. *Chlorophenols and their impurities in the Canadian environment*. Environmental Protection Service, Department of the Environment, Ottawa, Canada, Catalogue No. En46-3/81-2, 434 pp.

Joschek, H.I., and S.T. Miller. 1966. Photooxidation of phenol, cresols, and dihydroxybenzenes. *Journal of the American Chemical Society* **88**:3273–3281.

Jungclaus, G.A., V. Lopez-Avila, and R.A. Hites. 1978. Organic compounds in an industrial wastewater: a case study of their environmental impact. *Environmental Science and Technology* **12**:88–96.

Kaufman, D.D. 1976. Phenols. *In*: P.C. Kearney and D.D. Kaufman (Eds.), *Herbicides: chemistry, degradation and mode of action*. Vol. **2**, Marcel Dekker, New York, pp. 665–707.

Kaufman, D.D. 1978. Degradation of pentachlorophenol in soil and by soil microorganisms. *In*: K.R. Rao (Ed.), *Pentachlorophenol*. Environmental Science Research Series, Vol. **12**. Plenum Press, New York, pp. 29–39.

Kobayashi, K. 1978. Metabolism of pentachlorophenol in fishes. *In*: K.R. Rao (Ed.), *Pentachlorophenol*. Environmental Science Research Series, Vol. **12**. Plenum Press, New York, pp. 89–105.

Kobayashi, K., and H. Akitake. 1975. Studies on the metabolism of chlorophenols in fish. I. Absorption and excretion of PCP by goldfish. *Bulletin of the Japanese Society of Scientific Fisheries* **41**:87–92.

Kobayshi, K., and T. Kishino. 1980. Effect of pH on the toxicity and accumulation of pentachlorophenol in goldfish. *Bulletin of the Japanese Society of Scientific Fisheries* **46**:167–170.

Kuehl, D.W. 1981. Unusual polyhalogenated chemical residues identified in fish tissue from the environment. *Chemosphere* **10**:231–242.

Laity, J.L., I.G. Burstain, and B.R. Appel. 1973. Photochemical smog and the atmospheric reactions of solvents. *In*: R.W. Tess (Ed.), *Solvents theory and*

practice. Advances in Chemistry Series 124, American Chemical Society, Washington, D.C., pp. 95–112.
LeBlanc, G.A. 1980. Acute toxicity of priority pollutants to water flea (*Daphnia magna*). *Bulletin of Environmental Contamination and Toxicology* **24**: 684–691.
Lopez-Avila, V., and R.A. Hites. 1980. Organic compounds in an industrial wastewater. Their transport into sediments. *Environmental Science and Technology* **14**:1382–1390.
McLeese, D.W., V. Zitko, D.B. Sergeant, L. Burridge, and C.D. Metcalfe. 1981. Lethality and accumulation of alkylphenols in aquatic fauna. *Chemosphere* **10**:723–730.
Nagel, R., and K. Urich. 1981. Elimination and distribution of different substituted phenols by frog (*Rana temporaria*) and crayfish (*Astacus leptodactylus*). *Bulletin of Environmental Contamination and Toxicology* **26**:289–294.
Paasivirta, J., J. Sarkka, T. Leskijarvi, and A. Roos. 1980. Transportation and enrichment of chlorinated phenolic compounds in different aquatic food chains. *Chemosphere* **9**:441–456.
Pierce, Jr., R.H. 1978. *Fate and impact of pentachlorophenol in a freshwater ecosystem*. United States Environmental Protection Agency Report, Publication No. EPA-600/3-78-063, Athens, Georgia, 61 pp.
Pierce, Jr., R.H., C.R. Brent, H.P. Williams, and S.G. Reeves. 1977. Pentachlorophenol distribution in a freshwater ecosystem. *Bulletin of Environmental Contamination and Toxicology* **18**:251–258.
Piet, G.J., and F. De Grunt. 1975. *Organic chloro compounds in surface and drinking water of the Netherlands*. European Colloquium, Eur. 5196, Luxembourg 1974. Commission of the European Communities, Luxembourg, pp. 81–92.
Ramamoorthy, S., and G.G. Leppard. 1977. Fibrillar pectin and contact cation exchange at the root surface. *Journal of Theoretical Biology* **66**:527–540.
Reed, H.W.B. 1980. Alkyl phenols. *In*: *Encyclopedia of chemical terminology*. Vol. **2**. Wiley, Toronto, pp. 72–96.
Reiner, E.A, J. Chu, and E.J. Kirsch. 1978. Microbial metabolism of pentachlorophenol. *In*: K.R. Rao (Ed.), *Pentachlorophenol*. Environmental Science Research Series, Vol. **12**. Plenum Press, New York, pp. 67–81.
Saltzman, S., and S. Yariv. 1975. Infrared study of the sorption of phenol and *p*-nitrophenol by montmorillonite. *Soil Science Society of America Proceedings* **39**:474–479.
Schimmel, S.C., J.M. Patrick, Jr., and L.F. Faas. 1978. Effects of sodium pentachlorophenate on several estuarine animals: toxicity, uptake, and depuration. *Environmental Science Research* **11**:147–156.
Sheldon, L.S., and R.A. Hites. 1978. Organic compounds in the Delaware River. *Environmental Science and Technology* **12**:1188–1194.
Sittig, M. 1980. Priority toxic pollutants. *Health impacts and allowable limits*. Noyes Data Corporation, New Jersey, 370 pp.
Statistics Canada. 1981. *Manufacturers of industrial chemicals*. Manufacturing and Primary Industries Division, Catalogue 46-219, Ministry of Supply and Services, Ottawa, Canada.
Stom, D.J. *et al.* 1980. Methods of analyzing quinones in water and their application in studying the effects of hydrophytes on phenols. Part 5. Elimination of

carcinogenic amines from solutions under the action of *Nitella*. *Acta Hydrochimica et Hydrobiologica* **8**:241–245.
Thurman, C. 1982. Phenols. *In*: *Encyclopedia of chemical terminology*. Vol. **7**. Wiley, Toronto, pp. 373–384.
U.S. Environmental Protection Agency. 1980. *Ambient water quality criteria reports*. Office of Water Regulations and Standards, Washington, D.C.
U.S. International Trade Commission. 1975–1982. *Synthetic organic chemicals, U.S. production and sales*. U.S. Government Printing Office, Washington, D.C.
Veith, G.D., D.W. Kuehel, E.N. Leonard, F.A. Puglisi, and A.E. Lemke. 1979. Polychlorinated biphenyls and other organic chemical residues in fish from major watersheds of the United States. 1976. *Pesticides Monitoring Journal* **13**:1–11.
Versar. 1979. Water-related environmental fate of 129 priority pollutants. Vol. **II**. U.S. Environmental Protection Agency, Publication No. EPA-440/4-79-029b, Washington, D.C.
Virtanen, M.T., and M.L. Hattula. 1982. The fate of 2,4,6-trichlorophenol in an aquatic continuous-flow system. *Chemosphere* **11**:641–649.
Waggott, A., and A.B. Wheatland. 1978. Contribution of different sources to contamination of surface waters with specific persistent organic pollutants. *In*: O. Hutzinger, L.H. van Lelyveld, and B.C.J. Zoeteman (Eds.), *Aquatic pollutants: transformation and biological effects*. Proceedings of the Second International Symposium on Aquatic Pollutants, Amsterdam, The Netherlands, September 26–28, 1977, pp. 141–168.
Webb, P.W., and J.R. Brett. 1973. Effects of sublethal concentrations of sodium pentachlorophenate on growth rate, food conversion efficiency, and swimming performance in underyearling sockeye salmon (*Oncorhynchus nerka*). *Journal of the Fisheries Research Board of Canada* **30**:499–507.
Webb, R.G., *et al*. 1973. *Current practice in GC/MS analysis of organics in water*. Environmental Protection Technology Series EPA R2-73-277.
Wegman, R.C.C., and A.W.M. Hofstee. 1979. Chlorophenols in surface waters of the Netherlands (1976–1977). *Water Research* **13**:651–657.
Wegman, R.C.C., and H.H. van den Broek. 1983. Chlorophenols in river sediment in The Netherlands. *Water Research* **17**:227–230.
Wise, Jr., H.E., and P.D. Fahrenthold. 1981. Predicting priority pollutants from petrochemical processes. *Environmental Science and Technology* **15**: 1292–1304.
Wong, A.S., and D.C. Crosby. 1978. Photolysis of pentachlorophenol in water. *In*: K.R. Rao (Ed.), *Pentachlorophenol*. Environmental Science Research Series, Vol. **12**. Plenum Press, New York, 19–25p.
Yasuhara, A., H. Shiraishi, M. Tsujl, and T. Okuno. 1981. Analysis of organic substances in highly polluted river water by mass spectrometry. *Environmental Science and Technology* **15**:570–573.

9
Polychlorinated Biphenyls

Polychlorinated biphenyls (PCBs) are a group of chlorinated aromatic compounds with widespread applications. PCBs have been prepared synthetically since 1929 and most information on their manufacture under trade names and general characteristics are available from trade publications. The Monsanto Chemical Company, the sole manufacturer in the USA, has provided most information on their preparation and properties (Monsanto, 1972, 1974).

PCBs are produced synthetically by chlorinating the biphenyl (Figure 9.1) with anhydrous chlorine in the presence of iron filings or ferric hydroxide as a catalyst. The crude product is distilled with alkali to remove color, traces of HCl, and the catalyst, resulting in a mixture of chlorobiphenyls with a varying number of chlorine atoms per molecule. Chlorobiphenyls could also be synthesized from (i) arylation of aroyl peroxides, diazonium salts, phenyl hydrazines, and other aryl compounds, (ii) aryl condensation reactions, (iii) addition of chlorine to biphenyl systems, (iv) decarboxylation, and (v) dechlorination reactions.

All Aroclor products (Aroclor is the trade name of Monsanto's PCB) are characterized by a four-digit number. The first two digits stand for the type of molecule: 12 refers to chlorinated biphenyls and 54 to chlorinated terphenyls, though some 54 Aroclors could be a mixture of ter- and biphenyls. Aroclors 25- and 44- are mixtures of PCBs and polychlorinated terphenyls (75% and 60% PCB, respectively). The last two digits of the formula represent the percent weight of chlorine (Table 9.1).

Theoretically, 209 different chlorobiphenyls can exist based on the possible distribution of Cl atoms in the two rings of the biphenyl. However,

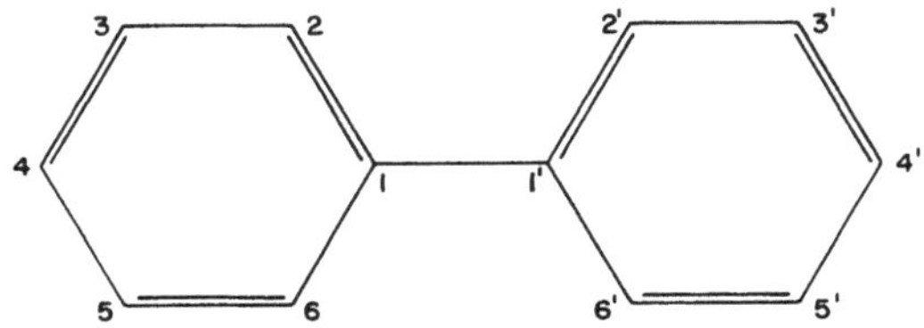

Figure 9.1. Numbering of the biphenyl rings.

mechanistic and statistical factors suggest that not all of these combinations exist. The amount of chlorine in chlorinated biphenyls ranges from ~18% to 71% (Table 9.2). As might be expected, Aroclor 1016 contains a predominance of Cl_2–Cl_4 biphenyls while Aroclor 1254 mainly contains Cl_4–Cl_6 biphenyls (Table 9.2).

Polychlorinated dibenzofurans (PCDFs) are the major identified toxic contaminants in Aroclors and Japanese Kaneclors, whereas European PCB products contain PCDFs and heptachloronaphthalenes as contaminants (Vos *et al.*, 1970). Laboratory studies have indicated the possibility of photochemical formations of CDFs and PCDFs as secondary products in commercial PCB mixtures (Roberts *et al.*, 1978). However, the effect of industrial use and environmental aging upon the concentration of PCDFs in commercial PCBs has not been studied.

Production, Uses, and Discharges

Total US production rose gradually during the 1960s, peaked at 38,600 metric tons in 1970, and decreased sharply in 1971 (Table 9.3). Exports followed a similar trend, reaching 6200 metric tons in 1970. Monsanto has

Table 9.1. Percent chlorine content of Aroclor formulations.

Aroclor	Percent Cl.	Average number of Cl/Aroclor molecule	Average mol. wt.
1221	20.5–21.5	1.15	192
1232	31.5–32.5	2.04	221
1242	42	3.10	261
1248	48	3.90	288
1254	54	4.96	327
1260	60	6.30	372
1262	61.5–62.5	6.80	389
1268	68	8.70	453

Source: Reprinted with permission from *The Chemistry of PCBs*, S. Hutzinger, S. Safe, and V. Zitko, 1974, 269 pp. CRC Press, Inc., Boca Raton, Florida.

Table 9.2. PCB component composition of some Aroclors.

PCB component	Mol. wt.	Percent chlorine	Presence (%) in Aroclor formulations				
			1016	1221	1242	1248	1254
$C_{12}H_{10}$	154.21	0	<0.1	11	<0.1	—	<0.1
$C_{12}H_9Cl$	188.65	18.79	1	51	1–3	—	<0.1
$C_{12}H_8Cl_2$	223.10	31.77	20	32	13–16	2	0.5
$C_{12}H_7Cl_3$	257.54	41.30	57	4	28–49	18	1
$C_{12}H_6Cl_4$	291.99	48.56	21	2	25–30	40	11–21
$C_{12}H_5Cl_5$	326.43	54.30	1	<0.5	8–22	36	48–49
$C_{12}H_4Cl_6$	360.88	58.93	<0.1	ND	1–4	4	23–34
$C_{12}H_3Cl_7$	395.32	62.77	ND	ND	<0.1	—	6
$C_{12}H_2Cl_8$	429.77	65.98	ND	ND	ND	—	ND
$C_{12}CHl_9$	464.21	68.73	—	—	—	—	—
$C_{12}Cl_{10}$	498.66	71.18	—	—	—	—	—

Sources: Reprinted with permission from *The Chemistry of PCBs*, S. Hutzinger, S. Safe, and V. Zitko, 1974, 269 pp. © CRC Press, Inc., Boca Raton, Florida.

been reluctant to disclose data for the remainder of the 1970s, but all US production has now ceased. This was a voluntary restriction, reflecting public concern for increasing residues in the environment. Although Japanese annual production peaked at ~11,800 metric tons, PCBs were apparently never manufactured in Canada. Little information is available on the production figures from other countries.

Uses

Owing to their chemical and thermal stability, inertness, and excellent dielectric properties, PCBs have found widespread industrial and commercial applications over the last 40 years. They are used as dielectric fluids

Table 9.3. Total US production and exports of PCB (metric tons).

Year	Production	Exports	Year	Production	Exports
1960	18,850	1700	1966	29,900	3100
1961	18,450	1900	1967	34,200	3700
1962	19,050	1650	1968	37,600	5100
1963	20,300	1650	1969	34,600	4800
1964	23,100	1900	1970	38,600	6200
1965	27,400	1900	1971	18,400	4500

Source: Reprinted with permission from *Chemical Engineering News,* vol. 49, page 15, Toxic Substances; Monsanto releases PCB data. © 1971 American Chemical Society.

in transformers and capacitors, as plasticizers in paints, plastics, sealants, resins, inks, printing, copy paper, and adhesives, and as components of hydraulic fluids in gas turbines and vacuum pumps. Highly chlorinated Aroclors mixed with chlordane, aldrin, and dieldrin may extend the kill-life of these compounds. It was reported that PCB alone had low toxicity to house flies but had a synergistic influence on the toxicity of dieldrin and DDT (Lichtenstein *et al.*, 1969). The effectiveness of these compounds decreased as the chlorinated level increased. For example, Aroclor 1221 increased the mortality of fruit flies from 59 to 93%, whereas Aroclor 1268 increased it to only 77% (Peakall and Lincer, 1970). Although some PCBs possess insecticidal and fungistatic activities, they were apparently never used as pesticides.

Estimates of the use of PCB for various industrial applications in the USA have been derived by Peakall and Lincer (1970) from the domestic sales figures released by Monsanto for the period 1957–1972 (Table 9.4). In 1960, about 11,300 metric tons of Aroclor were used in transformers and capacitors and about 29% of the total on other uses as plasticizers and in hydraulic fluids. In 1970, the latter fraction increased to about 45%. After voluntary restrictions by Monsanto, this figure dropped to 23% and total PCB sales dropped to 18.4 metric tons (Table 9.3).

Discharges

Although PCBs constitute a small fraction of the hazardous chemical load in the environment, they receive special mention in the U.S. Toxic Substances Control Act (TSCA) of 1977. This reflects their environmental behavior, which in turn forced discontinuance of their use in many areas. Between 1929 and 1977, about 6.36×10^8 metric tons of PCBs were produced in the USA; about 3.41×10^8 metric tons are still in use, of which about 22% (7.4×10^7 metric tons) are in the electric utility industry (Miller, 1982). For this reason, EPA continues to regulate electrical utilities containing PCB concentrations (i) over 500 mg L^{-1} and (ii) 50–500 mg L^{-1}. Concentrations <50 mg L^{-1} are not regulated. Approximately 10% of the electrical utility transformers are contaminated with PCB in the 50–500 mg L^{-1} range. The utility is responsible for locating the faulty, contaminated transformers, which is time-consuming and expensive. This includes frequent testing of transformer oil for PCB content and eventual draining and disposal or decontamination. The utilities are currently investigating methods for better and cost-beneficial reduction to <50 mg L^{-1} while saving the oil, which could range from 200–600 gallons for each transformer. There are about 20 million transformers in the USA (Miller, 1982). Many are located in high-rise buildings, originally installed for fire safety reasons. There are no methods currently available to forewarn of an incipient transformer failure.

Table 9.4. Some suggested uses of PCBs.

Base material	Type of Aroclor used and its percent content	End use
Polyvinyl chloride	1248, 1254, 1260 (7–8%)	Secondary plasticizers to increase flame retardance and chemical resistance
Polyvinyl acetate	1221, 1232, 1242 (11%)	Improved quick-track and fiber-tear properties
Polyester resins	1260 (10–20%)	Stronger fiberglass; reinforced resins and economical fire retardants
Polystyrene	1221 (2%)	Plasticizer
Epoxy resins	1221, 1248 (20%)	Increased resistance to oxidation and chemical attack; better adhesive properties
Styrene-butadiene co-polymer	1254 (8%)	Better chemical resistance
Neoprene	1268 (40%) (1.5%)	Fire retardant injection moldings
Crepe rubber	1262 (5–50%)	Plasticizer in paints
Nitrocellulose lacquers	1262 (7%)	Co-plasticizer
Ethylene vinyl acetate	1254 (41%)	Pressure-sensitive adhesives
Chlorinated rubber	1254 (5–10%)	Enhances resistance, flame retardance, electrical insulation properties
Varnish	1260 (25% of oil)	Improved water and alkali resistance
Wax	1262 (5%)	Improved moisture and flame resistance

Source: Monsanto Technical Bulletin O/PL 306.

An estimated 2.8 million capacitors are in use in the USA. About 2000 of them rupture every year, causing spillage into the environment. As in the case of transformers, there are no tools to predict capacitor failure. Soil samples have to be tested for contamination, and the solid PCB-containing material must be shredded prior to incineration. The utility industry considers it to be cost-ineffective to recover the part/parts of the contaminated capacitors. Portable screening devices such as x-ray fluorescence meters for transformer oil, acoustical detectors to detect the ultrasonic sound from a faltering capacitor, and an infrared scanner to measure the temperature-rise of the faltering capacitor have been studied in the field (Miller, 1982).

Behavior in Natual Waters

Aroclors possess properties different from their individual chlorobiphenyl compounds. For example, the individual compounds are solids at room temperature, whereas the Aroclors are either mobile oils (Aroclor 1221, 1232, 1242, and 1248), viscous liquids (Aroclor 1254), or sticky resins (Aroclor 1260 and 1262). PCBs are sparingly soluble in water, and the solubility decreases with increasing chlorine content (Table 9.5). Consequently, the aqueous solubilities of some Aroclors are also variable as follows: 1242, 200 $\mu g\ L^{-1}$; 1248, 100 $\mu g\ L^{-1}$; 1254, 40 $\mu g\ L^{-1}$; and 1260, 25 $\mu g\ L^{-1}$. Bias in these results owing to selective solubilization of lower chlorinated biphenyls (LCBPs) should be taken into account in considering the environmental behavior of specific PCBs and their toxicity to organisms. The relative pattern of partitioning into tissues often reflects the differential water solubilities of the compounds rather than preferential uptake.

Table 9.5. Aqueous solubility of chlorobiphenyls (mg L^{-1}).

Compound	Solubility	Compound	Solubility
Monochlorobiphenyls		Pentachlorobiphenyls	
2-	5.9	2,2′,3,4,5′-	0.022
3-	3.5	2,2′,4,5,5′-	0.031
4-	1.19	Hexachlorobiphenyl	
Dichlorobiphenyls		2,2′,4,4′,5,5′-	0.0088
2,4-	1.40	Octachlorobiphenyl	
2,2′-	1.50	2,2′,3,3′,4,4′,5,5′-	0.0070
2,4′-	1.88	Decachlorobiphenyl	0.015
4,4′-	0.08	4,4′-Dichlorophenyl	
		in Tween 80*	
Trichlorobiphenyls		(0.1% aqueous solution)	5.9
2,4,4′-	0.085	in Tween 80*	
		(0.1% aqueous solution)	≥10.0
2′,3,4-	0.078	in humic acid extract	0.07
Tetrachlorobiphenyls			
2,2′,5,5′-	0.046		
2,2′,3,3′-	0.034		
2,2′,3,5′-	0.170		
2,2′,4,4′-	0.068		
2,3′,4,4′-	0.058		
2,3′,4′,5-	0.041		
3,3′,4,4′-	0.175		

*Surfactant.

Source: Reprinted with permission from *The Chemistry of PCBs*, S. Hutzinger, S. Safe, and V. Zitko, 1974, 269 pp. © CRC Press, Inc., Boca Raton, Florida.

Sorption

PCBs are often associated with bottom sediments in aquatic systems and are transported over a considerable distance from the discharge source. Bottom sediments of the Housatonic River in Connecticut along a 20-km section contained Aroclors 1248, 1254, and 1260 in the ratio 1:0.6:1.9 (Sawhney *et al.*, 1981). Since residues were highly correlated with clay and organic matter, these clay-organic complexes were largely responsible for retention and transport of PCBs. Similarly, the primary sedimentation process of sewage treatment removes 50–70% of the PCB load from sewage (McIntyre *et al.*, 1981a; Shannon *et al.*, 1976). At present, the guidelines for sludge disposal in some countries include only heavy metals without any concern for organic substances. Disposal of sludges on land and sea may contaminate and accumulate PCB in plants and marine fauna.

Chemical conditioning and pressure filtration of sewage sludges prior to disposal result in high recoveries of solids as sludge cakes. The treatment does not seem to interfere in the analysis of PCBs, and about 83% of Aroclor 1260 of the digested sludge was found in the sludge cake (McIntyre *et al.*, 1981b). However, any residual polyelectrolyte used as a conditioner may increase the water solubility of nonpolar compounds such as PCB and organo-chlorine insecticides. This must be considered in determining the behavior and fate of these chemicals subsequent to disposal on land.

Volatilization

Vapor pressures and vaporization rates significantly influence the behavior of PCBs in water. As might be expected, LCPB (lower chlorinated biphenyl) components of Aroclors volatilize at a relatively rapid rate and may introduce bias into vaporization data. It is clear that interfacial reactions such as sorption onto suspended solids and subsequent sedimentation and volatilization of otherwise easily biodegradable LCBP from the water column will dominate in the environmental dynamics of PCBs in natural waters. This, combined with little chemical breakdown owing to low residence time in water as a result of poor solubility, have resulted in accumulation in other phases such as biota and sediment.

Breakdown studies of Aroclor 1260 in water samples collected from the field showed only evaporational losses and no metabolic activity for 12 weeks (Oloffs *et al.*, 1972). The evaporational loss at concentrations less than water solubility was due to water evaporation carrying the PCB with it. At concentrations > water solubility, the rate of loss was dependent upon the distribution of undissolved PCB in the water column and its vapor pressure. In the absence of solubility data, it was assumed that Aroclor 1260 accumulated at the water-air interface, leading to evaporation. Absence of any change in the chromatograms over the 12 weeks suggested that no metabolism occurred.

Chemical Transformations

Owing to their chemical inertness, PCBs are resistant to most chemical degradation reactions in the environment except photochemical breakdown. First experimental evidence of this process was obtained in 1971 when a hexachlorobiphenyl photodegraded at 310 nm to yield products formed by dechlorination, molecular rearrangement, and condensation reactions (Safe and Hutzinger, 1971). This study was followed by several other investigations showing the photochemical reactivity of pure chlorobiphenyls as well as the commercial PCB formulations under laboratory ultraviolet irradiation sources (>290 nm) and natural sunlight. Andersson *et al.* (1973) studied the photolysis of hexachlorobiphenyl in methanol and identified a series of 0-methoxy PCBs and methoxy-substituted chlorodibenzofurans among the 80 compounds formed (Figure 9.2). They showed that extrapolation of some earlier results to the natural environment was not realistic since the energy sources used (254 nm) were higher than that encountered in solar radiation (290–450 nm). Secondly, high concentrations of organic solvents were employed under laboratory conditions, quite unnatural to environmental conditions. In aqueous suspension, 2,5-dichlorobiphenyl and 2,5,2′,5′-tetrachlorobiphenyls yielded low quantities (~0.2%) of 2-chlorodibenzofuran. In addition, the higher molecular weight polychlorinated dibenzofurans could be the primary photochemical product, undergoing photoreduction in water.

Ultraviolet sources of >290 nm are being increasingly used to simulate natural sunlight. Irradiations have been conducted in gas, solution, and solid phases to create results comparable to natural environmental conditions. Hustert and Korte (1972) showed that gas-phase irradiations did not affect a hexachlorobiphenyl compound but formed polar products from a tetrachlorobiphenyl and water. Irradiation in the gas phase is especially relevant to LCBPs that are relatively volatile.

The biphenyl molecule has two absorption maxima; the main band at $\lambda_{max} = 202$ nm ($\varepsilon = 44{,}000$) and the second band, called the k band, at $\lambda_{max} = 242$ nm ($\varepsilon = 17{,}000$). Chlorine substitution in a biphenyl molecule

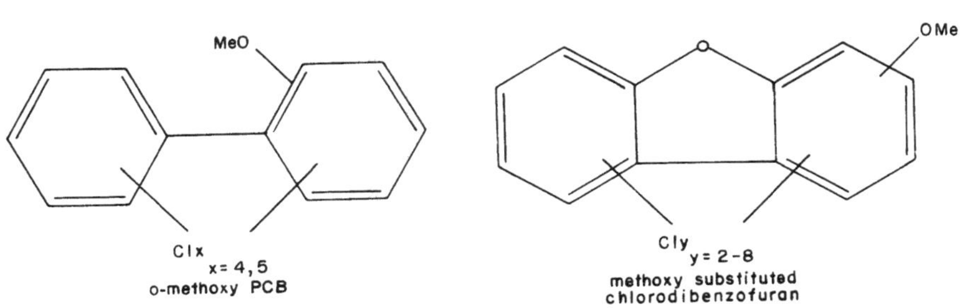

Figure 9.2. Chemical structure of methoxy-substituted PCB and CDF compounds.

has a marked bathochromic shift on the k band with a greater shift occurring for the *para* substituent (~13 nm) than for the *meta* substituent (~6 nm). For the more highly substituted chlorinated biphenyl isomers, both the main and k bands are shifted toward the visible region with increasing Cl substitution. Thus, almost all chlorobiphenyls absorb in the region >280 nm; compounds with <2 Cl atoms at *ortho* positions absorb 240–250 nm and compounds with >2 Cl atoms at >270 nm. Higher chlorinated biphenyls photodegrade faster than LCBPs. The mechanism for the photochemical dechlorination of 4,4′-dichlorophenyl is illustrated in Figure 9.3. The following figure (9.4) gives the photodegradation products of chlorinated biphenyls in isopropanol.

Metabolic Transformations

The metabolism of PCB in animal tissues is characterized by the disappearance or reduction in the concentration of LCBPs. This has been observed in rats, birds, cows, and carp. Fish seem to have relatively low PCB-metabolizing activity. In fact, prolonged dietary exposure with Aroclor 1254 (>200 days) and additional control feed for >200 days was required for the disappearance of the first two peaks in Atlantic salmon (Hutzinger *et al.*, 1974). Actively metabolizing rats consistently retain a greater proportion of HCBPs of the standard Aroclor 1254. Predamaging the rat liver with CCl_4 failed to produce any changes in dose PCB composition,

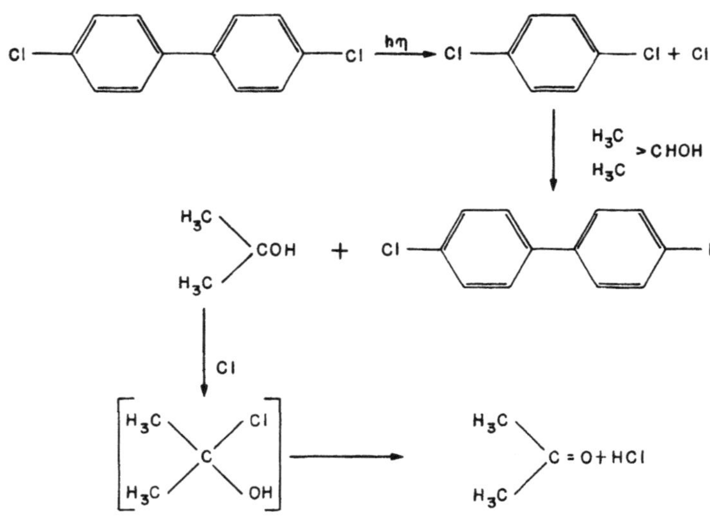

Figure 9.3. Photochemical degradation of 4,4′-DCB. (Reprinted with permission from *The Chemistry of PCBs*, S. Hutzinger, S. Safe, and V. Zitko, 1974. 269 pp. © CRC Press, Inc., Boca Raton, Florida.)

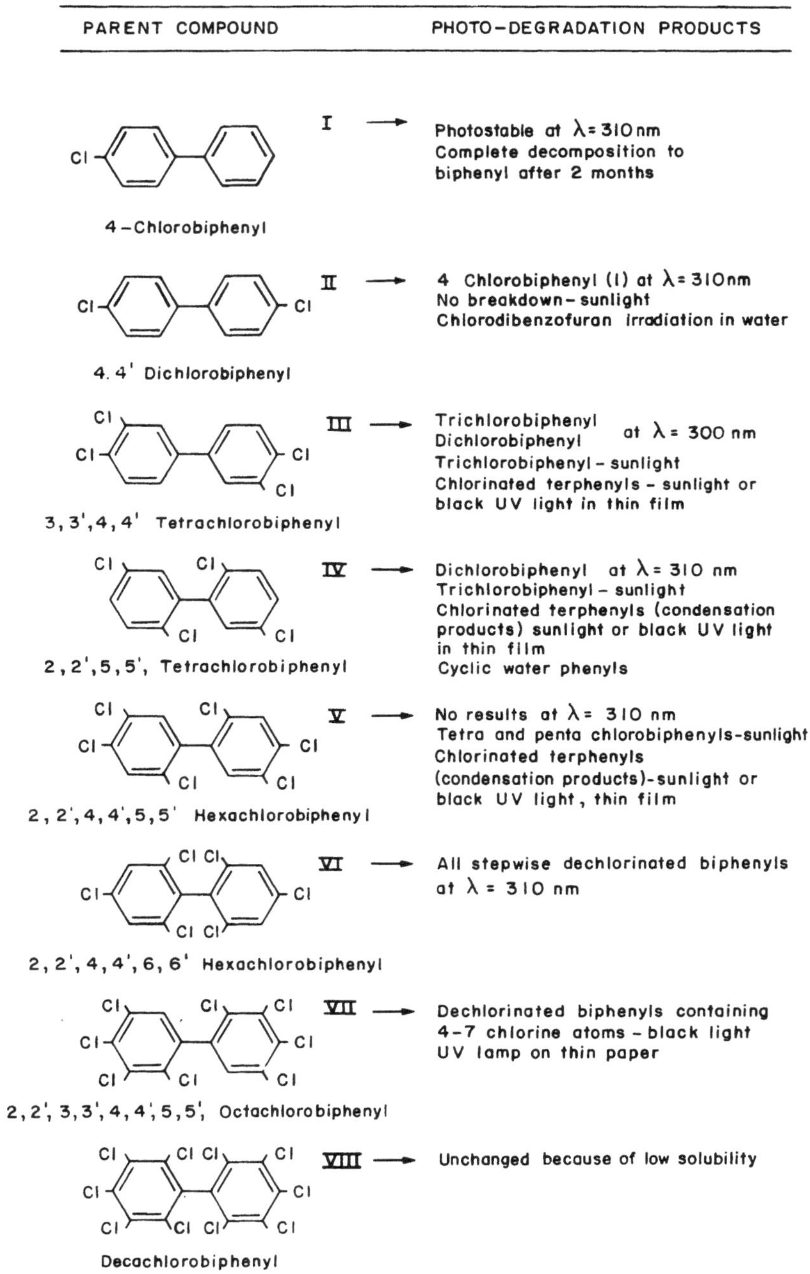

Figure 9.4. Photochemical degradation products of PCB. (Reprinted with permission from *The Chemistry of PCBs*, S. Hutzinger, S. Safe, and V. Zitko, 1974, 269 pp. © CRC Press, Inc., Boca Raton, Florida.)

thus confirming preferential metabolism of PCB components with liver (or comparable organ) as the primary active site.

PCB analyses in environmental samples reveal an isomer pattern similar to that of Aroclor 1254 or 1254/1260 (Task Force on PCB, 1976), although North American sales figures indicate the primary input of PCB is LCBP, particularly Aroclor 1242. This indicates the LCBPs are less persistent in the environment than HCBPs, and many recent studies confirm this observation. Reports from the Monsanto Company showed the order of bacterial degradation in an activated sludge test unit to be biphenyl>Aroclor 1221>Aroclor 1016>Aroclor 1254. It was evident from these studies that mono- to tetrachloro-isomers could be degraded.

A considerable variety of biota are capable of metabolizing LCBP up to 6 Cl atoms into polar metabolites. Conjugated and free forms of PCB metabolites have been identified in feces and urine of animals. Present evidence indicates that hydroxylation with the formation of an arene oxide intermediary is the primary metabolic mechanism for the breakdown of PCBs. However, direct hydroxylation of PCBs yielding a single monohydroxylated product without the arene oxide intermediary has also been reported (Gardner *et al.*, 1976). Studies on the metabolism of a related compound (biphenyl) by liver microsomal preparations from 11 species of animals showed hydroxylation yielding 2- and 4-hydroxy biphenyls. Species differences with different enzyme systems might account for the variation in results.

Accumulation by wild guillemot of certain PCBs with 4,4′-chlorine substitutions was attributed to the bird's inability to hydroxylate the PCBs with 4,4′ positions occupied by Cl atoms (Jansson *et al.*, 1975). Studies on species influence on the metabolism of specific chlorobiphenyls both *in vitro* and *in vivo* would be of considerable value in understanding the differences in PCB breakdown and accumulation.

Isomerization and dechlorination reactions have been implicated in the metabolism of HCBPs (McKinney, 1976; Hutzinger *et al.*, 1974). However, identification of potentially toxic dibenzofuran structures in some metabolites, and lower ratios of PCBs to polychlorinated dibenzofurans in liver compared to adipose tissue, have caused concern regarding the metabolic formation and accumulation of PCDFs in the liver (Kuratsune *et al.*, 1976). Further work in this area is essential. Figure 9.5 lists the various metabolites identified in mammalian metabolism involving different species.

Metabolic Breakdown by Microorganisms

A number of commercial PCB formulations have been tested for their degradability by microbes in aerobic activated sludge in a semicontinuous operation. Similar to animal metabolic results, mono-, di-, and trichlorobiphenyls are significantly biodegraded and volatilized, whereas PCBs with

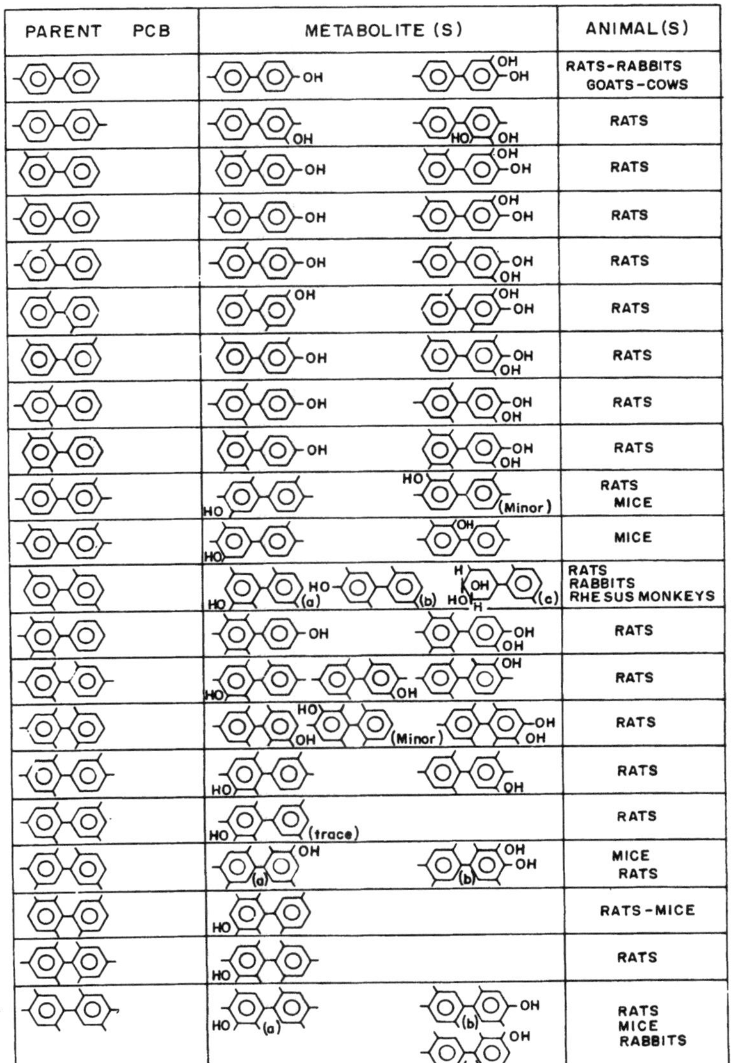

Figure 9.5. Metabolites identified in mammalian metabolism involving different species. (From Roberts *et al.*, 1978.)

≥5 Cl atoms/molecule tend to sorb to suspended particulates and sediments, bioaccumulate because of poor water solubility, and resist biodegradation (Clark, 1979; Tulp *et al.*, 1978). Recent work has shown that Aroclor 1221 and Aroclor 1232 are the only mixtures showing significant biodegradation (Tabak *et al.*, 1981). The new mixtures 1016 and 1242 showed some breakdown at 5 mg L^{-1} and low activity at 10 mg L^{-1} level. Aroclors 1248 and 1260 exhibited virtually no breakdown at both concentrations. The availability of C-H bonds in PCB determines the extent of hydroxylation and in turn the biodegradation.

Baxter *et al.* (1975) reported a faster degradation rate for commercial PCB mixtures than their single components. Furthermore, the addition of biphenyl to the substrate enhanced the biodegradation of certain PCBs; this indicated the possible cometabolism in the degradation of HCBPs. Furakawa and Matsumura (1976) showed that the bacteria *Alkaligenes* sp. degraded PCBs in two major steps: (i) production of a yellow colored intermediate absorbing at $\lambda \cong 400$ nm and (ii) its breakdown into chlorobenzoic acid. It was reported that *Pseudomonas* 7509, originally isolated from the activated sludge, degraded Aroclor 1221 10 times more rapidly than the sewage, reflecting the preference of *Pseudomonas* 7509 for Aroclor 1221 over other organic substrates (Liu, 1981) It was claimed that *Pseudomonas* 7509 has an excellent activity between 15 and 35°C and still maintained enough activity to oxidize Aroclor 1221 at 4°C, making it suitable for cold climate operations. Presence of N and P nutrients did not retard the microbial degradation.

Emulsification of PCB mixtures by sodium lignosulfonate greatly enhances the microbial degradation of PCB mixtures from Aroclor 1221 (low chlorinated) to 1254 (higher chlorinated). This is due to the increase in surface area of the substrate, which is the limiting factor in the biodegradation process. Also, by being inert to microbial attack, lignosulfonate provides a stable emulsion. The critical importance of interface in biodegradation was observed in oil breakdown by microorganisms (Marshall, 1976). It was shown that the solubility of PCB is not as significant as the interfaces on microbial activity. Cometabolism as shown by the enhanced degradation of Aroclor 1254 when Aroclor 1221 was added to the growth medium was in agreement with the findings of Baxter *et al.* (1975). It is imprecise to predict the biodegradability of a mixture from pure components since the physico-chemical properties are different (for example, depression of melting point of the mixture compared with the components.

Residues

Air, Water, and Sediments

Accurate measurements of PCB in the atmosphere have been hampered by analytical and technical problems, including insufficient collection of volatile fractions and insufficient distinction between particulate and vapor phase fractions. In addition, no collection equipment is uniformly accepted. Despite these difficulties, it is safe to conclude that much of the transport and deposition of PCBs in North America and Europe is atmospheric. Eisenreich *et al.* (1981) suggested that atmospheric inputs to Lake Michigan ranged from 2500 to 9000 kg yr^{-1}, accounting for ~60% of total deposition. Although this latter percentage may represent an upper limit for atmospheric deposition in industrial areas, there is little doubt that the detectable levels of PCBs found in Arctic lakes are almost entirely due to atmospheric transport. The same may be said about transport across and deposition into the Atlantic Ocean (Doskey and Andren, 1981).

Total PCB concentrations in air and precipitation have generally declined in recent years, reflecting the ban on production in many countries. Eisenreich *et al.* (1981) reported that the average concentration of airborne PCBs for Lake Superior was 1.5 ng m^{-3} in 1978 and 1.0 ng m^{-3} in 1979. Nevertheless, total atmospheric burdens over major industrial centers have shown a smaller change in recent years. These values can be compared with averages of 119 and 14.9 ng L^{-1} for rain water in Chicago and rural areas of the UK, respectively (Murphy and Rzeszutko, 1977; Wells and Johnstone, 1978). In the continental USA, vapor concentrations generally range from 1 to 10 ng L^{-1} (Richard and Junk, 1981), whereas the corresponding range for British coastal waters was <0.2–.08 ng L^{-1} (Dawson and Riley, 1977). Residues of 25 to 3200 mg L^{-1} were found in rainfall at Fort Edward (USA), owing to the disposal and volatilization of PCB-containing wastes (Brinkman *et al.*, 1980).

PCBs are only slightly soluble, resulting in low dissolved levels in water (Table 9.6). Maximum residues seldom exceed 2 ng L^{-1} and may fall below 0.2 ng L^{-1} in off-shore marine areas. Unfiltered water samples, containing particulate matter, often bear much higher residues (Table 9.6). Concentrations >100 ng L^{-1} have been reported for both marine and fresh waters. Recent data have also shown that the probability of finding PCBs in ground water is comparable to that of surface waters (Page, 1981). This poses a new and long-term hazard to potable water supplies in industrialized areas.

Sediments are the primary sink for PCBs and will continue to be a major source of contamination to the food chain for many years to come. Although concentrations in some major lakes, such as St. Clair, have declined in recent years, industrial and municipal sources still contribute significantly to the total PCB burden in most industrial zone waters. Frank *et al.* (1981), for example, reported sediment residues of 10–20 μg kg^{-1} dry weight in Lake Michigan near Chicago, Milwaukee, and Green Bay. By contrast, the whole lake average was only 9.7 μg kg^{-1}. Comparable trends have been reported in recent years for other major lakes such as Huron, St. Clair, Erie, and Superior. However, in the vicinity of waste outfalls, residues may range from 2000 to >500,000 μg kg^{-1} (Elder *et al.*, 1981.)

Inshore marine areas of industrial zones are also often highly contaminated, resulting in the transfer of significant amounts of PCBs to fish, marine mammals, and birds. Sediment levels in the Southern California

Table 9.6. Concentration of PCBs in marine and fresh waters.

Location	Average	Range (ng L^{-1})	Year
	Dissolved		
English Channel[1]	0.19	(0.15–0.30)	1974
Irish Sea[2]	0.5	(<0.2–1.0)	1974
Mediterranean Sea[3]	2.0	(<0.2–8.6)	1975
Grand River (Canada)[4]	5.8	(ND–100)	1975–1976
Grand River (Canada)[4]	3.7	(ND–100)	1976–1977
Saugeen River (Canada)[4]	0.2	(ND–3)	1975–1976
Saugeen River (Canada)[4]	0.3	(ND–5)	1976–1977
	Suspended Solids ($\mu g\ kg^{-1}$)		
Inland waters (19 states, USA)[4]	—	(ND–4000)	1971–1972
Dority Reservoir (USA)[5]	99	(70–130)	1978
Tiber Estuary (Mediterranean)[6]	297	(9–1000)	1976
	135	(ND–380)	1977
Baltic Sea[7]	5.0	(0.3–139)	1977
Brisbane Estuary (Australia)[8]	—	(ND–50)	—
Niagara River (Canada)[9]	961	(—)	1979–80

Sources: [1]Crump-Wiesner *et al.* (1974); [2]Dawson and Riley (1977); [3]Elder and Villeneuve (1977); [4]Frank (1981); [5]Brinkman *et al.* (1980); [6]Puccetti and Leoni (1980); [7]Ehrhardt (1981); [8]Shaw and Connell (1980); [9]Warry and Chan (1981).
ND, not detected.
—, no data.

Bight (Los Angeles) generally exceeded 1000 $\mu g\ kg^{-1}$ and in some areas reached 10,000 $\mu g\ kg^{-1}$ (Young and Heesen, 1978). Similar, or slightly lower values, have been reported for the harbors of most major cities and industrial zone estuaries around the world. Although data are limited, offshore sediments, such as those in the Mediterranean and Baltic Seas, usually bear much lower residues, <5 $\mu g\ kg^{-1}$ (Basturk *et al.*, 1980).

Aquatic Plants and Invertebrates

PCBs are rapidly sorbed from water by aquatic plants, and CFs, though variable, generally fall within a range of 1×10^4–5×10^4. Accordingly, since dissolved PCB levels in water are low, residues in natural populations of algae seldom exceed 1 mg kg^{-1} wet weight. Attached species may also mobilize PCBs directly from sediments. It has also been shown that the rate of sorption depends on the quantity of chlorine in the different PCB homologs (Mrozek and Leidy, 1981). These differences are reflected throughout the food chain and may significantly effect total residues in fish; for example,

CFs for Aroclor 1016 and Aroclor 1260 in fathead minnows varied from 42,500 to 194,000 (Veith *et al.*, 1979).

PCBs are a significant and widespread contaminant of both marine and fresh-water invertebrates. Goldberg *et al.* (1978) reported concentrations of >5000 $\mu g\ kg^{-1}$ dry weight for mussels on the Atlantic and Pacific coasts of the USA. Comparable or slightly lower residues have been found in benthic invertebrates inhabiting coastal waters of many other industrialized nations. In the Medway Estuary (UK), shrimp *Crangon vulgaris* had whole body burdens ranging from trace to 275 $\mu g\ kg^{-1}$ wet weight, whereas residues in several species of polychaetes and crabs from the Brisbane River Estuary (Australia) reached 520 $\mu g\ kg^{-1}$ wet weight (Shaw and Connell, 1980; van den Broek, 1979). Since PCB levels in water are often low compared with those in sediments, zooplankton are more likely to carry lower body burdens than benthic species. For example, mixed microplankton collections from the Mediterranean yielded average residues of 65 $\mu g\ kg^{-1}$ dry weight (Fowler and Elder, 1980–1981). However, Linko *et al.* (1979) reported peak concentrations of 3300 and 720 $\mu g\ kg^{-1}$ wet weight for collections made in the Baltic Sea (Finland) during 1974 and 1976, respectively. This was due to unusually large seasonal lipid levels in the tissues of the plankton. Such variation implies that monitoring programs involving invertebrates should account for seasonal differences in residues.

Fish

PCBs are a widespread contaminant of fish tissues. Although residues have generally declined in recent years, there are still many examples of significant adulteration of tissues. Based on 1977 data, residues in American shad increased from an average of 2.0 to 6.1 mg kg^{-1} wet weight during the annual migration up the Hudson River, USA (Pastel *et al.*, 1980), whereas other nonmigratory species from the same river had burdens of 1.8–3.8 mg kg^{-1} (Skea *et al.*, 1979). Bluefin tuna, a carnivorous highly migratory species collected from the North Aegean Sea (1975–1979), carried average residues of 2.6 mg kg^{-1} compared with 0.5–0.7 mg kg^{-1} for smaller, more sedentary fish (Kilikidis *et al.*, 1981). The decline in levels over the last few years has been dramatic in some instances but less so in others. For example, concentrations in brown trout from the River Rhone (Switzerland) averaged 7.2 and 0.2 mg kg^{-1} in 1972 and 1975, respectively (Schweizer and Tarradellas, 1980). By contrast, smallmouth bass from Lake Erie yielded mean residues of 0.3 mg kg^{-1} in 1968 and 1.7 mg kg^{-1} in 1975 (Frank *et al.*, 1978).

Seasonal variations in PCB levels can be quite marked in fish. Edgren *et al.* (1981) reported a three- to seven-fold seasonal difference in residues in perch and roach collected from a bay on the Baltic coast (Figure 9.6). Similarly, residues in whiting from the Medway Estuary (UK) peaked at

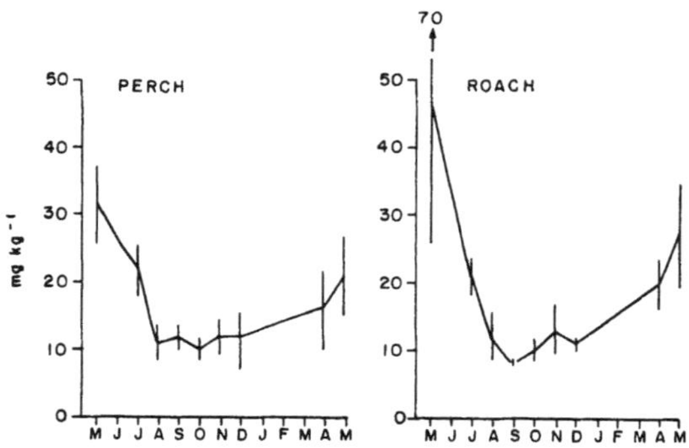

Figure 9.6. Average concentration (± 95% confidence limits) of PCB in extractable fat from perch and roach collected from Hamnefjarden, Sweden (Reprinted with permission from *Chemosphere*, vol. 10, M. Edgren, M. Olsson, and L. Reutergardh, A one year study of the seasonal variations of DDT and PCB levels in fish from heated and unheated areas near a nuclear power plant, ©1981, Pergamon Press, Ltd.).

0.16 mg kg^{-1} in April, declining to 0.01 mg kg^{-1} by October (van den Broek, 1979). Such variability stems from the interplay of nutritional, reproductive, and activity cycles and may have a highly significant impact on the usefulness of guidelines for safe consumption by humans. For example, it is possible that fish would meet safe standards (generally 2 mg kg^{-1}) at one time of the year, yet be unsafe for human consumption at another time. Monitoring programs must clearly account for seasonal variability in residue levels.

Maximum concentrations in fish are usually associated with fat deposits. The liver is another site of accumulation, though levels in the gonads are variable and often low. Residues are generally lowest in muscle and, by and large, parallel the amount of fat in the tissue. Overall, measurements of concentration in fat are a good tool for monitoring PCB contamination and should be conducted in conjunction with edible muscle tissue determinations.

Food is probably a more important route of uptake in fish than water under natural conditions. Retention from the diet is generally high, >50%, but decreases with the amount of dietary-PCB (Leatherland and Sonstegard, 1980). Uptake from water is also rapid, at least under laboratory conditions; for example, exposure of fathead minnows for 32 days to contaminated water resulted in a CF 194,000 for Aroclor 1260, while the corresponding values for A-1254, A-1248, and A-1016 were 100,000, 70,500, and 42,500, respectively (Veith *et al.*, 1979). The half-life of PCBs in fish is highly variable, depending on the homolog under investigation and species and growth rate of fish. Roberts *et al.* (1978) quoted values ranging from 3 to >16 weeks for a variety of species and conditions. These lengthy periods are one of the reasons why residues increase with the age, size, and trophic position of the fish.

Toxicity

Aquatic Plants, Invertebrates, and Fish

PCBs generally inhibit the growth of aquatic plants at concentrations of 10–100 $\mu g\ L^{-1}$, but reductions in photosynthesis and carbon uptake in sensitive species may occur at 0.1–1.0 $\mu g\ L^{-1}$ (Fisher and Wurster, 1973). Comparably low concentrations have been known to alter the community structure and species composition of natural algal populations. Toxicity increases as the temperature tolerance limit for each species is approached; thus, plants growing under suboptimal conditions may be more vulnerable to PCB-induced stress than those growing under optimal conditions. This is one of the reasons why there are regional differences in sensitivity among algal populations (Fisher *et al.*, 1973). Toxicity depends on species and the homolog under investigation. Aroclor 1242 appears to be particularly toxic to fresh-water algae (Glooschenko and Glooschenko, 1975). It is also known that PCBs interact synergistically with pesticides and arsenic in their toxic effect on algae. This implies that the methods of disposal of mixed PCB/pesticide wastes might consider potential effects on aquatic plants and the remainder of the food chain.

Marine and fresh-water invertebrates are also highly sensitive to PCBs. Nimmo *et al.* (1975) recorded toxicities of Aroclor 1254 to several species of shrimp and oysters of 0.1–12.5 $\mu g\ L^{-1}$ in testing that lasted 30 weeks. Using Aroclor 1016, Hansen *et al.* (1974) gave 96-h LC_{50}s of 10.5–12.5 $\mu g\ L^{-1}$ for two species of a shrimp, whereas Stahl (1979) found no acute effect of Aroclor 1254 on hermit crabs at concentrations of up to 30 $\mu g\ L^{-1}$. As with aquatic plants, toxicity differs considerably among the different Aroclors. Nebeker and Puglishi (1974) found that the relative toxicity (3-week LC_{50}) of Aroclor to the cladoceran *Daphnia magna* was as follows: Aroclor $1248 < 1254 < 1260 < 1262 < 1242 < 1232 < 1221 < 1268$. It has also been shown that juvenile/immature stages of many invertebrate species are more sensitive to PCBs than adults. This implies that there is probably a seasonal cycle in the sensitivity of invertebrate populations to PCB-containing wastes under natural conditions.

Most species and life stages of fish are sensitive to PCBs. DeFoe *et al.* (1978) calculated that the 30-day LC_{50} in fathead minnows exposed to Aroclor 1260 was 3.3 $\mu g\ L^{-1}$ compared with 4.7 $\mu g\ L^{-1}$ for Aroclor 1248. Although reproduction occurred at concentrations as high as 3 $\mu g\ L^{-1}$ (A-1248), there was a 20% reduction in the standing crop of second-generation fish exposed to much lower levels (0.4 $\mu g\ L^{-1}$). This was mainly due to the death of larvae soon after hatching. Enhanced sensitivity of juvenile stages is widespread among fish. Nebeker *et al.* (1974) reported a 96-h LC_{50} of 15 $\mu g\ L^{-1}$ for newly hatched fathead minnows exposed to Aroclor 1242 and 300 $\mu g\ L^{-1}$ for 3-month old minnows. Thus the decision to discharge PCB-containing wastes into natural waters should consider the susceptibility of sensitive life stages of indigenous species of fish.

PCBs have been implicated in reducing eggshell thickness in water fowl in laboratory experiments. Although a comparable response is likely to occur under natural conditions, it has not been possible to separate the effects of DDT/DDE on eggshell thickness from those of PCBs. It is known, however, that PCB residues in bird eggs have recently declined over wide geographic areas (Haseltine *et al.*, 1981). This implies that the potential threat to eggshell thickness has also probably declined. Ingestion of PCBs may produce generalized edema and hypopericardium in birds. Other pathological changes include enlargement and damage to the kidneys and livers, internal hemorrhage, and splenic atrophy. Exposure to nontoxic levels may result in increased susceptibility to viral infection.

Human Health

PCBs induce a wide range of pathological symptoms in the liver of experimental mammals, including adenofibrosis and development of neoplastic nodules and carcinomas. Although these symptoms are generally comparable to those associated with chlorinated pesticides, the level of carcinogenic activity induced by PCBs is relatively low (Sittig, 1980). Consequently, several epidemiological studies have been unable to draw statistically significant correlations between levels of occupational exposure and cancer mortality (Brown and Jones, 1981). Apart from their carcinogenic properties, PCBs are known mutagens and teratogens. Human experience with PCBs stems largely from an incidence in Japan in 1968 where contaminated cooking oil was accidently ingested (Kurzel and Centrulo, 1981). Over 1000 victims exhibited symptoms such as nausea, headache, diarrhea, and acne. Of the 13 infants born to exposed women, 11 showed retarded growth and 2 were still-born. These latter findings are consistent with experimental studies which have shown that PCBs are rapidly transferred to fetuses and that nursing infants are at greater risk than their mothers when the mothers are exposed to lipophilic toxins (Kodama and Ota, 1980).

References

Andersson, K., A. Norstrom, C. Rappe, B. Rasmuson, and H. Swahlin. 1973. *Photochemical degradation of PCB's.* National Meeting of the American Chemical Society, August 26–31, Chicago, Illinois.

Basturk, O., M. Dogan, I. Salihoglu, and T. Balkas. 1980. DDT, DDE, and PCB residues in fish, crustaceans and sediments from the eastern Mediterranean coast of Turkey. *Marine Pollution Bulletin* **11**:191–195.

Baxter, R.A., P.E. Gilbert, R.A. Lidgett, J.H. Mainprize, and H.A. Vodden. 1975. The degradation of polychlorinated biphenyls by microorganisms. *The Science of the Total Environment* **4**:53–61.

Brinkman, M., K. Fogelman, J. Hoeflein, T. Lindh, M. Pastel, W.C. Trench, and D.A. Aikens. 1980. Distribution of polychlorinated biphenyls in the Fort Edward, New York, water system. *Environmental Management* **4**:511–520.

Brown, D.P., and M. Jones. 1981. Mortality and industrial hygiene study of workers exposed to polychlorinated biphenyls. *Archives of Environmental Health* **36**: 120–129.

Chemical and Engineering News. 1971. Monsanto releases PCB data. **49**:15.

Clark, R.R. 1979. Degradation of polychlorinated biphenyls by mixed microbial cultures. *Applied and Environmental Microbiology* **37**:680–689.

Crump-Wiesner, H.J., H.R. Feltz, and M.L. Yates. 1974. A study of the distribution of polychlorinated biphenyls in the aquatic environment. *Pesticides Monitoring Journal* **8**:157–161.

Dawson, R., and J.P. Riley. 1977. Chlorine-containing pesticides and polychlorinated biphenyls in British coastal waters. *Estuarine and Coastal Marine Science* **4**:55–69.

Defoe, D.L., G.D. Vieth, and R.W. Carlson. 1978. Effects of Aroclor 1248 and 1260 on the fathead minnow (*Pimephales promelas*). *Journal Fisheries Research Board of Canada* **35**:997–1002.

Doskey, P.V., and A.W. Andren. 1981. Concentration of airborne PCBs over Lake Michigan. *Journal of Great Lakes Research* **7**:15–20.

Edgren, M., M. Olsson, and L. Reutergardh. 1981. A one year study of the seasonal variations of DDT and PCB levels in fish from heated and unheated areas near a nuclear power plant. *Chemosphere* **10**:447–452.

Ehrhardt, M. 1981. Organic substances in the Baltic Sea. *Marine Pollution Bulletin* **12**:210–213.

Eisenreich, S.J., B.B. Looney, and J.D. Thornton. 1981. Airborne organic contaminants in the Great Lakes ecosystem. *Environmental Science and Technology* **15**:30–38.

Elder, D.L., and J.P. Villeneuve. 1977. Polychlorinated biphenyls in the Mediterranean Sea. *Marine Pollution Bulletin* **8**:19–22.

Elder, V.A., B.L. Proctor, and R.A. Hites. 1981. Organic compounds found near dump sites in Niagara Falls, New York. *Environmental Science and Technology* **15**:1237–1243.

Fisher, N.S., and C.F. Wurster. 1973. Individual and combined effects of temperature and polychlorinated biphenyls on the growth of three species of phytoplankton. *Environmental Pollution* **5**:205–212.

Fisher, N.S., L.B. Graham, E.J. Carpenter, and C.F. Wurster. 1973. Geographic differences in phytoplankton sensitivity to PCBs. Nature **241**:548–549.

Fowler, S.W., and D.L. Elder. 1980–1981. Chlorinated hydrocarbons in pelagic organisms from the open Mediterranean Sea. *Marine Environmental Research* **4**:87–96.

Frank, R. 1981. Pesticides and PCB in the Grand and Saugeen river basins. *Journal of Great Lakes Research* **7**:440–454.

Frank, R., H.E. Braun, M. Holdrinet, D.P. Dodge, and S.J. Nepszy. 1978. Residues of organochlorine insecticides and polychlorinated biphenyls in fish from Lakes

Saint Clair and Erie, Canada—1968–76. *Pesticides Monitoring Journal* **12**: 69–80.

Frank, R., R.L. Thomas, H.E. Braun, D.L. Gross, and T.T. Davies. 1981. Organochlorine insecticides and PCB in surficial sediments of Lake Michigan (1975). *Journal of Great Lakes Research* **7**:42–50.

Furukawa, K., and F. Matsumura. 1976. Microbial metabolism of polychlorinated biphenyls. Studies on the relative degradability of polychlorinated biphenyl components by *Alkaligenes* sp. *Journal of Agricultural Food Chemistry* **24**: 251–256.

Gardner, A.M., H.R. Righter, and J.A.G. Roach. 1976. Excretion of hydroxylate polychlorinated biphenyl metabolites in cow's milk. *Journal of the Association of Official Analytical Chemists* **59**:273–277.

Glooschenko, V., and W. Glooschenko. 1975. Effect of polychlorinated biphenyl compounds on growth of Great Lakes phytoplankton. *Canadian Journal of Botany* **53**:653–659.

Goldberg, E.D., V.T. Bowen, J.W. Farrington, G. Harvey, J.H. Martin, P.L. Parker, R.W. Risebrough, W. Robertson, E. Schneider, and E.Gamble. 1978. The mussel watch. *Environmental Conservation* **5**:101–126.

Hansen, D.J., P.R. Parrish, and J. Forester. 1974. Aroclor 1016: toxicity to and uptake by estuarine animals. *Environmental Research* **7**:363–373.

Haseltine, S.D., G.H. Heinz, W.L. Reichel, and J.F. Moore. 1981. Organochlorine and metal residues in eggs of waterfowl nesting on islands in Lake Michigan off Door County, Wisconsin, 1977–78. *Pesticides Monitoring Journal* **15**:90–97.

Hustert, K., and F. Korte. 1972. Synthesis of PCB's in reaction with UV light. *Chemosphere* **1**:7–10.

Hutzinger, S., S. Safe, and V. Zitko. 1974. *The Chemistry of PCBs*. CRC Press, Boca Raton, Florida, 269 pp.

Jansson, B., S. Jensen, M. Olsson, L. Renberg, G. Sundstrom, and R. Vaz. 1975. Identification by GC-MS of phenolic metabolites of PCB and p,p′-DDE isolated from Baltic guillemot and seal. *Ambio* **4**:93–97.

Kilikidis, S.D., J.E. Psomas, A.P. Kamarianos, and A.G. Panetsos. 1981. Monitoring of DDT, PCBs, and other organochlorine compounds in marine organisms from the North Aegean Sea. *Bulletin of Environmental Contamination and Toxicology* **26**:496–501.

Kodama, H., and H. Ota. 1980. Transfer of polychlorinated biphenyls to infants from their mothers. *Archives of Environmental Health* **35**:95–100.

Kuratsune, M., Y. Masuda, J. Nagayama. 1976. *Some recent findings concerning Yusho*. Proceedings of the National Conference on Polychlorinated Biphenyl, November 19–21, 1975, Chicago, Illinois. U.S. Environmental Protection Agency, Publication No. EPA-560/6-75-004, pp. 14–29.

Kurzel, R.B., and C.L. Centrulo, 1981. The effect of environmental pollutants on human reproduction, including birth defects. *Environmental Science and Technology* **15**:626–640.

Leatherland, J.F., and R.A. Sonstegard. 1980. Effect of dietary Mirex and PCBs in combination with food deprivation and testosterone administration on thyroid activity and bioaccumulation of organochlorines in rainbow trout *Salmo gairdneri* Richardson. *Journal of Fish Diseases* **3**:115–124.

Lichtenstein, E.P., K.R. Schulz, T.W. Fuhremann, and T.T. Liang. 1969. Biological interaction between plasticizers and insecticides. *Journal of Economic Entomology* **62**:761–765.

Linko, R.R., P. Rantamaki, K. Rainio, and K. Urpo. 1979. Polychlorinated biphenyls in plankton from the Turku archipelago. *Bulletin of Environmental Contamination and Toxicology* **23**:145–152.

Liu, D. 1981. Biodegradation of Aroclor 1221 type PCBs in sewage wastewater. *Bulletin of Environmental Contamination and Toxicology* **27**:695–703.

Marshall, K.C. 1976. *Interfaces in microbial ecology*. Harvard University Press, Cambridge, Massachusetts.

McIntyre, A.E., R. Perry, J.N. Lester. 1981a. The behaviour of polychlorinated biphenyls and organo-chlorine insecticides in primary mechanical wastewater treatment. *Environmental Pollution* **2**:223–233.

McIntyre, A.E., J.N. Lester, and R. Perry. 1981b. The influence of chemical conditioning and dewatering on the distribution of polychlorinated biphenyls and organochlorine insecticides in sewage sludges. *Environmental Pollution* **2**: 309–320.

McKinney, J.D. 1976. *Toxicology of selected symmetrical hexachlorobiphenyl isomers; correlating biological effects with chemical structure.* Proceedings of the National Conference on Polychlorinated Biphenyl, November 19–21, 1975, Chicago, Illinois. U.S. Environmental Protection Agency Publication No. EPA-560/6-75-004, pp. 73–76.

Miller, S. 1982. The persistent PCB problem. *Environmental Science and Technology* **16**:98A–99A.

Monsanto Industrial Chemical Corporation. 1972. *Presentation to the Interdepartmental Task Force on PCB's*. Washington, D.C., 1972.

Monsanto Industrial Chemical Corporation. 1974. *PCBs*. Aroclor Technical Bulletin O/PL 306A, St. Louis, Missouri, 20 pp.

Monsanto Technical Bulletin O/PL—306., Monsanto Company, St. Louis, Missouri, U.S.A.

Mrozek, Jr., E., and R.B. Leidy. 1981. Investigation of selective uptake of polychlorinated biphenyls by *Spartina alterniflora* Loisel. *Bulletin of Environmental Contamination and Toxicology* **27**:481–488.

Murphy, T.J., and C.P. Rzeszutko. 1977. Precipitation inputs of PCBs to Lake Michigan. *Journal of Great Lakes Research* **3**:305–312.

Nebeker, A.V., and F.A. Puglisi. 1974. Effect of polychlorinated biphenyls (PCB's) on survival and reproduction of *Daphnia, Gammarus* and *Tanytarsus*. *Transactions American Fisheries Society* **103**:722–728.

Nebeker, A.V., R.A. Puglisi, and D.L. Defoe. 1974. Effect of polychlorinated biphenyl compounds on survival and reproduction of the fathead minnow and flagfish. *Transactions of the American Fisheries Society* **103**:562–568.

Nimmo, D.R., D.J. Hansen, J.A. Couch, N.R. Cooley, P.R. Parrish, and J.I. Lowe. 1975. Toxicity of Aroclor 1254 and its physiological activity in several estuarine organisms. *Archives of Environmental Contamination and Toxicology* **3**: 22–39.

Oloffs, P.C., L.J. Albright, and S.Y. Szeto. 1972. Fate and behavior of fine chlorinated hydrocarbons in three natural waters. *Canadian Journal of Microbiology* **18**:1393–1398.

Page, W.G. 1981. Comparison of groundwater and surface water for patterns and levels of contamination by toxic substances. *Environmental Science and Technology* **15**:1475–1481.

Pastel, M., B. Bush, and J.S. Kim. 1980. Accumulation of polychlorinated biphenyls in American shad during their migration in the Hudson River, spring 1977. *Pesticides Monitoring Journal* **14**:11–22.

Peakall, D.B., and J.L. Lincer. 1970. Polychlorinated biphenyls. Another long-life widespread chemical in the environment. *Bio-Science* **20**:958–964.

Puccetti, G., and V. Leoni. 1980. PCB and HCB in the sediments and waters of the Tiber estuary. *Marine Pollution Bulletin* **11**:22–25.

Richard, J.J., and G.A. Junk. 1981. Polychlorinated biphenyls in effluents from combustion of coal/refuse. *Environmental Science and Technology* **15**: 1095–1100.

Roberts, J.R., D.W. Rodgers, J.R. Bailey, and M.A. Rorke. 1978. *Polychlorinated biphenyls: biological criteria for an assessment of their effects on environmental quality.* Associate Committee on Scientific Criteria for Environmental Quality, National Research Council of Canada Publication No. NRCC 16077, 172. pp.

Safe, S., and O. Hutzinger. 1971. PCB's: photolysis of 2,4,5,2′,4′,6′-hexachlorobiphenyl. *Nature* **232**:641–642.

Sawhney, B.L., C.R. Frink, and W. Glowa. 1981. PCBs in the Housatonic River: determination and distribution. *Journal of Environmental Quality* **10**:444–448.

Schweizer, C., and J. Tarradellas. 1980. Etat des recherches sur les biphényles polychlorés en Suisse. *Chimia* **34**:507–517.

Shannon, E.E., F.J. Ludwig, and I. Valdemanis. 1976. *Polychlorinated biphenyls in municipal wastewaters.* Environment Canada Research Report No. 49.

Shaw, G.R., and D.W. Connell. 1980. Polychlorinated biphenyls in the Brisbane River estuary, Australia. *Marine Pollution Bulletin* **11**:356–358.

Sittig, M. 1980. *Priority toxic pollutants. Health impacts and allowable limits.* Noyes Data Corporation, New Jersey, 370 pp.

Skea, J.C., H.A. Simonin, H.J. Dean, J.R. Colquhoun, J.J. Spagnoli, and G.D. Veith. 1979. Bioaccumulation of Aroclor 1016 in Hudson River fish. *Bulletin of Environmental Contamination and Toxicology* **22**:332–336.

Stahl, Jr., R.G. 1979. Effect of a PCB (Aroclor 1254) on the striped hermit crab, *Clibanarius vittatus* (Anomura: Diogenidae) in static bioassays. *Bulletin of Environmental Contamination and Toxicology* **23**:91–94.

Tabak, H.H., S.A. Quave, C.I. Mashni, E.F. Barth. 1981. Biodegradability studies with organic priority pollutant compounds. *Journal Water Pollution Control Federation* **53**:1503–1518.

Task Force on PCB. 1976. *Background to the regulation of polychlorinated biphenyls (PCB) in Canada.* Environmental Canada and Health and Welfare Canada, Joint Report No. 76-1, April 1, 1976.

Tulp, M.Th.M., R. Schmitz, and O. Hutzinger. 1978. The bacterial metabolism of 4,4′-dichlorobiphenyl, and its suppression by alternative carbon sources. *Chemosphere* **1**:103–108.

van den Broek, W.L.F. 1979. Seasonal levels of chlorinated hydrocarbons and heavy metals in fish and brown shrimps from the Medway estuary, Kent. *Environmental Pollution* **19**:21–38.

Veith, G.D., D.L. DeFoe, and B.V. Bergstedt. 1979. Measuring and estimating the bioconcentration factor of chemicals in fish. *Journal of the Fisheries Research Board of Canada* **36**:1040–1048.

Vos, J.G., J.H. Koeman, H.L. van der Maas, M.C. ten Noever de Brauw, and R.H. de Vos. 1970. Identification and toxicological evaluation of chlorinated dibenzofuran and chlorinated naphthalene in two commercial polychlorinated biphenyls. *Food and Cosmetics Toxicology* **8**:625–633.

Warry, N.D., and C.H. Chan. 1981. Organic contaminants in the suspended sediments of the Niagara River. *Journal of Great Lakes Research* **7**:394–403.

Wells, D.E., and S.J. Johnstone. 1978. The occurrence of organochlorine residues in rainwater. *Water, Air, and Soil Pollution* **9**:271–280.

Young, D.R., and T.C. Heesen. 1978. DDT, PCB, and chlorinated benzenes in the marine ecosystem off Southern California. *In*: R.L. Jolley (Ed.), *Water chlorination, environmental impact and health effects*, Vol. **2**, pp. 267–290.

10
Polychlorinated Dibenzo-*p*-Dioxins (PCDD)

There are 75 homologs and isomers within the polychlorinated dibenzo-*p*-dioxin group (Table 10.1), exhibiting the following characteristics:

(i) a relatively stable aromatic nucleus;
(ii) many isomers have parallel toxic properties;
(iii) increasing halogen content in the two rings increases their environmental stability, lipophilicity, thermal stability, and resistance to acids, bases, oxidants, and reductants; and
(iv) property (iii) has contributed to the widespread presence of PCDDs in the environment.

PCDDs have been detected in trichlorophenol (TCP), tetrachlorophenol (TeCP), and pentachlorophenol (PCP) and in chlorophenol derivatives such as 2,4-dichlorophenoxyacetic acid (2,4-D) and 2,4,5-trichlorophenoxyacetic acid (2,4,5-T). Since combustion of organic materials and seepage from chemical dumps also apparently generate detectable residues in the environment, PCDDs are now suspected to be ubiquitous contaminants in both aquatic and terrestrial ecosystems. However, it is not possible to estimate accurately the amounts of PCDDs discharged to the environment or to measure residues except in cases of extreme ambient contamination, and thus the overall threat posed by PCDDs has not been determined at the present time.

Production, Uses, and Discharges

PCDDs are not produced commercially and have no direct use. They are inadvertently formed during the production of 2,4,5-trichlorophenol from

Table 10.1. A list of PCDD isomers.*

Monochloro-(2)		Tetrachloro-(22)	
1,		1,2,3,4;	1,2,6,9;
2,		1,2,3,6;	1,2,7,8;
		1,2,3,7;	1,2,7,9;
Dichloro-(10)		1,2,3,8;	1,2,8,9;
		1,2,3,9;	1,3,6,8;
1,2;	1,8;	1,2,4,6;	1,3,6,9;
1,3;	1,9;	1,2,4,7;	1,3,7,8;
1,4;	2,3;	1,2,4,8;	1,3,7,9;
1,6;	2,7;	1,2,4,9;	1,4,6,9;
1,7;	2,8;	1,2,6,7;	1,4,7,8;
		1,2,6,8;	2,3,7,8;
Trichloro-(14)			
		Pentachoro-(14)	
1,2,3;	1,3,7;		
1,2,4;	1,3,8;	1,2,3,4,6;	1,2,3,8,9;
1,2,6;	1,3,9;	1,2,3,4,7;	1,2,4,6,7;
1,2,7;	1,4,6;	1,2,3,6,7;	1,2,4,6,8;
1,2,8;	1,4,7;	1,2,3,6,8;	1,2,4,6,9;
1,2,9;	1,7,8;	1,2,3,6,9;	1,2,4,7,8;
1,3,6;	2,3,7;	1,2,3,7,8;	1,2,4,7,9;
		1,2,3,7,9;	1,2,4,8,9;
Hexachloro-(10)			
		Heptachloro-(2)	
1,2,3,4,6,7;			
1,2,3,4,6,8;		1,2,3,4,6,7,8;	
1,2,3,4,6,9;		1,2,3,4,6,7,9;	
1,2,3,4,7,8;			
1,2,3,6,7,8;		Octachloro-(1)	
1,2,3,6,7,9;			
1,2,3,6,8,9;		1,2,3,4,6,7,8,9;	
1,2,3,7,8,9;			
1,2,4,6,7,9;			
1,2,4,6,8,9;			

Source: National Research Council of Canada (1981a).
*The numbers in parentheses indicate the number of isomers for that group of PCDD.

1,2,4,5-tetrachlorobenzene. Since 2,4,5-TCP is used in the production of numerous herbicides and preservatives, PCDDs have been reported to be a contaminant of 2,4,5-trichlorophenoxy acetic acid, 2,4,5-T esters, 2,4-D, clopen, and silvex. PCDDs are also released into the atmosphere and industrial wastes, and by combustion of vegetation treated with 2,4,5-T.

Total inadvertent production of PCDD is extremely difficult to estimate. The National Research Council of Canada recently speculated that as much

Table 10.2. Estimated input (kg) of PCDD into the Canadian environment from major identified sources.

Source	British Columbia	Prairies	Ontario	Quebec	Maritimes	Total
Pentachlorophenol usage	570	360	350	110	130	1520
2,4-D production usage	0.3	8.9	<0.1	<0.1	<0.1	9.2
Precipitated fly ash	<0.1	<0.1	1.6	5.1	<0.1	6.7
Airborne fly ash	<0.1	1.5	2.2	2.2	0.6	6.5
Total	570	370	353	117	131	1542

Source: National Research Council of Canada (1981a).

as 1.5 metric tons could be entering the Canadian environment each year (Table 10.2). Of this total, 0.6 metric tons originated in British Columbia from the heavy use of chlorinated phenols in the forestry and wood-preserving industries (Table 10.2). Moderately large amounts were also emitted in the prairie provinces and Ontario, whereas annual production in Quebec and the maritimes was only 0.11–0.13 tons. The production and use of 2,4-D was a significant source of PCDD only in the prairie provinces. It also appears that precipitated and airborne fly-ash is not a primary source in most of Canada. Maximum rates of input occur in heavily populated areas (Ontario, Quebec) and are <0.002 metric tons per year in British Columbia, the prairies, and the maritimes (Table 10.2).

Significant amounts of PCDD are probably transported across international borders (National Research Council of Canada, 1981a). One example is the movement of 2,3,7,8-tetrachlorodibenzo-*p*-dioxin from the Hooker Chemical Plant (Niagara Falls, USA), down the Niagara River to Lake Ontario, where it has been implicated in the adulteration of potable water and tissues of aquatic organisms. Total US production data do not appear to be available. It is known, however, that approximately 2 million acres in the USA have been treated for weed control with about 6800 metric tons of either 2,4,5-T, 2,4-D, or a combination of the two. Assuming an average TCDD content of 0.008 mg kg^{-1}, approximately 54 metric tons of TCDD have been released in the USA from this source alone. No information is available on the quantities of PCDD originating from chemical dumps and land fill sites.

PCDDs enter surface waters primarily through liquid discharges from several different industries. These include wood-processing plants, wood protection and preservation plants, kraft pulp mills, leather tanneries, sewage treatment plants, and chemical manufacturing plants. Each of these industries either uses or produces PCDD-contaminated material that is eventually discharged to the environment. In Canada, chlorinated phenols are probably the major source of PCDD in surface waters (National Research Council of Canada, 1981a). Since chlorinated phenols occur at detectable levels in water, sediments, and biota throughout southern Canada, it is likely that PCDDs are also present throughout the aquatic ecosystem. Herbicides such as 2,4,5-T and 2,4-D are no longer used in large amounts in Canada and thus make only a small contribution to total PCDD burden in lakes and rivers.

The amount of PCDD entering surface waters from combustion processes is extremely difficult to estimate. It is known that fly ash from municipal incinerators contains all of the higher homologs (TCDD-O_8CDD) at variable concentrations (Table 10.3). Such material may eventually find its way into surface waters, and it is logical to assume that a correlation would exist between concentrations in fish and the amount of combustion occurring along the shores of lakes, rivers, and estuaries (Crummett *et al.*, 1981). However, the importance of such inputs relative to other sources has yet to be determined.

Behavior in Natural Waters

Sorption

The majority of available data on the fate of PCDD pertains only to 2,3,7,8-tretrachlorodibenzo-*p*-dioxin (TCDD). Since the data are generally not quantitative, interpretation and extrapolation to natural environmental conditions has to be viewed with some caution. Hence, only a qualitative concept of the behavior of PCDD in the environment is presented. PCDDs tend to be rapidly and strongly sorbed by most soils (Firestone, 1977; Ward and Matsumura, 1978), suggesting relative immobility in most matrices. Using partition coefficient data for an octanol/water mixture and its relationship to the organic matter and its content in sediment and soils,

Table 10.3. Average concentrations (μg kg^{-1}) of PCDD in fly ash from municipal incinerators.

	T_4CDD	P_5CDD	H_6CDD	H_7CDD	O_8CDD	Total
Municipal incinerators (Netherlands)[1]	54	182	326	288	106	956
Municipal incinerators (Ontario)[2]	14	23	26	15	6	84

Sources: [1]Lustenhouwer *et al.* (1980); [2]National Research Council of Canada (1981a).

sorption of organic compounds can be predicted. The following equations show this relationship (Karickhoff *et al.*, 1979; Chiou *et al.*, 1979).

$$\log K_{oc} = 1.00 \log K_{ow} - 0.21 \quad (1)$$

$$\text{when } K_{om} = 1.8\ K_{oc} \quad (2)$$

and

$$K_s = \frac{K_{om} \cdot \%\ om}{100} \quad (3)$$

where ow = octanol/water; oc = organic carbon content; om = organic matter.

Using the above equation, 2,3,7,8-TCDD (with a log P (octanol/water) of 6.146 (Kenaga, 1980)) has an estimated log K_{om} value of 6.2041. Since K_s is dependent on percent organic matter, the value of log K_s can vary from 4.2041 to 6.2041 for soils with organic matter ranging from 1% to 100% of soil content. However, this relationship has not been confirmed by extensive analysis of a variety of soil types. Isensee and Jones (1975) reported soil to water equilibria for soils of differing organic matter. The K_{om} values calculated from these data show dependency on organic content and vary from 1×10^5 to 6×10^5. Kinetic data on the rate of these sorption equilibria are not available. It is safe to assume that the sorption rate constant will be governed by diffusion of the compound in the medium as well at the interface and characterized by Fick's law. Using this theoretical assumption and the equations derived by Frost and Pearson (1961), the sorption rate constant has been estimated (National Research Council of Canada, 1981b). A 5×10^{-3} cm diameter sediment particle of specific gravity 2.7 was calculated to have a sorption rate constant of ~1000 ml g^{-1} day^{-1}. It should be noted here that sorption is highly dependent upon the sediment particle size. For the commonly found range of 1–10% organic matter in sediments, the sorption rate constants are expected to fall between 10 and 100 ml g^{-1} day^{-1}. Most compounds studied seem to fit within this range. For example, the desorption rate of PCB (Aroclor 1254) is slow with a minimum half-life of 90 years (Halter and Johnson, 1977). The sorption rate constant was calculated to be 10 g of water g^{-1} sediment day^{-1} from the partition coefficient value of 6×10^4. This shows that sorption will continue for several years and an equilibrium will be reached only after ~55 years (Marshall and Roberts, 1977).

Anomalous movement of 2,3,7,8-TCDD deep into soils has been reported (Cavallaro *et al.*, 1979). Presence of fissures and excessive rainfall probably caused this downward flow of the compound that is otherwise highly immobile. Since the aqueous solubility of 2,3,7,8-TCDD is low (0.2 μg L^{-1}), it is unlikely that the movement of the dissolved phase could have caused the flow.

Simple deterministic models have been used recently to screen chemical persistence in modeling the fate of organic chemicals. The model in Table

Table 10.4. Definition of the standardized conditions of the two aquatic systems.

Parameter	Pond	Lake
Latitude (°N)	45	45
Temperature (°C)	20	20
Water		
Surface area (hectare)	0.37	66.0
Mean depth (m)	1.00	15.0
Volume (L)	3.7×10^6	1.0×10^{10}
Suspended soilds (mg L^{-1})	50	5.0
Attenuation[1]	high*	low†
Sediment		
Weight (kg)	7.3×10^4	2.9×10^7
Effective depth (cm)	1.0	1.0
Organic matter (%)	10	10
Biota		
Number of fish	185	5×10^5
Average weight (g)	100	100

Source: National Research Council of Canada (1981b).
*Nonpollutant light absorption.
†Distilled water.

10.4 consists of four compartments including water, suspended solids, sediments, and biota. Two standard aquatic systems, a highly eutrophic and a large oligotrophic lake, were used to evaluate the relative importance of the various processes in the overall fate of the organic compound (Table 10.5).

The actual pollutant loading was kept at concentrations <0.01 times its solubility in water so that the various removal processes would likely follow first-order kinetics. At equilibrium conditions, the sediment/suspended solids pool was the main sink for 2,3,7,8-TCDD and determined its persistence (Table 10.5; Figure 10.1). Therefore, under natural conditions, residence time would depend on the availability of suspended solids and organic-rich sediments. Reduction of the organic content would lead to enhanced volatilization and reduced residence time and, in fact, lowering the organic matter in the pond from 10% to 1% reduced residence time from 7 years to 0.5 years (Table 10.5). It was also concluded that at equilibrium, sediments and suspended solids may be the best monitoring matrices for 2,3,7,8-TCDD. Although residues in fish bordered on the detection limit, such measurements were a reliable initial indicator, considering the longer time taken for the sediments to reach equilibrium (Figure 10.1).

PCDDs exhibit widely varying water solubilities, octanol/water partitioning coefficients, and soil partition coefficients (Table 10.6). The variation could be as much as 1×10^5. Thus, when a mixture of PCDD is

Table 10.5. Fate of TCDD at equilibrium condition under constant load.

	Pond	Lake
Input concentration ($\mu g\ L^{-1}\ d^{-1}$)	0.74	2×10^{2}
Input volume (L)	10	10
Load ($\mu g\ d^{-1}$)	7.4	2×10^{3}
Loading concentration to water ($\mu g\ L^{-1}\ d^{-1}$)	2×10^{-6}	2×10^{-7}
Approximate time to equilibrium of slowest compartment	(sediment)	(sediment)
days	93,000	110,000
years	255	300
Concentration in:		
Water ($pg\ L^{-1}$)	1.8	2.5
Suspended solids ($\mu g\ kg^{-1}$)	2.8	4.2
Sediments ($\mu g\ kg^{-1}$)	0.28	0.42
Fish ($ng\ kg^{-1}$)	61	91
Fraction (%) of total load in:		
Water	0.03	0.21
Suspended solids	2.47	1.69
Sediments	97.49	98.06
Fish	0.01	0.04
Fraction degraded or lost by:		
Volatilization	100	100
Photolysis	0	0
Biodegradation, fish	0	0
Approximate residence time (d)	2868	6288
Half-life of system clearance (d)	1987	4358

Source: National Research Council of Canada (1981a).

Table 10.6. Relative partition coefficients of PCDDs.

Homolog	Relative lipophilicity*	Relative BCF†	Relative soil partition coefficient‡	Relative aqueous solubility§
OCDD	700	300	700	0.001
HCDD (hepta)	100	70	100	0.007
HCDD (hexa)	30	20	30	0.04
PCDD	5	4	5	0.2
TeCDD	1	1	1	1.0
TCDD	0.2	0.2	0.2	5
DCDD	0.04	0.06	0.04	30
MCDD	0.007	0.01	0.007	100

Source: National Research Council of Canada (1981a).

*Derived from relative $k_{octanol/water}(K_{ow})$.

†Log BCF $= -0.70 + 0.85 \log K_{ow}$.

‡Derived from $\log K_{oc} = 1.0 \log K_{ow} - 0.21$;

§Log $C_w = 0.7178 - 0.9874 \log K_{ow}$ where C_w = aqueous solubility of the liquid PCDD.

Figure 10.1. Equilibration and depuration of 2,3,7,8-TCDD in various compartments of a model aquatic ecosystem. (From National Research Council of Canada, 1981b.)

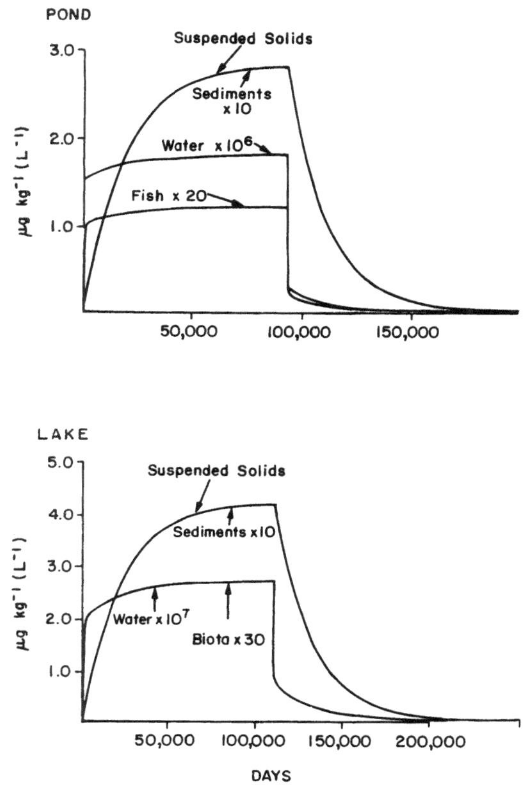

released into the ecosystem, there will be extensive shifts in the behavior pattern to matrices. In fact, saturation of the sorbent surface or dominance of other fate processes other than sorption could override the differences in homolog pattern. Higher PCDDs tend to associate with abiotic matrices such as suspended solids or organic sediments (Table 10.6). Corrections must be made for differences in their affinities to accumulate in order to evaluate meaningful patterns of ambient levels of PCDD loadings. For example, presence of higher PCDDS does not necessarily suggest a higher loading rate for the PCDDs but higher affinity (National Research Council of Canada, 1981a).

Volatilization

The rate constant for volatilization of 2,3,7,8-TCDD from water can be predicted using the general formulas of Liss and Slater (1974), Mackay (1978), and Southworth (1979). The rate of volatilization is given by the following equation:

$$K_v = \frac{1}{d\sqrt{M}[3.1407 \times 10^{-4} + (2.0413 \times 10^{-4}\ T_k\ S/P)]}$$

where d = water depth (cm); M = molecular weight (321.87); T_k = temperature in °kelvin, 298; P = vapor pressure in torr, 1.7×10^{-6}; S = solubility in water in mole L^{-1}, 6.2×10^{-10}; and k_v = rate of volatilization in cm d^{-1}, 1.7×10^2. The half-life of 2,3,7,8-TCDD was calculated to be 6 minutes in a 1 cm deep pool and about 600 minutes in a 1 m pool (National Research Council of Canada, 1981a). This agrees with the observation by Ward and Matsumura (1978) that water loss was correlated with the loss of TCDD. Volatilization rates of other low solubility compounds (2.79×10^2 for Aroclor 1242 and 3.45×10^3 for Aroclor 1260, Mackay and Wolkoff, 1974) are consonant with this estimate for TCDD. The fate modeling experiments in Tables 10.3 and 10.4 and Figure 10.1 showed that the primary removal process for 2,3,7,8-TCDD was volatilization.

Photolysis

PCDDs seem to be resistant to photolytic degradation under conditions relevant to environmental situations. In aqueous suspensions and on wet or dry soil, the photodecomposition of 2,3,7,8-TCDD was negligible (Crosby *et al.*, 1971). TCDD did degrade rapidly in alcohol solution under artifical light and sunlight and the rate of decomposition was dependent upon the degree of chlorination (Crosby *et al.*, 1971). In addition, thin films of pure TCDD on glass plates were stable in sunlight for at least 14 days (Crosby *et al.*, 1971). Similar layers of the herbicide Agent Orange (5 mg cm^{-2}) containing 15 mg L^{-1} of TCDD exposed to natural sunlight on glass plates, leaves, or soil lost almost all of the TCDD during a single day mainly owing to photochemical dechlorination. The three requirements to degrade the otherwise persistent TCDD are (i) dissolution in a light-transmitting film, (ii) the presence of an organic hydrogen donor such as a solvent or pesticide, and (iii) ultraviolet light. The levels of herbicide application used by Crosby *et al.* (1971) were much greater than any that ever have been used on fields (5 mg cm^{-2} = 500 kg $hectare^{-1}$) (Crosby and Wong, 1977). Daily variations in light intensity and the differences in initial concentrations of TCDD could account for any difference in degradation rates between various amounts of herbicide. Based on their studies, Crosby and Wong (1977) suggested that environmental residues of TCDD would often be lower than previously expected.

Dobbs and Grant (1979) showed that the chlorine atom in the *meta* position in OCDD in the presence of hexane is more labile to sunlight than the *ortho* position. Hexa-CDD and hepta-CDD containing the same number of *meta* chlorines but fewer *ortho* chlorines degraded with relative ease; in

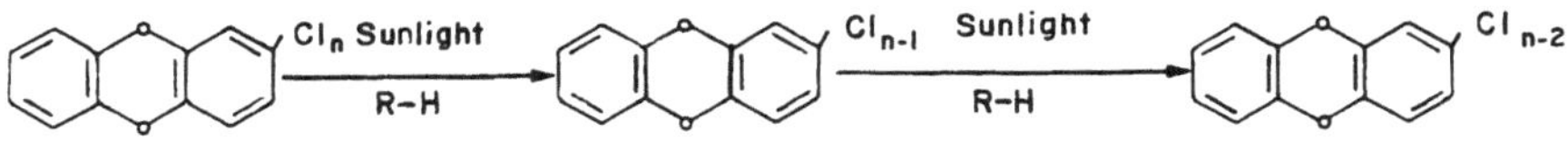

Figure 10.2. Step-wise photolytic degradation of PCDDs. (From National Research Council of Canada, 1981a.)

addition, 2,3,7,8-TCDD was apparently the most photolabile PCDD. The proposed pathway of degradation was dechlorination to lower chlorinated homologs as transient intermediates that undergo subsequent photolysis to lower homologues (Figure 10.2). The lower chlorinated PCDDs, including 2,3,7,8-TCDD, eventually degrade to unidentified products (Stehl *et al.*, 1973).

Under laboratory conditions and in the absence of hydrogen donors, photolysis is an insignificant fate process in soils (Plimmer, 1978) or inert surfaces (Helling *et al.*, 1973). No quantitative studies have been carried out in water or on the role of fulvic acid as a potential hydrogen source. Addition of surfactant to an aqueous suspension of 2,3,7,8-TCDD resulted in significant photolysis (Plimmer *et al.*, 1973). The same compound was reported to be photosusceptible on silica gel in the absence of hydrogen donors (Parlar *et al.*, 1978). These observations should be taken into account in the photolytic degradation of PCDDs.

Biotransformation

PCDDs are highly resistant to microbial degradation. Matsumura and Benezet (1973) found only 5 out of 100 microbial strains capable of degrading TCDD. Presence of lake sediment and nutrients promoted the microbial metabolism of 2,3,7,8-TCDD, releasing small amounts of metabolites to the water column (Ward and Matsumura, 1978). Studies performed by the US Air Force suggested the half-life for TCDD was 230–320 days depending on the bacteria present (Young *et al.*, 1976; Hay, 1981). However, a study of TCDD in the soil of Seveso, Italy, found a much longer half-life, in excess of 10 years (DiDomenico *et al.*, 1980a). The clearance of 2,3,7,8-TCDD in mammals is slow, with a half-life of 13–30 days (Van Miller *et al.*, 1976). TCDD is hydroxylated in rat, possibly as a means of detoxification (Poiger and Schlatter, 1979). Using the bile-loop cannulae in rats, Ramsey *et al.* (1979) showed that ^{14}C-radioactivity in bile was more polar than 2,3,7,8-TCDD and postulated that they were glucuronide conjugate metabolites whereas Guenthner *et al.* (1979) proposed that 2,3,7,8-TCDD was catabolized to arene oxides-TCDD intermediates by P_1-450 (P-488) enzyme system. The intermediates bind covalently to nucleophilic macromolecules at such a high rate that further metabolism is

Table 10.7. Estimated half-life for clearance of 2,3,7,8-TCDD in various mammalian species.

Species	Dose (μg kg-bw^{-1}.d^{-1})*	Duration of dose	Half-life (days)
Rat	0.5–1.5	42 days	11–21
Rat	50	Single oral	17.4 ± 5.6
Rat	1	Single oral	31 ± 6 (21–39)
Rat	0.01, 0.1, 1	5 d wk^{-1} for 7 wk	29.7 (16–35)
Guinea pig	0.5	Single i.p.†	30.2 ± 5.8
Guinea pig			22–43
Hamster	650	Single i.p. ^{3}H-TCDD	12 ± 2
		Single i.p. ^{14}C-TCDD	10.8 ± 2.4
		Single oral	15.0 ± 2.5

Source: National Research Council of Canada (1981a).
*kg-bw = kilogram body weight
†i.p. = intraperitoneal.

undetectable. This strong binding of arene oxides of TCDD to proteins is possibly the cause of the extreme toxicity of TCDD. Further work in this area on TCDD, and PCDDs in general, will improve our understanding of their mode of toxic action. Table 10.7 lists the half-lives of 2,3,7,8-TCDD in various mammalian species.

Table 10.8 summarizes the predicted partition, transfer, and degradation constants generated from fate modeling experiments. It can be seen that a large amount of useful information can be generated from a few predictive parameters. Although not tested by experiments, the information provides a useful baseline for understanding the fate of chemicals in the environment.

For meaningful and safe monitoring of PCDDs in the aqueous environment, the chosen indicator matrices must include good accumulators like organic sediments, organic soils, suspended solids, older fatty fish, piscivorous animals, as well as good synthetic sorbents such as charcoal filters. The results should be adjusted for the different accumulation patterns of PCDDs to avoid bias. Care must be taken to assure that the indicators have not reached the saturation limit. Various feeding patterns, physico-chemical removal processes, and degradation processes operating in an aquatic system that should be considered in the choice of indicator matrix and interpretation of results are given in Figure 10.3.

Table 10.8. Summary of the dynamics of 2,3,7,8-TCDD in the aqueous environment.

Volatilization (25°C) (cm^{-1} day^{-1})	1.7×10^2
Photolysis (45°N, June) (cm^{-1} day^{-1})	
Attenuation: high	0
low	0
Sorption	
Partition	
(100% organic, K_{om})	1.6×10^6
(10% organic, K_s)	1.6×10^5
Transfer	
(g water (g sediment)$^{-1}$ day^{-1})	
(100%, organic)	1.0×10^2
(10%, organic)	1.0×10^1
Bioaccumulation	
Partition (K_b)	3.3×10^4
Transfer (day^{-1} $g^{0.2}$)	
uptake (k_{wu})	1.9×10^2
clearance (k_{wc}	5.9×10^{-3}
Fraction degraded	0

Source: National Research Council of Canada (1981b).

Residues

Air, Water, and Sediments

PCDDs rarely occur at detectable levels in the atmosphere, reflecting their low concentration in other substrates. However, the incineration of municipal and industrial wastes and coal and volatilization of chlorinated pesticides may produce detectable residues at site-specific points. Hutzinger *et al.* (1981) reported that fly ash extract contained total –PCDDs of 4200 ng m^{-3}. TCDD occurred at 130 ng m^{-3}, whereas total-PCDF was 3600 ng m^{-3}. Much lower levels of TCDD were found in dust from metropolitan areas of Michigan and, in a Dow Chemical plant, dust contained TCDD levels of 1–4 ng g^{-1} (Kriebel, 1981). Although the importance of atmospheric release of PCDD by incineration and volatilization in determining levels in surface waters is not known, large-scale accidental industrial stack emissions may produce significant atmospheric contamination over wide geographic areas. For example, the release of TCDD at Seveso (Milan, Italy) during July 1976 resulted in atmospheric dust residues of 0.06–2.1 ng g^{-1} (Figure 10.4). Since these values were recorded in 1977 and 1978, much higher levels were probably present at the time of the accident. It was also

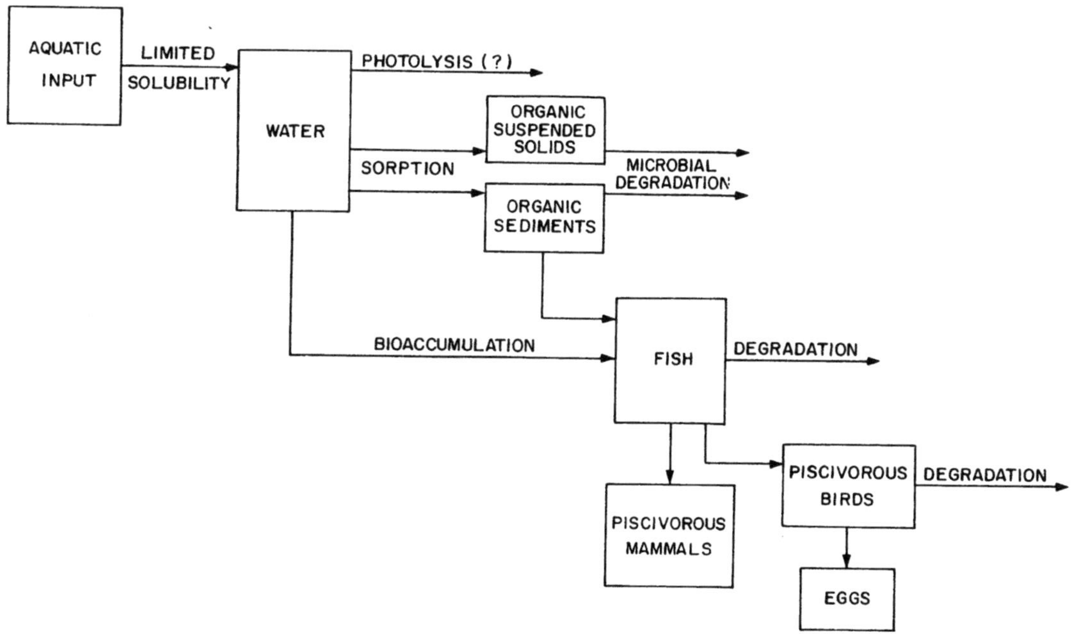

Figure 10.3. Step-wise fate processes predominating bioaccumulation. (From National Research Council of Canada, 1981a.)

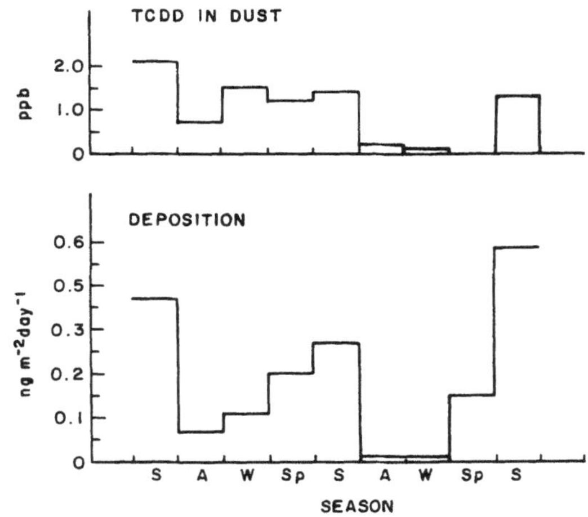

Figure 10.4. Seasonal variations in 2,3,7,8-TCDD levels in dust and rate of deposition of 2,3,7,8-TCDD at Seveso. Sp, spring; S, summer; A, autumn; W, winter. (From DiDomenico *et al.*, 1980a.)

estimated that TCDD fallout ranged from <0.04 to 0.79 ng m^{-2} day^{-1} (Figure 10.4). These rates produced significant contamination of soil and presumably water, though there are no data to confirm this latter point.

PCDDs do not accumulate significantly in water and consequently residues are generally below detectable limits, except in cases of extreme ambient pollution. For example, concentrations in Lake Ontario and Lake Erie are generally <1 pg L^{-1}, despite the input of substantial amounts of PCDD-bearing wastes from various chemical plants (National Research Council of Canada, 1981a). By contrast, less toxic derivatives of TCDD are found in relatively high concentrations in industrial zone waters. Lake *et al.* (1981) reported that the average concentration of 2,4,8-trichlorobenzofuran in suspended sediments was 0.25 ng L^{-1} in an industrialized region of Narragansett Bay (USA) and ~0.01 ng L^{-1} for a rural area.

Although little information is available, sediments probably act as a sink for PCDD. Their half-life in such substrates is long (>1 year), creating the possibility of long-term adulteration of the food chain in aquatic systems (Lake *et al.*, 1981). PCDDs also move downward through soils, thereby increasing the chance of contamination of ground water (DiDomenico *et al.*, 1980b). At the Seveso site, TCDD residues in the upper 7 cm of soil exceeded 1000 μg m^{-2} near the discharge point and decreased to <0.75 μg m^{-2} at 2 km from this area (DiDomenico *et al.*, 1980c). Soil samples collected near a Dow Chemical plant (USA) contained 1–120 μg TCDD kg^{-1}, compared with <0.03 and <0.005 μg kg^{-1} for urban and rural soils, respectively (Kriebel, 1981). Soil near a plant manufacturing trichlorophenol contained TCDD residues of up to 559 μg kg^{-1} (Van Ness *et al.*, 1980). Such data imply that surface run-off, particularly during periods of heavy, wet precipitation, may be a significant source of TCDD in natural waters.

Aquatic Invertebrates and Fish

Detectable levels of PCDD can often be related to discharge from specific industries and/or exposure to dioxin-containing pesticides. Following the heavy use of 2,4,5-T in South Vietnam, fish from the interior rivers of Dong Nai and Sai Gon contained average TCDD residues of 70–810 ng kg^{-1} wet weight (Baughman and Meselson, 1973). Detectable levels were also found in invertebrates (420 ng kg^{-1}) and in fish from the sea coast (80 ng kg^{-1}). Generally lower concentrations (<5–55 ng kg^{-1}) occurred in five species of fish from an industrial area of Michigan (USA), but one species yielded average residues of 157 ng kg^{-1} (Table 10.9). Comparable levels were found in piscivorous salmonids collected from Lake Ontario (Table 10.9); however, omnivorous species contained average residues of <10 ng kg^{-1}, whereas concentrations in fish from Lake Erie were generally below detectable limits. Apart from 2,3,7,8-TCDD, other PCDDs and derivatives occur in various aquatic and semiaquatic species (Table 10.10). The significance of such residues has not been established at this time.

Table 10.9. Concentration (ng kg^{-1} wet weight) of 2,3,7,8-TCDD in fish collected from the Tittabawasssee and Grand Rivers (Michigan), Lake Ontario, and Lake Erie.

Species	Average	Range
	Michigan Rivers[1]	
Channel catfish	157	28–695
Carp	55	20–153
Yellow perch	14	10–20
Smallmouth bass	8	7–8
White sucker	11	4–21
	Lake Ontario[2]	
Chinook salmon	35	26–39
Coho salmon	22	19–31
Lake trout	65	41–107
Rainbow trout	20	9–32
Brown trout	44	8–162
White perch	21	5–45
White sucker	7	<3–18
Brown bullhead	4	2–5
	Lake Erie[2]	
Coho salmon	1.4	<1–1.9
Walleye	0.6	<0.2–2.6
Smallmouth bass	0.5	<0.2–1.6

Sources: [1]Harless *et al.* (1982); [2]National Research Council of Canada (1981a).

Table 10.10. Concentrations (ng kg^{-1}) of various PCDDs in herring gulls and fish.

Homolog	Herring gull (Saginaw Bay, Lake Huron)	Unidentified fish (Tittabawasse River, Michigan)	Carp (Niagara River, USA)	Carp (Niagara River, Lake Ontario)
2,3,7,8-TCDD	70–160	81	417	53
other TCDD	—	ND–100	—	—
PCDD	—	31	33	9
HCDD (hexa)	88–20	ND–90	36	24
HCDD (hepta)	—	ND–44	100	12
OCDD	19–28	ND–150	158	19
2,3,7,8-TCDF	15–16	—	5	6
2,3,4,7,8-/ 1,3,4,8,9-P_5CDF	26–48	—	80	66
other PCDFs	2	—	—	9
HCDF (hexa)	40–59	—	203	135
HCDF (hepta)	17–18	—	304	44
OCDD	3–5	—	388	6

Source: National Research Council of Canada (1981a).
ND, not detected.
—, not measured.

Because sensitive analytical methods have only recently been developed, few data are available on long-term trends in PCDD residues. Ogilvie (1981) did, however, report that TCDD in herring gull eggs, collected from Lake Ontario in 1971 and then frozen, had an average burden of 0.800 μg kg^{-1} compared with 0.060 μg kg^{-1} in 1980. Although this decline largely reflects improvement in the control of waste discharges and use of TCDD-containing chemicals, uncontrolled seepage from chemical dumps continues to be a significant source of TCDD. For example, 3-month-old spottail shiners caught near the Love Canal in 1981 carried average residues of 59 ng kg^{-1}, the highest ever recorded for this species in the Niagara River (Canadian Environmental Control Newsletter, 1982).

CFs for TCDD in fish, ranging from 5400 to 35,500, are only slightly lower than those recorded for PCBs, DDT, and its metabolites but greater than those of other pesticides (Kenaga, 1980). Appreciable uptake from water occurs at concentrations as low as 0.001 μg L^{-1}, though no detectable uptake was found at concentrations <0.001 μg L^{-1} (Miller *et al.*, 1979). Fish and invertebrates may also accumulate TCDD directly from food (Matsumura and Benezet, 1973). However, the relative importance of the two routes of potential contamination have not been adequately documented. The half-life of TCDD in fish tissues is extremely long, ranging from at least 114 to 317 days (Miller *et al.*, 1979). This in turn implies that whole body residues in fish should be an indicator of the presence of TCDD in natural waters and a tool for detecting levels of TCDD at ecosystems <0.001 μg L^{-1}.

Toxicity

One of the most toxic environmental contaminants known is 2,3,7,8-TCDD. Helder (1980) reported that concentrations as low as 0.1 ng L^{-1} retarded egg development and growth of larval pike. Equally low levels resulted in 50% mortality of yolk-bearing fry within 96 hours whereas 10 ng L^{-1} caused almost 100% mortality. Similarly, consumption by rainbow trout of diets containing <2.3 mg kg^{-1} TCDD caused an average mortality of 88% in 71 days of exposure (Hawkes and Norris, 1977). Symptoms of chronic/sublethal intoxication by TCDD are numerous and include pathological changes in liver, such as focal necrosis and nuclear enlargement. These changes, together with induction of the enzyme aryl hydrocarbon hydroxylase, are similar to those observed in mammalian laboratory animals (Dewse, 1976). Other symptoms include a decrease in food consumption and weight gain, and an increase in the incidence of diseased fins. Many of the remaining homologs and derivatives of the PCDD group are significantly less toxic to fish than 2,3,7,8-TCDD. The 48-h LC_{50}'s of tetrachlorodibenzofuran to sheepshead minnows and the cladoceran *Daphnia magna* were >3.2 and 1.7 mg L^{-1}, respectively (Heitmuller *et al.*, 1981; LeBlanc, 1980). Dichlorobenzofuran also has low acute toxicity, possibly reflecting poor absorption from the gut and rapid excretion of conjugated hydroxy derivatives (Zitko *et al.*, 1973).

TCDD is extremely toxic to mammals, exhibiting acute, chronic, and subchronic effects. The LD_{50} in guinea pigs is only 2 μg kg^{-1} bw. The liver is a major target organ during acute exposure, and there appears to be little or no metabolism of TCDD in this organ. Dewse (1976) reported that TCDD was 30,000 times more potent an inducer of aryl hydrocarbon hydroxylase (AHH) than 3-methyl-cholanthrene. AHH is strongly implicated in yielding chemical intermediates that are carcinogenic to the host organism. Similarly, Hay (1981) reported that repeated doses of 5 μg kg^{-1} to pregnant mice produced cleft palate in 65% of fetuses. Symptoms of sublethal/chronic intoxication in humans include chloracne, alopecia, hematurea, peripheral numbness, and weight loss, occasionally culminating in extreme emaciation. Industrial exposure to TCDD may produce metabolic disturbances such as elevated levels of lipids with abnormalities in the lipoprotein spectrum (Pazderova-Vejluklovė *et al.*, 1981). There is also focal damage to the nervous system and development of mild liver lesions. Most individuals exposed to TCDD have apparently completely recovered from intoxication.

References

Baughman, R., and M. Meselson. 1973. An analytical method for detecting TCDD (dioxin): levels of TCDD in samples from Vietnam. *Environmental Health Perspectives* **5**:27–35.

Canadian Environmental Control Newsletter. 1982. **213**:1791–1798.

Cavallaro, A., G. Tebaldi, G. DeFelice, G. Colli, and R. Gualdi. 1979. Indagine sperimentale sulla penetrazione di TCDD nella zona A di Seveso. *Boll. Chim. d'Unione Ital. Lab. Prov. Sci.* **5**:489–541.

Chiou, L.J., L.J. Peters, and V.H. Freed. 1979. A physical concept of soil-water equilibra for non-ionic organic compounds. *Science* **206**:831–832.

Crosby, D.G., and A.S. Wong. 1977. Environmental degradation of 2,3,7,8-tetrachlorodibenzo-*p*-dioxin (TCDD). *Science* **195**:1337–1338.

Crosby, D.G., A.S. Wong, J.P. Plimmer, and E.A. Woolson. 1971. *Photodecomposition of chlorinated dibenzo-p*-dioxins. *Science* **173**:748–749.

Crummett, W.B., R.R. Bumb, L.L. Lamparski, N.H. Mahle, T.J. Nestrick, and L.W. Whiting. 1981. Environmental chlorinated dioxins from combustion—the trace chemistries of fire hypothesis. *In*: O. Hutzinger, R.W. Frei, E. Merian, and F. Pocchiari (Eds.), *Impact of chlorinated dioxins and related compounds on the environment*. Pergamon Press, Oxford, pp. 253–263.

Dewse, C.D. 1976. Dangers of T.C.D.D. *The Lancet* **2**:363.

DiDomenico, A., V. Silano, G. Viviano, and G. Zapponi. 1980a. Accidental release of 2,3,7,8-tetrachlorodibenzo-*p*-dioxin (TCDD) at Séveso, Italy. V. Environmental persistence of TCDD in soil. *Ecotoxicology and Environmental Safety* **4**:339–345.

DiDomenico, A.D., V. Silano, G. Viviano, and G. Zapponi. 1980b. Accidental release of 2,3,7,8-tetrachlorodibenzo-*p*-dioxin (TCDD) at Séveso, Italy. IV. Vertical distribution of TCDD in soil. *Ecotoxicology and Environmental Safety* **4**:327–338.

DiDomenico, A.D., V. Silano, G. Viviano, and G. Zapponi. 1980c. Accidental release of 2,3,7,8-tetrachlorodibenzo-*p*-dioxin (TCDD) at Séveso, Italy. II. TCDD distribution in the soil surface layer. *Ecotoxicology and Environmental Safety* **4**:298–320.

Dobbs, A.J., and C. Grant. 1979. Photolysis of highly chlorinated dibenzo-*p*-dioxins by sunlight. *Nature* **278**:163–165.

Firestone, D. 1977. Chemistry and analysis of pentachlorophenol and its contaminants. *FDA By-lines* **2**:57–89.

Frost, A.A., and R.G. Pearson. 1961. *Kinetics and mechanism: a study of homogeneous chemical reactions*. John Wiley and Sons, New York.

Guenthner, T.M., J.M. Fysh, and D.W. Nebert. 1979. 2,3,7,8-tetrachlorodibenzo-*p*-dioxin: covalent binding of reactive metabolic intermediates principally to protein *in vitro*. *Pharmacology* **19**:12–22.

Halter, M.T., and H.E. Johnson. 1977. A model system to study the desorption and biological availability of PCB in hydrosoils. *In*: F.L. Mayer and J.L. Hamelink (Eds.), Aquatic toxicology and hazard evaluation. *ASTM STP* **634**:178–195.

Harless, R.L., E.O. Oswald, R.G. Lewis, A.E. Dupuy, Jr., D.D. McDaniel, and H. Tai. 1982. Determination of 2,3,7,8-tetrachlorodibenzo-*p*-dioxin in fresh water fish. *Chemosphere* **11**:193–198.

Hawkes, C.L., and L.A. Norris. 1977. Chronic oral toxicity of 2,3,7,8-tetrachlorodibenzo-*p*-dioxin (TCDD) to rainbow trout. *Transactions of the American Fisheries Society* **106**:641–645.

Hay, A. 1981. Chlorinated dioxins and the environment. *Nature* **289**:351–352.

Heitmuller, P.T., T.A. Hollister, and P.R. Parrish. 1981. Acute toxicity of 54 industrial chemicals to sheepshead minnows (*Cyprinodon variegatus*). *Bulletin of Environmental Contamination and Toxicology* **27**:596–604.

Helder, T. 1980. Effects of 2,3,7,8-tetrachlorodibenzo-*p*-dioxin (TCDD) on early life stages of the pike (*Esox lucius* L.) *The Science of the Total Environment* **14**:255–264.

Helling, C.S., A.R. Isensee, E.A. Woolson, P.D.J. Ensor, G.E. Jones, J.R. Plimmer, and P.C. Kearney. 1973. Chlorodioxins in pesticides, soils and plants. *Journal of Environmental Quality* **2**:171–178.

Hutzinger, O., K. Olie, J.W.A. Lustenhouwer, A.B. Okey, S. Bandiera, and S. Safe. 1981. Polychlorinated dibenzo-*p*-dioxins and dibenzofurans: a bioanalytical approach. *Chemosphere* **10**:19–25.

Isensee, A.R., and G.E. Jones. 1975. Distribution of 2,3,7,8-tetrachlorodibenzo-*p*-dioxin (TCDD) in an aquatic model ecosystem. *Environmental Science and Technology* **9**:668–672.

Karickhoff, S.W., D.S. Brown, and T.A. Scott. 1979. Sorption of hydrophobic pollutants on natural sediments. *Water Research* **13**:241–248.

Kenaga, E.E. 1980. Correlation of bioconcentration factors of chemicals in aquatic and terrestrial organisms with their physical and chemical properties. *Environmental Science and Technology* **14**:553–556.

Kriebel, D. 1981. The dioxins: toxic and still troublesome. *Environment* **23**:6–13.

Lake, J.L., P.F. Rogerson, and C.B. Norwood. 1981. A polychlorinated dibenzofuran and related compounds in an estuarine ecosystem. *Environmental Science and Technology* **5**:549–553.

LeBlanc, G.A. 1980. Acute toxicity of priority pollutants to water flea (*Daphnia magna*). *Bulletin of Environmental Contamination and Toxicology* **24**: 684–491.

Liss, P.S., and P.G. Slater. 1974. Flux of gases across the air-sea interface. *Nature* **247**:181.

Lustenhouwer, J.W.A., K. Olie, and O. Hutzinger. 1980. Chlorinated dibenzo-*p*-dioxins and related compounds in incinerator effluents: a review of measurements and mechanisms of formation. *Chemosphere* **9**:501–522.

Mackay, D. 1978. Volatilization of pollutants from water. *In*: O. Hutzinger, I.H. Van Lelyveld, and B.C.J. Zoeteman (Eds.), *Aquatic pollutants: transformation and biological effects*. Pergamon Press, Toronto, pp. 175–185.

Mackey, D., and A.W. Wolkoff. 1974. Rate of evaporation of low solubility contaminants from water bodies to atmosphere. *Environmental Science and Technology* **8**:611–614.

Marshall, W.K., and J.R. Roberts. 1977. Simulation modelling of the distribution of pesticides in ponds. *In*: J.R. Roberts, R. Greenhalgh, and W.K. Marshall (Eds.), *Proceedings of a symposium on fenitrothion. The long-term effects of its use in forest ecosystems*. National Research Council of Canada, Publication No. 16073.

Matsumura, F., and H.J. Benezet. 1973. Studies on the bioaccumulation and microbial degradation of 2,3,7,8-tetrachlorodibenzo-*p*-dioxin. *Environmental Health Perspectives* **5**:253–258.

Miller, R.A., L.A. Norris, and B.R. Loper. 1979. The response of coho salmon and guppies to 2,3,7,8-tetrachlorodibenzo-*p*-dioxin (TCDD) in water. *Transactions of the American Fisheries Society* **108**:401–407.

National Research Council of Canada. 1981a. *Polychlorinated dibenzo-*p*-dioxins: Criteria for their effects on man and his environment.* Associate Committee on Scientific Criteria for Environmental Quality, National Research Council of Canada Publication No. 18574, Ottawa, 251 pp.

National Research Council of Canada. 1981b. *A screen for the relative persistence of organic chemicals in aquatic ecosystems—an analysis of the role of simple computer models.* Associate Committee on Scientific Criteria for Environmental Quality, National Research Council of Canada Publication No. 18570. Ottawa, 302 pp.

Ogilvie, D. 1981. Dioxin found in the Great Lakes basin. *Ambio* **10**:38–39.

Parlar, H., S. Gaeb, and I. Gevefugi. 1978. Photo-chemical reactions of 2,3,7,8-tetrachlorodibenzo-*p*-dioxin (TCDD) adsorbed on a silica gel. *Pergamon Series on Environmental Science* **1**:465–466.

Pazderova-Vejłupkova, J., M. Nemcova, J. Pícková, L. Jirásek, and E. Lukás. 1981. The development and prognosis of chronic intoxication by tetrachlorodibenzo-*p*-dioxin in men. *Archives of Environmental Health* **36**:5–11.

Plimmer, J.R. 1978. Photolysis of TCDD and trifluratin on silica and soil. *Bulletin of Environmental Contamination and Toxicology* **20**:87–92.

Plimmer, J.R., U.I. Klingebeil, D.C. Crosby, and A.D. Wong. 1973. Photochemistry of dibenzo-*p*-dioxins. *Advances in Chemistry Series* **120**:44–54.

Poiger, H., and Ch. Schlatter. 1979. Biological degradation of TCDD in rats. *Nature* **281**:706–707.

Ramsey, J.C., J.G. Hefirer, R.J. Karbowski, W.H. Braun, and P.J. Gehring. 1979. The *in vivo* bio-transformation of 2,3,7,8-tetrachlorodibenzo-*p*-dioxin (TCDD) in the rat. *Toxicology and Applied Pharmacology* **48**:A162.

Southworth, G.R. 1979. Transport and transformation of anthracene in natural waters. *In*: L.L. Marking and R.A. Kimerle (Eds.), *Aquatic toxicology. ASTM STP* **667**:359–380.

Stehl, R.H., R.R. Papenfuss, R.A. Bredeweg, and R.W. Roberts. 1973. The stability of pentachlorophenol and chlorinated dioxins to sunlight, heat, and combustion. *Advances in Chemistry Series* **120**:119–125.

Van Miller, J.P., R.J. Marlar, and J.R. Allen. 1976. Tissue distribution and excretion of tritiated tetrachlorodibenzo-*p*-dioxin in non-human primates and rats. *Food and Cosmetic Toxicology* **14**:31–34.

Van Ness, G.F., J.G. Solch, M.L. Taylor, and T.O. Tiernan. 1980. Tetrachlorodibenzo-*p*-dioxins in chemical wastes, aqueous effluents and soils. *Chemosphere* **9**:553–563.

Ward, C.T., and F. Matsumura. 1978. Fate of 2,3,7,8-tetrachlorodibenzo-*p*-dioxin (TCDD) in a model aquatic environment. *Archives of Environmental Contamination and Toxicology* **7**:349–357.

Young, A.L., C.E. Thalken, E.L. Arnold, J.M. Cupello, and L.G. Cockerham. 1976.

*Fate of 2,3,7,8-tetrachlorodibenzo-*p*-dioxin (TCDD) in the environment: summary and decontamination recommendations*. United States Air Force Academy Report No. TR 76-18, 41 pp.

Zitko, V., D.J. Wildish, O. Hutzinger, and P.M.K. Choi. 1973. Acute and chronic oral toxicity of chlorinated dibenzofurans to salmonid fishes. *Environmental Health Perspectives* **5**:187–189.

11
Prioritization and Hazard Assessment of Chemicals

Synthetic organic chemicals are increasing in number and diversity every year. This creates a need to assess their potential environmental hazard, in addition to screening chemicals that are already in use but have not been tested adequately. Complete experimental assessment, with all conceivable tests, is almost impossible considering the necessary labor and resources. Our knowledge of various physico-chemical and biological processes operating in the environment, conditions that tend to slow down or limit the processes, and the interaction among processes have grown greatly in the last decade. This allows the extrapolation of laboratory findings to natural environmental situations. Decisions should be made to prevent needless experimentation and to ensure results from critical tests. The basic process in any hazard evaluation will be to predict Expected Environmental Concentration (EEC) from the environmental characteristic behavior of the compound and the use pattern. If the EEC is less than the experimentally determined no-effect level for appropriate environmental organisms, then the chemical should be considered acceptable. Estimation of environmental exposure can be made for a localized situation where the source input(s) or the receiving system can be identified. In many systems "the benchmark approach" was used to match the properties of a new chemical with a homolog or analog to predict its environmental distribution, e.g. "DDT-like materials."

The persistence and fate of a pollutant depend as much on environmental factors as on its physico-chemical properties. Hence, it is important that testing protocols be designed to maximize the applicability of the end results to different environmental conditions. It is necessary to decide how closely models should simulate the natural environments. Too many descriptive details can hinder its use for assessment of general patterns of pollutant's fate (Thomann, 1981). Mathematical models can potentially be used to predict the fate of a chemical in any well-characterized environmental situation.

The resolution in prediction does not necessarily increase linearly with the refinements made on the relation describing a given environmental process (O'Neill, 1971; Miller *et al.*, 1976). In the case of screening models, there is an optimum point above which an increase in the complexity of the model does not increase the usefulness of its predictions as basic indicators of behavior patterns (Roberts *et al.*, 1981). Complex models can mask some very basic problems inherent to our data base. There are fundamental dangers in this situation where we can create elegant scenarios and complex tables of numbers that do not possess well-established or reasonable confidence limits (Roberts *et al.*, 1981). Thus, a less-detailed model is preferable to permit the user to recognize easily its limitations and make scientific adjustments accordingly.

Environmental Distribution

Development of the Decision Tree

Neely (1979) developed a decision tree (Figure 11.1) that is an extension of several previous studies on compartmental analysis. The profile gives an estimated distribution of the chemical in air, water, and soil. Decisions on environmental assessment can be made by comparing this profile with the intended use pattern. The "YES" output from boxes K, L, and M indicate the need for further studies on degradation. Description of the lettered boxes is as follows:

A. Use Pattern

Information on application pattern of the chemical, sphere of entry (atmosphere, hydrosphere, etc.), and the estimate of rate of environmental release is required.

B. Confined Use

If the use pattern indicates no entry into the environment leading to negligible EEC, no further testing is required.

C. Polymer

If the chemical is an insoluble polymer, then interactions associated with the solid waste disposal as landfill or incineration should be investigated.

D. Ionic Material

Interaction with suspended solids, sediments, and transformation is possible and should be assessed.

E. Partitioning Pattern

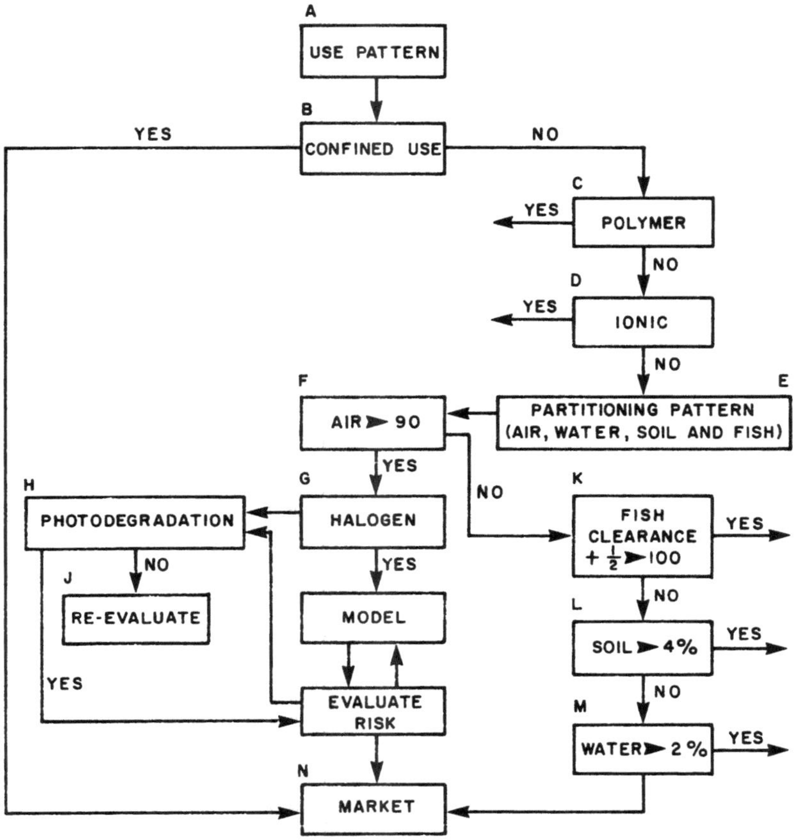

Figure 11.1. Decision tree for choosing appropriate testing. (From Neely, 1979.)

Distribution profile of the chemical in air, water, fish, and sediment is calculated.

The chemical is introduced to the water compartment (Figure 11.2) at a fixed rate of 0.15 g hr^{-1} for a 30-day period, followed by a 30-day clearance phase. The half-life for clearance from the fish biomass is estimated and the percent distribution of the chemical in all the compartments at the end of the 30-day clearance period is determined.

From a knowledge of solubilities and vapor pressures of a wide range of chemicals from toluene to DDT, Neely (1978) developed a set of four regression equations that account for the results in a statistically significant manner.

$$\text{Percent of chemical in air} = -0.247\,(1/H) + 7.9 \log S + 100.6 \quad (1)$$

$$\text{Percent of chemical in water} = 0.054\,(1/H) + 1.32 \quad (2)$$

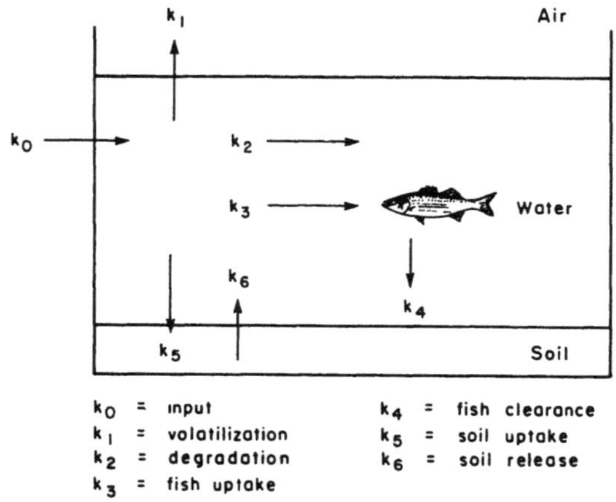

Figure 11.2. Dynamics of a chemical in an aquatic ecosystem. (Neely, 1980.)

$$\text{Percent of chemical in sediment/soil} = 0.194\ (1/H) - 7.65 \log S - 1.93 \tag{3}$$

$$\text{and } \log (t_{1/2}) = 0.0027\ (1/H) - 0.282 \log S + 1.08 \tag{4}$$

where

$$H\ (\text{mm Hg}\cdot\text{mole}^{-1}) = \frac{\text{vapor pressure} \times \text{molecular weight}}{\text{solutility (mg L}^{-1})}$$

$$S\ (\text{mM liter}^{-1}) = \frac{\text{solubility (mg L}^{-1})}{\text{molecular weight}} \quad \text{and}$$

$t_{1/2}$ (hours) = half-life for clearance from fish in this ecosystem.

The chemicals tested with their relevant properties are given in Table 11.1.

Table 11.2 presents the actual and predicted distributions from the regression equations of the tested chemicals. The agreement seems to be reasonably good.

F. Air

Based on the results from E, either go to G (air >90%) or proceed to K for other interactions.

G. Halogen

If the chemical is a halogenated compound, use any suitable model (Crutzen *et al.*, 1978; National Academy of Sciences, 1976; Neely, 1977) to estimate the mass transfer of chlorine from the troposphere to the stratosphere. This is to evaluate the relative risk caused by halogen atoms in the

Table 11.1. Chemicals tested in the simulated aquatic ecosystem and their properties.

Chemical	Molecular weight	Vapor pressure (mm Hg)	Water solubility (mg L^{-1})
Toluene	92	30	470
p-Dichlorobenzene	147	1	79
Trichlorobenzene	180	0.5	30
Hexachlorobenzene	285	10^{-5}	0.035
Diphenyl	154	9.7×10^{-3}	7.5
Trichlorobiphenyl	256	1.5×10^{-3}	0.05
Tetrachlorobiphenyl	291	4.9×10^{-4}	0.05
Pentachlorobiphenyl	325	7.7×10^{-3}	0.01
DDT	350	10^{-7}	1.2×10^{-3}
Perchloroethylene	166	14	150

Source: Neely (1980).

destruction of atmospheric ozone. If the volatile chemical is free from halogen, proceed to box H.

H. Photodegradation

Here the photodegradation potential of chemicals containing C—H or C≡C bonds that are susceptible to attack by hydroxyl radical in the troposphere will be estimated. Such degradation will eliminate the tropo-

Table 11.2. Actual and predicted distribution of chemicals tested in the simulated ecosystem.

Chemical	Water, %	Soil, %	Air, %	$t_{1/2}$ from fish (hours)
Toluene	0.9 (1.33)	0.4 (~0)	98.6 (~100)	10 (7.6)
p-Dichlorobenzene	1.24 (1.31)	1.28 (0.24)	97.5 (98)	15 (14)
Trichlorobenzene	1.33 (1.34)	2.06 (4.09)	96 (94)	17 (20)
Hexachlorobenzene	3.57 (1.98)	39.4 (31)	56 (68)	162 (164)
Diphenyl	2.27 (1.59)	5.4 (9)	92.2 (89)	27 (29)
Trichlorobiphenyl	1.38 (1.33)	15.2 (26)	83 (71)	96 (134)
Tetrachlorobiphenyl	1.5 (1.34)	17 (27)	81 (71)	104 (139)
Pentachlorobiphenyl	1.5 (1.34)	21 (33)	77 (65)	229 (226)
DDT	1.26 (3.17)	67.5 (46.5)	28 (49)	915 (517)
Perchloroethylene	1 (1.32)	1 (~0)	98 (100)	14 (12)

Source: Neely (1980).

spheric buildup of the parent chemical. If there is no potential problem, proceed to N.

J. Reevaluation

Once a material is assigned to this box, a continuous reevaluation considering search for other degradation pathways must be conducted.

K. Fish Clearance

Values of $t_{1/2} > 100$ hours indicate a potential bioconcentration problem in the aquatic environment. This decision is arbitrary and based on the bench-mark concept comparing with chemicals that are known to bioconcentrate and have $t_{1/2} > 100$ hours (Table 11.2). Consequently, further tests on the types of degradation products and possible bioaccumulation in the aquatic system should be investigated. When $t_{1/2} < 100$ hours, proceed to L.

L. Soil/Sediment

Using the bench-mark approach, if the amount of the chemical in soil/sediment is ≥4%, degradation should be examined. If not, proceed to M.

M. Water

If the concentration of the chemical in the aqueous phase is >2%, transformation should be studied.

If no long-term environmental problem based on the use pattern is anticipated from this approach, then the production of the chemical should be either continued or permitted if it is a new chemical. It should be pointed out that this model is designed to assess environmental as opposed to human health hazard. It assumes that the industrial hygiene and animal toxicity studies indicate a reasonable balance in the risk-benefit analysis. It should be concluded that the testing should continue to insure that the EEC resulting from the use is below the no-effect level. If the EEC is close to no-adverse biological effect level, then further characterization of the ecosystem, for example the receiving water body, is required. It includes surface area, depth, hydrological characteristics, temperature, salinity, trophic status, suspended solid concentration, and percent organic content of the bottom sediment.

Case Studies

Kepone, a pesticide, was accidentally discharged into the James River, Virginia (USA). Performing analysis with the decision tree, Neely (1980) identified Box E as the most sensitive compartment, and the internal profile predicted that bioconcentration ($t_{1/2} > 100$ hours) (Table 11.3) and sorption-desorption from bottom sediment will determine the distribution and longevity of the problem, respectively. This analysis was based on physical properties and did not include any degradative mechanisms since they were not available. Further tests on its degradation (Dawson *et al.*, 1978) also supported the importance of bioconcentration and confirmed preliminary

Table 11.3. The partioning profile of case study chemicals generated from box E of the decision tree analysis.

Chemical	% of Chemical in the Soil	Air	Water	$t_{1/2}$ for clearance from fish (hours)
Kepone	62	23	14	231
Mirex	37	60	1.4	320
Chlorpyrifos	74	8.5	18	335
Monochlorobenzene	~0	~100	1.34	8

Source: Neely (1979).

analysis. In the kepone incident, the decision tree was capable of focusing quickly on the key areas for further testing.

Mirex, a chlorinated hydrocarbon, behaved similarly to kepone except for one important difference. Mirex has a relatively high volatility rate similar to DDT and thus circulated widely through the atmosphere. The third chemical studied was chlorpyrifos (0-0-diethyl 3,5,6-trichlor -2 pyridyl phosphorothioate), used as an insecticide. Initial analysis showed $t_{1/2} = 335$ hours, but using data for aqueous hydrolysis, metabolism by fish, and photodegradation in air and water, a much faster fish clearance time (<100 hours) was calculated. The analysis concluded that chlorpyrifos is not persistent in aquatic ecosystems. The only precaution is that the initial application rate must be adjusted below the acute toxicity level for fish. The last chemical tested, monochlorobenzene, showed a different profile from the rest. Its residence phase is the atmosphere with no potential for bioconcentration (Table 11.3). Data on its rapid photodegradation in the troposphere (Dilling *et al.*, 1976) were used in the model to predict that >99% of MCB will be degraded in the lower troposphere with very little entering the stratosphere to damage the ozone layer. In conclusion, using physical and chemical properties, it is possible to predict with reasonable accuracy the resident phase of a chemical. Based on this information, the next steps should include the determination of the magnitude of bioconcentration effect, biodegradation rates in sediment and water, and acute and chronic effects on various target organisms. This information should be fed into the cycle and the process repeated until the investigator is satisfied that the EEC is below the no-effect level. Once this is reached, no further testing is required.

It should be pointed out here that for any prioritization scheme, only a limited set of critical tests is performed to arrive at the EEC. Estimation of EEC is a difficult process, controlled by identifying correctly and accurately the source inputs for the chemical. It may be accomplished in a localized aquatic situation but for most other systems, a bench-mark approach is used to predict the environmental distribution and persistence of the chemical. It should be pointed out that the decision tree model proposed in Figure 11.1 is designed to assess the environmental hazard and not the human health hazard. Although the results are given in numbers (percents), they are not absolute and provide only a relative ranking of testing priorities among compartments. This profile allows the investigator a visualization of the environmental distribution of the chemical. Further environmental tests can be designed to verify the prediction.

It is important at this stage to outline the limitations of Neely's decision tree model (1979) and identify gaps in the information base for several compartments, if not for all. It is hoped that the user will be aware of the limitations and restraints and the fact that the applicability of the end-results from this model will depend upon the information available for the various compartments for either specific scenario or a general assessment.

Use Pattern. This step is important in any assessment program, and the lack of reliable information on the amount of the chemical released into the environment is the single largest source of error (Branson, 1980). The current or expected pattern of usage (for chemical in use or new chemical under development, respectively), disposal, and release to the environment should be determined as accurately as possible. Efforts should be devoted to identifying special exposure situations, if any. Missing information should be categorized and documented for a chemical that is likely to be accepted. Extra assurance should be guaranteed by gathering data on gross amount used from (i) direct production estimates, (ii) indirect production estimates, and (iii) product usage estimated by calculating the content of the given chemical in product(s) and, in turn, the total per capita usage. Care must be exercised to account adequately for the multiple-use materials and to allow reasonable increases in amounts for growth factors and new applications (Duthie, 1977). In addition, time pattern for usage and regularity of usage should be assessed in determining the potential for build-up of the chemical in the environment. Geographical patterns for heavy usage in certain areas leading to higher concentrations of the chemical than the national average should be recognized. More importantly, periodic reviewing of usage and disposal pattern is essential in the decision-making process.

Confined Use. The term "confined use" must be well defined with established bench-mark histories of products that are in a similar use pattern before reaching a conclusion that the product is truly confined and sealed away from environmental exposure.

Polymer. Possible depolymerization processes (thermal and photochemical) and their kinetics should be investigated before concluding that the polymer is highly inert.

Ionic Material. The kinetics of sorption-desorption processes of the ionic compound with the solid substrates under a variety of environmental conditions should be determined to arrive at the equilibrium concentration of the ionic compound in the free form. Complex/chelate formation of the ionic compound with the anions (organic and inorganic), micro-, and macro-solutes should be investigated to determine the extent of alteration in speciation and their behavior in water. Chelate formation with ligands containing donor atoms such as N, S, and Se, or even Cl^- in some cases ($HgCl_2$) can modify the hydrophilic ionic compound into a strong lipid-soluble covalent compound.

Partitioning Pattern. The determination of accurate values for solubility and vapor pressure of organic chemicals in water is difficult. Reported literature values for solubility and vapor pressure for several organic chemicals vary by several fold (Versar, 1979). This will seriously bias the distribution pattern of the chemical. In addition, sorption to surfaces of the container materials as well as volatilizational loss of the chemical will lead to lower exposure concentrations in water. Hence, these fate processes have to be accounted for and the system is mass balanced to arrive at the true exposure concentration for uptake at the interface of biotic and abiotic materials.

Recently reported nonlinear dependence of toxicity on log K_p (octanol/water) (Hansch, 1980a) as well as the role of hydrophobicity in the sequestering of chemicals by proteins and other materials must be considered in the toxicity evaluation based on log K_p (octanol/water).

Air. Compounds other than halogenated ones such as NO_x and SO_x can likely cause damage to the ozone layer; hence, they also should be included for investigation under this box.

Clearance of Compounds from Fish. For screening purposes, the benchmark compounds with the following characteristics should be chosen:

(i) The compounds chosen must include the various man-made categories (industrial, agricultural, etc.) to represent a matrix of susceptibility to fate processes.

(ii) Many industrial chemical products are not pure single compounds. Instead, they are mixtures of isomers, homologs, varied polymeric chain length, as well as impurities or by-products. When formulated mixtures (or complex technical grade material) are to be used, comparison of the testing mixture (or technical grade compound) and the major active ingredient should be considered to rule out unexpected additive or synergistic effects.

(iii) Within each class of chemical reactivity represented, compounds with different sorptive capacities (high or low log K_p (octanol/water) must be included as bench-mark compounds. This is important in identifying the partitioning compartment and determining the extent of the degradation of the sorbed chemical. For example, consider two hypothetical compounds that are readily degraded only by anaerobic bacteria. If one compound is highly hydrophilic and the other highly hydrophobic, then the latter compound will partition into the sediment and will undergo anaerobic microbial degradation. Thus, it will be less persistent in water than the hydrophilic compound.

(iv) The bench-mark compounds must be well characterized for their fate processes operative under field situations.

At present, the available bench-mark compounds cannot satisfy these criteria owing to insufficient number of well-characterized compounds (Roberts *et al.*, 1981). It is necessary to develop and validate test methods for measuring rates of degradation, evaporational loss, and bioaccumulation of chemicals in water and the results expressed as measured rate constants. It is assumed that pollutant concentrations are always below their solubility limits and not in enough concentrations to saturate all competing compartments. For comparative purposes, all the processes are assumed to follow the simple first-order or pseudo-first-order rate kinetics. One of the long-term objectives of any screening program using models must be to examine the possibility of more complex kinetics operating in the environment as a function of time and concentration. This might reveal the saturation point of the compartment, including the simple process such as sorption.

In real situations, the EEC and No-Effects Concentration (NEC) are not constant, although they appear constant in figures. Varying usage and disposal pattern and accompanying fate processes contribute to the fluctuation in EEC. Varying EECs and NECs result in a fluctuating risk level. If estimates of these variations are available, the probability of occurrence of extremely high EEC and extremely low NEC can be calculated. This will refine the risk estimation and help to estimate the frequency at which a given risk level is exceeded in any given fixed time period such as summer season during which the flow is low and the temperature is high (Dickson *et al.*, 1979).

Prioritization for Detailed Assessment

Initial Chemical Selection

The names of chemicals are fed into the initial lists for assessment via several routes; on an *ad hoc* basis from within EPA, other agencies, public petitions, and in future from chemical industries (Muir, 1980).

Abbreviation of the List

Chemicals that are not in commercial production or those that are already being regulated such as food additives, pesticides, and drugs are eliminated from the initial list. Also, some chemicals are eliminated on the basis of production volume, environmental release, occupational exposure, and general human exposure. Finally, the list is further condensed by considering the potential biological activity and the need for health and environmental effects testing. Thus scoring in the assessment process assists in the application of the limited analytical resources to probable high-priority chemicals.

Preliminary Evaluation

Once a chemical is selected for assessment, a chemical hazard information profile (CHIP) is compiled through review of available literature, existing documentation from the Toxic Substances Control Act (TSCA) files, and computer information systems. The profile is a summary of known effects and exposure information (Figure 11.3). After scrutiny, the chemical is assigned either a low priority, a need for more information, or high priority to warrant further assessment.

Restricted access to production and use information will limit the development of sound estimates of exposure and could weaken the assessment process. Also, the method of exposure evaluation favors chemicals of multiple routes of exposure over chemicals having an extremely high value for only one type of exposure. Property-effect relationships, toxicity models, and scoring systems are valuable when used at their specific level; however, setting priorities for testing, which is a very complex task, still involves almost entirely subjective judgments of experienced researchers.

The transport and transformation of chemicals provide valuable information on their exposure concentration used in risk analysis. The chemical on entry into the environment will be influenced by its physico-chemical properties and the transport processes to concentrate in a particular compartment (Figure 11.4). Biotic and abiotic transformation processes could degrade the chemical.

The net concentration (exposure concentration) of the toxic chemical in the environment is given by equation (5)

$$\frac{dc}{dt} = \Sigma \frac{k_i C}{1 + K_s} \tag{5}$$

where C = original concentration of the chemical, K = partition coefficient between two phases, k_i = rate constant for various fate processes, and s = mass of the sorbent in a particular compartment. Neely *et al.* (1974) first reported a semiempirical relationship between log P_{ow} (octanol/water

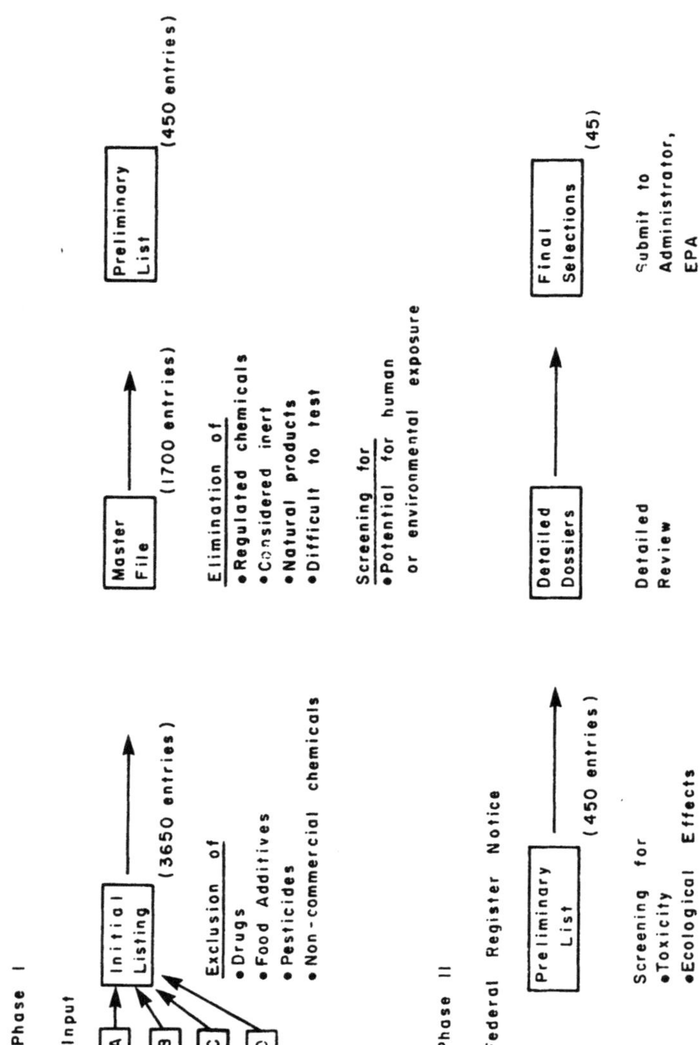

Figure 11.3. Schematics of the prioritization process. (From Stephenson, 1980.)

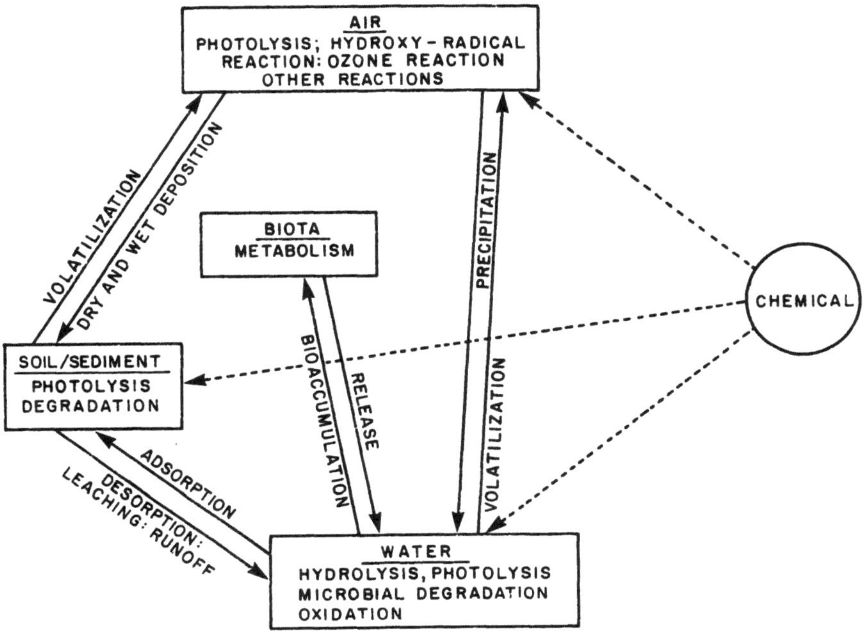

Figure 11.4. Schematic illustration of transport and fate processes of a chemical in the environment. (Haque *et al.*, 1980.)

mixture) and the bioconcentration factor (BCF) for many organic chemicals (equation 6)

$$\log(\text{BCF}) = 0.524 \log[P] + 0.124 \tag{6}$$

Laboratory studies have shown a similar relationship for ecological magnification $[E \cdot M.]$ (Lu and Metcalf, 1975) (equation 7).

$$\log[E \cdot M.] = 0.7285 + 0.6335 \log P_{ow} \tag{7}$$

It should be noted that these correlations are semiquantitative and may fail for low P values or for chemicals whose speciation is not the same in both phases.

Screening of Toxic Chemicals

Information from transport and fate processes can be utilized to screen toxic chemicals as to their environmental behavior using (i) the bench-mark concept and (ii) the structure-activity relationship.

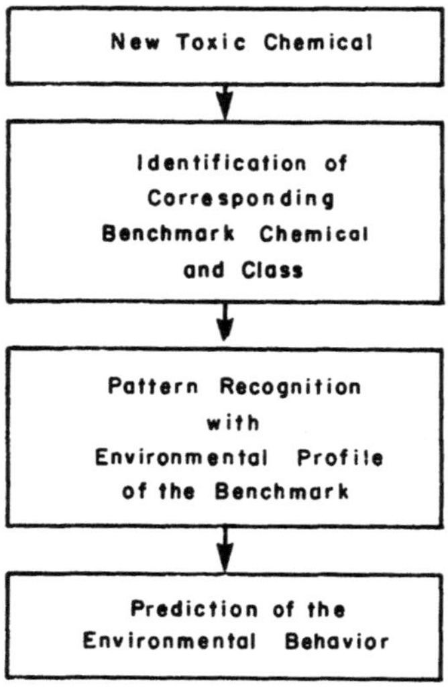

Figure 11.5. Schematic illustration of the bench-mark concept. (From Haque *et al.*, 1980.)

Bench-Mark Concept

This concept involves the selection of one or more bench-mark chemicals from important classes of toxic chemicals and measurement of their essential environmental parameters and physico-chemical properties. This information is then integrated to build an environmental profile (Haque *et al.*, 1980) for the class of chemical. The new chemical is compared with the corresponding bench mark for structural similarity and, using pattern recognition methods, the behavior of the new chemical can be predicted (Figure 11.5).

The environmental parameters and physico-chemical properties included in the bench-mark concept are aqueous solubility, vapor pressure, aqueous hydrolysis, sorption-desorption, biotic and abiotic transformations, and partitioning measured under standard conditions. Chemicals with high log P_{ow} and high environmental persistence value will be flagged for their potential toxicity through bioaccumulation and relatively long periods of exposure to humans and the environment, respectively. Figure 11.6 shows a hypothetical environmental profile of a chemical that could be used to predict its behavior with time. Further development is needed for field data, validation, and extension of the concept to effects and toxicity pattern.

This concept is based on the correlation of variation of certain biological activities of chemicals with functional substituents on the chemical (Hansch, 1980b). Hansch demonstrated that toxicity of a series of chemicals may be correlated to their log P_{ow} and electronic properties of the substituent. In general,

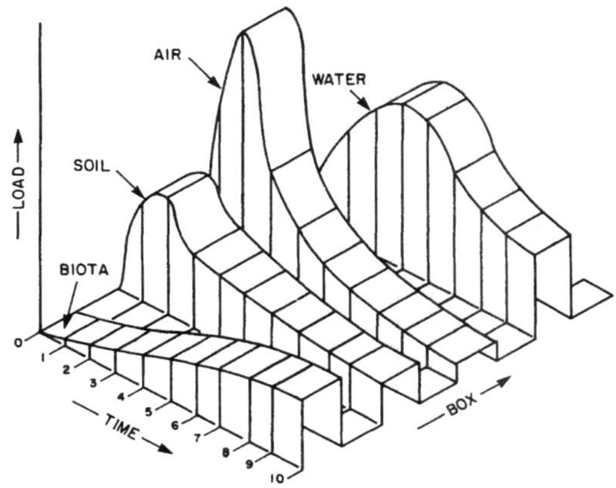

Figure 11.6. Hypothetic environmental profile of a chemical. (From Haque *et al.*, 1980.)

$$\log C = a \, . \log P_{ow} + b . \log P_{ow}^{2} + k \quad (8)$$

where P_{ow} = partition coefficient in octanol/water, C = concentration of the chemical required to produce certain toxic effects, and a, b, and K are constants. The constant K represents electronic parameters. The structure-activity concept is restricted to a series of chemicals of common basic skeletal structure with varying functional groups and substituents. Relatively little work has been carried out extending the structure-activity concept to correlating structure with environmental degradation parameters.

The predictive methodology for assessing the fate of a chemical is schematically represented in Figure 11.7. The data came from laboratory test methods and environmental measurements. The procedure is evaluated for accuracy in predicting concentrations as a function of time and location of the field site. A useful validation procedure is a field test where the chemical is applied at a constant known rate to a simple compartment, air, water, or soil, for which all of the environmental properties are either known or can be measured during the exposure period. The concentration of chemical is monitored as a function of time and location within or without the test plot. The input rate can be altered, and the concentration with time and location are followed. In effect, the individual fate processes are integrated. Field data are compared with the laboratory tests under comparable conditions to evaluate the agreement between prediction and observation. In most cases, lack of matching conditions in the field and laboratory leads to deviation between predicted and observed values.

Since the fate processes controlling the concentrations of chemicals in the environment are dynamic, kinetic and equilibrium-rate constants do describe the contribution of various processes to the net change and rate of change in its concentration. Since chemicals are usually present in low concentrations in the environment, the kinetics are usually first order. The set of simple first-

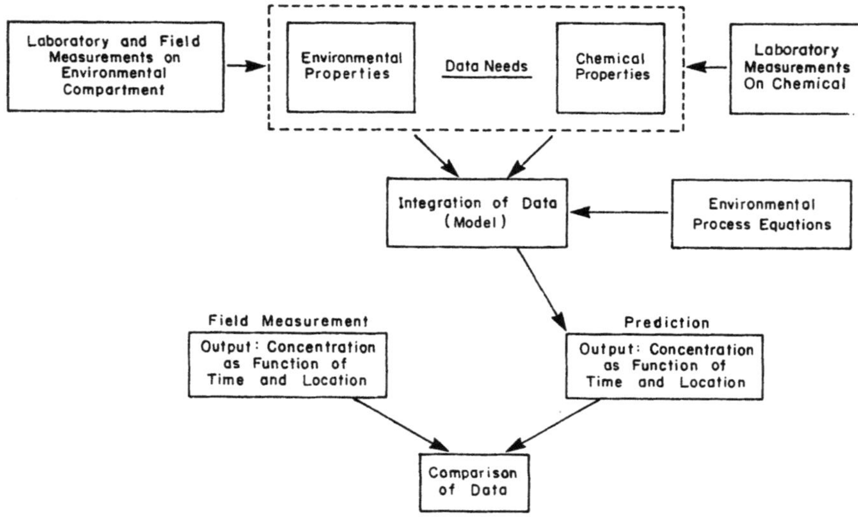

Figure 11.7. Illustrative scheme for predictive methodology, development, processing, and validation. (From Mill, 1980.)

order processes can quickly evaluate the contribution from each fate process to the total rate of loss of chemical. It serves as a screening tool for preliminary assessment and to decide what further tests are needed. It can take into account variability in conditions and also provide an estimate of persistence. If we agree that a limited number of discrete chemical, physical, and biological processes control EEC, then it is appropriate to evaluate the validity of their predictive capabilities.

Predictive Capability of Transport and Fate Processes

Sorption

Many organic chemicals, especially the nonpolar and water-insoluble ones, sorb strongly to sediments/suspended solids/soils. This determines the fraction available for other fate processes, and, in effect, sorption buffers the EEC in the aqueous phase. The conventional sorption equilibrium constant has been modified to account for the correlation with the organic content of the sorbent. For an organic content value of unity, K_s (sorption constant) is equivalent to K_{ow} (partition coefficient in octanol/water mixture). For screening purposes, one concentration of chemical and two loadings of sediment (low and high) will provide a reliable estimate of K_s. It is important that some accurately detectable amount of the chemical be present in both phases. Prior estimate of K_s using the relationship between solubility and K_{oc}

and calculation of the equilibrium amounts presented in a certain volume of water, about half the concentration of chemical in each phase, can be easily arranged (Smith *et al.*, 1978).

Predictive Capability. Estimation of K_{sc} from solubility would suffice as a screening test for making decisions for further detailed measurements. Mill (1980) developed a regression equation:

$$\log K_{sc} = -0.782 \log [C] - 0.27 \quad (9)$$

where $[C]$ = concentration of the chemical in moles L^{-1}.

A regression line has been plotted using K_{sc} and the aqueous solubility of a variety of chemicals (accurately measured). One can estimate K_{sc} within a power of ten for most nonpolar chemicals, which is sufficiently accurate for screening most chemicals.

Hydrolysis

Hydrolysis of organic compounds usually results in the substitution of a group by a hydroxyl group. In water, the reaction is catalyzed by H^+, OH^-, and H_3O^+ ions. In moist soils, labile complexes of calcium and copper act as catalysts. Sorption tends to increase a chemical's reactivity to H^+ or OH^- radicals. Mabey and Mill (1978) reviewed the kinetic data for hydrolysis of a variety of organic chemicals in aquatic systems and the chemical characteristics of most fresh-water systems. These data could be used to calculate the persistence (half-life for hydrolysis) of chemicals under natural environmental conditions.

Predictive Capability. Screening and detailed test methods to calculate various hydrolytic kinetic parameters and their temperature dependence are available (Mill *et al.*, 1978). A significant data base on semiempirical structure-activity relationships for most kinds of hydrolytic reactions are available in literature (Wells, 1963; Taft, 1956). The data base is large enough to predict a rate constant within a factor of 2 or 3 for a specific chemical within a family of structurally related chemicals.

Volatilization

Volatilizational loss of chemicals from water to air is an important fate process for chemicals with low aqueous solubility and low polarity. Many chemicals, despite their low vapor pressure, can volatilize rapidly owing to their high activity coefficients in solution. Volatilizational loss from surfaces is a significant transport process. Volatilization of organic chemicals from the soil surface is complicated by other variables. There is no simple laboratory measurement that will reliably extrapolate itself to the field for the soil situation.

Predictive Capability. For chemicals with H_c (Henry's law constant) >1000 torr M^{-1}, a reasonable estimate of the ratio $k^c_{vw}/k^{O_2}_{vw}$ can be arrived from the following ratio (Smith *et al.*, 1978):

$$\frac{k^c_{vw}}{k^{O_2}_{vw}} = \frac{D^{o_2}}{D^C} \tag{10}$$

where the left hand term is the volatilization rate quotient for the chemical and oxygen and the right hand term is the ratio of molecular diameters for O_2 and the chemical.

For chemicals with $H_c<1000$ torr M^{-1}, no satisfactory predictive procedure is available for volatility.

Oxidation

The essential oxidation rate steps in the environment involve reactive species such as free radicals ($RO_2^{\cdot}$, $RO^{\cdot}$, $HO^{\cdot}$), ozone, and $^1O_2$ (singlet oxygen). For predictive purposes, it is important to identify the important oxidants and their concentrations in the environmental compartments. Values of many oxidation rate constants are known reliably (Hendry *et al.*, 1974). In the atmosphere, oxidation by the $OH^{\cdot}$ radical is significant, whereas ozone is important in the oxidation of some olefins and possibly some sulfur or phosphorus compounds. The $RO_2^{\cdot}$ radical can be important in sunlight photolysis in natural waters. Competitive kinetic techniques could be used to evaluate the relative loss of two chemicals, one of which is a standard of known reactivity to a specific oxidant (Mill *et al.*, 1978). Azo compounds could be used to generate $RO_2^{\cdot}$ radical in water, nitrous acid to form the $HO^{\cdot}$ radical in air, and a dye to generate $^1O_2$ in water.

Predictive Capability. Several empirical structure-activity relationships for oxidation by $RO_2^{\cdot}$ (polyani relation), $HO^{\cdot}$ (Hammett sigma-rho relations), and $^1O_2$ have been developed. A large data base is available to estimate reliably the rate constants for new chemicals within a factor of 3–5. Precise evaluation of rate constants for highly susceptible compounds such as phenols, aromatic amines, alkyl sulfides, and electron-rich olefins and dienes is needed. Oxidation of saturated alkyl compounds including alkanes, haloalkanes, esters, and ketones is slow in water and air. Most chemicals excepting the above-mentioned compounds are readily oxidized by $HO^{\cdot}$ radicals.

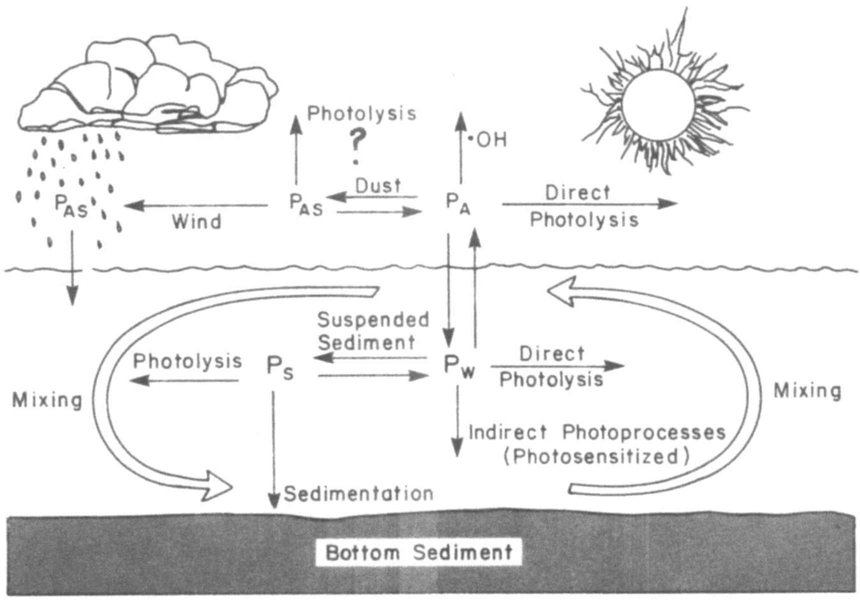

Figure 11.8. Environmental processes and their effect on photolysis of pollutants. (From Zepp, 1980.)

Photolysis

Figure 11.8 shows the two general types of photolytic processes and the influence of various processes on the photolytic rate of aquatic pollutants. A simple way of determining the photolytic rate of a chemical is to expose it in aqueous solution or in a thin layer to outdoor sunlight and measure the rate of disappearance. Simultaneously, photolysis of another chemical of known quantum yield (ϕ) with a similar spectral range should be monitored. This method avoids the determination of quantum yield or spectral analysis.

Another method of determining environmental photolysis is to measure ϕ at a single wavelength (λ) in the laboratory. Sunlight intensity (I_λ) data as a function of time of day, season, and latitude are available in the literature. The rate constant in sunlight $k_{p(s)}$ is given by equation (11):

$$k_{p(s)} = \phi \Sigma I_\lambda \varepsilon_\lambda \tag{11}$$

and the half-life in sunlight is given by

$$(t_{1/2})_{(s)} = \frac{\ln 2}{k_{p(s)}} \tag{12}$$

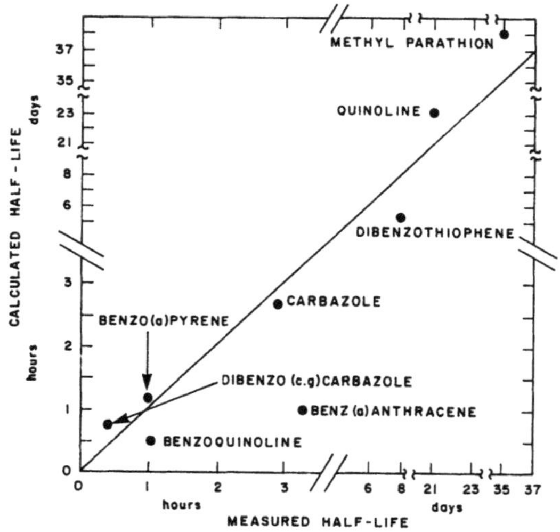

Figure 11.9. Measured and calculated half-lives for direct photolysis. (From Mill, 1980.)

Both computer and hand methods are available to sum the products of $\varepsilon_\lambda I_\lambda$ over a wavelength range to plot the photolytic half-life of a chemical in air or water with time and latitude (Mabey *et al.*, 1979). Comparison of the measured and calculated half-lives for direct photolysis in sunlight for eight selected chemicals in water using the above method gave excellent agreement with a factor of two (Figure 11.9).

Predictive Capability. For screening purposes, equation (11) may be used to calculate the upper limit of k_p by assuming $\phi = 1$. If the calculated rate constant is small relative to other fate processes, no additional photolytic measurements are needed. Prediction of ϕ from structure-activity relationship is still empirical. Caution must be used in extrapolating quantum yield from one solvent to the other.

Biodegradation

Biological transformations include a variety of reactions such as oxidation, reduction, hydrolysis, and rearrangement. Since they are difficult to separate from each other, biological transformation is treated as another single discrete fate process. Although laboratory studies using single organism cultures with optimum nutrient conditions interpret the more complex mixed culture situations, they represent extreme simplifications. Several procedures are available to measure kinetics in mixed-culture systems. In the simplest one, the microbial population is kept constant and the chemical is present in low concentration. This condition is similar to the natural water and soil environments. The organisms are usually collected from the environmental compartment of interest and acclimated to the chemical over several weeks in the presence of nutrients so that the necessary enzyme apparatus is in place. No structure-activity relationship for biodegradation has been developed except to identify some structures that are inherently recalcitrant to biotransformation. Some halogenated compounds are thermochemically limited in oxidative biodegradation.

Impact on Biological Species

The preliminary list of target chemicals is evaluated for biological activity factors of carcinogenicity, mutagenicity, teratogenicity, acute toxicity, other toxic effects (such as reproductive damage or organ-specific toxicity), bioaccumulation, and ecological disturbances. Scores are assigned for each of the above seven factors either numerically from 0, 1, 2 to 3, or a letter score such as x, xx, and xxx. The magnitude of the former score indicates the degree to which the particular effect is confirmed. For example, a score of 3 for carcinogenicity means that the chemical is a well-established carcinogen in humans or experimental animals. On the other hand, the magnitude of the letter score indicates the degree of concern and need for further testing. For example, a score of xxx for carcinogenicity means that the chemical is a strongly suspected carcinogen but has not been confirmed by adequate testing. Consensus should be reached on the type of scale, not necessarily the magnitude of the agreed scale. From the exposure and biological activity scoring procedures, analyses and single biological effect tests are performed. The cumulative totaling of the scores and multiple exposure yields a guide to select the priority list of chemicals for further testing and evaluation of specific toxicological effects. The priority in biological impact relates primarily to direct interactions of the chemical with receptor of significance and importance and type of effect. Wherever the biological data are few or nil, chemicals are placed by analogies into various groups of varied levels of concern.

The necessary toxicological information about a chemical are as follows (Murphy, 1980):

(i) Chemical and physical properties.
(ii) Metabolism: (a) Determination of the activity of the parent compound and metabolite and their deposition kinetics, (b) comparison of the type of metabolites in plants and animals, and (c) partitioning of a chemical or metabolite to determine biomagnification.
(iii) Acute Toxicity—LD_{50}: This is a useful screening test to determine the sensitive and resistant organisms. If the former is part of the food chain, it can cause serious consequences for organisms at higher trophic levels. On the other hand, the latter resistant organism could provide a medium for subtle bioaccumulation.

(iv) Repeated Short-Term Studies: This involves the administration of the chemical in the diet, drinking water, or inhaled air for usually 3 months in rodents. A broad biological test program can nearly identify all types of effects excepting cancer or heritable mutation.
(v) Long-Term Studies: Through properly designed long-term exposure (lifetime exposure in rodents and major fraction of lifetime in other species), evaluation of carcinogenicity, multigeneration reproduction effects, and delayed chronic effects can be examined.
(vi) Special Studies: This category could include toxico-kinetic studies, genetic toxicity studies, synergism by other chemicals, and dose-response analyses. The difficulties in extrapolating toxicological data from laboratory mammals to humans are small compared with those in assessing and evaluating the toxicological hazards of chemicals to biota in the environment.

Fate studies identify the type of compounds and their exposure concentration at different points in the ecosystem and the organism(s) at greatest risk of exposure. The toxicant released into the environment can reach humans through a number of pathways (Figure 11.10). Although greatest immediate attention should be paid to human health, ultimate human welfare may depend on the overall quality of the ecosystem. Future environmental hazard assessment should be based on comparative toxicology.

Assessment of Hazards

In the last decade, several hazard assessment programs were applied throughout the world to evaluate new and in-use chemicals. Most existing programs use the tier concept where the tests are arranged in definite steps in which scope and priority of additional testing can be identified. Basically, in any hazard assessment process, investigations on observed biological effects and Expected Exposure Concentration (EEC) are conducted in parallel (Figure 11.11). Theoretically, every chemical has a concentration at which it has no adverse effect on survival, growth, or reproduction. This concentration is called No Observed Effect Concentration (NOEC) and is determined from full-life cycle chronic toxicity testing of fish, macroinvertebrates, or plant species for aquatic hazard determination. Similarly, there is a highest EEC resulting from the normal manufacture and use pattern of the chemical. Any hazard evaluation program accurately determines the two concentrations to evaluate the margin of safety for the chemical. It can be seen that the accuracy and statistical reliability of the data increase along the x-axis with continued test procedures (Figure 11.11). Some chemicals, through synergism or antagonism, may exhibit environmental toxicity different from clean water laboratory effect (Figures 11.12 and 11.13). As the evaluation procedure continues with more refined tests, the estimates on the fate and effects became more accurate to distinguish

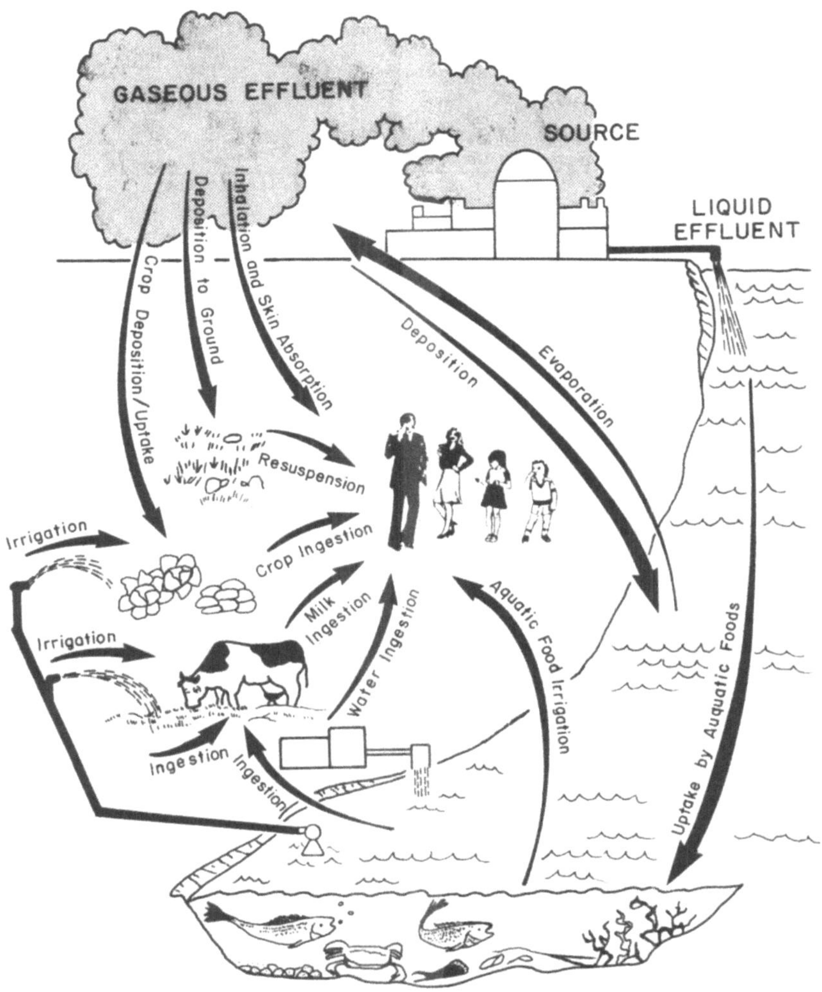

Figure 11.10. Possible exposure pathways of a chemical. (From Moghissi *et al.*, 1980.)

clearly between EEC and NOEC with high degree of confidence. Various hazard evaluation programs have their tier concept, each tier becoming increasingly complex, time consuming, and expensive. The tier concept is useful since it identifies the data requirements and outlines the most direct path for accurate estimates of realistic biological effects and fate determinations. The effectiveness of the tier concept depends upon the implementation of decision criteria regarding pass/ban options, regulation, and need for additional information. Otherwise the entire program degenerates to cataloging of tests and results in no clear direction whether the process has

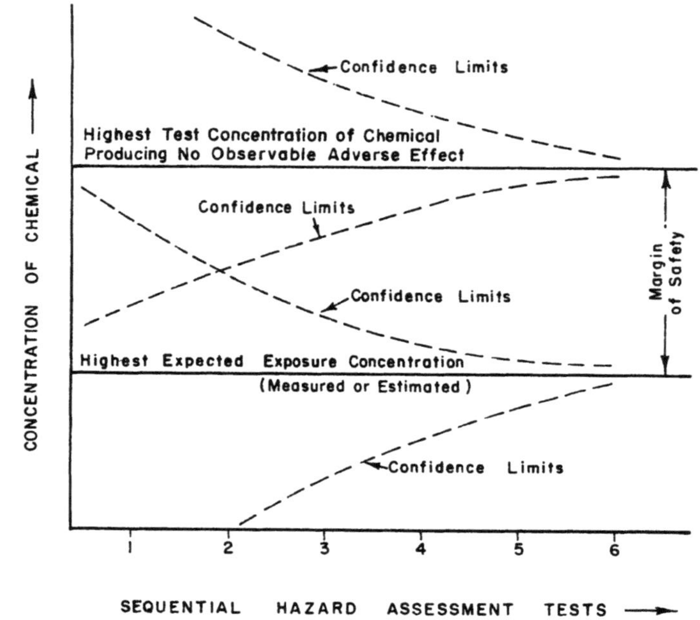

Figure 11.11. Sequential hazard assessment of a chemical. Larger gap between effect and exposure lines means higher margin of safety of the chemical. (From Branson, 1980.)

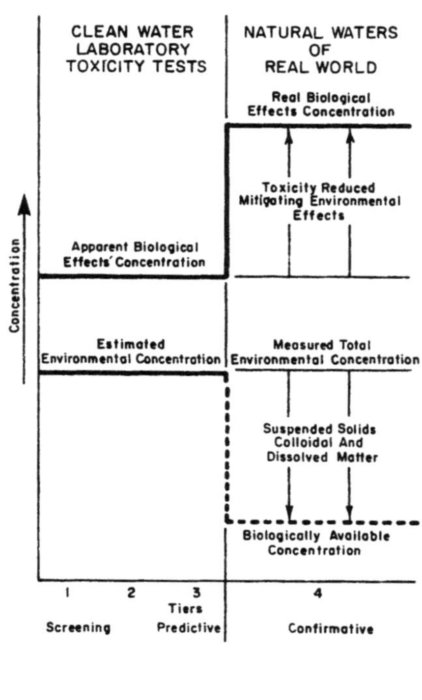

Figure 11.12. Environmental toxicity much less than laboratory data, leading to greater margin of safety. (Kimerle, 1980.)

Figure 11.13. Environmental toxicity greater than laboratory toxicity, leading to a narrower margin of safety than anticipated. (Kimerle, 1980.)

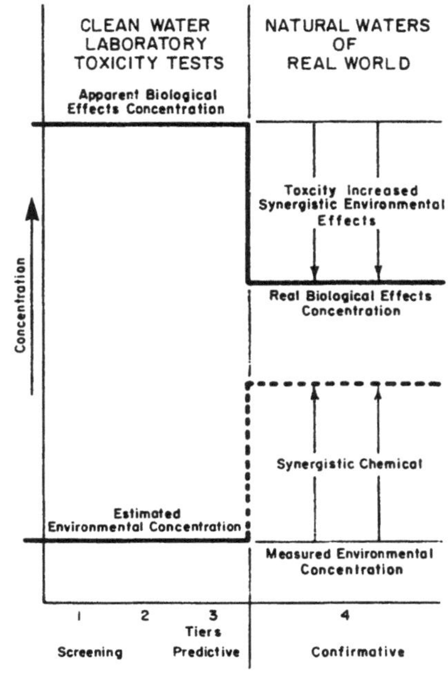

achieved the needed goals or more testing is required. Table 11.4 lists some of the typical hazard assessment programs available in the literature and the intent of each program.

To further examine the 13 hazard evaluation programs listed above, comparative testing on three hypothetical materials were performed. Realistic physical, chemical, and fate data along with toxicity data for mammalian and aquatic life were created. Potentials for carcinogenic, teratogenic, and mutagenic effects were not included to simplify the comparative evaluations. Material 1 was relatively nontoxic with low water solubility and little effect on aquatic life. Material 2 was a relatively toxic material, highly water soluble with an ability to chelate metals. Material 3 was a highly toxic, readily soluble pesticide degrading to a more toxic and persistant intermediate. The results of this comparative evaluation are summarized in Table 11.5.

The relatively rigid specifications of high volume use or potential for exposure of large environmental areas directs the decision to obtain advanced level or higher testing regardless of inherent nontoxic nature of the test substance. Examination of early phases of these programs shows wide variation on implications from acute toxicity testings (Table 11.6). These criteria were taken out of context from one or more programs.

All programs agree on the need for basic chemical/physical characterization data, estimates of use, and volume distribution. In early tiers,

Table 11.4. Hazard assessment programs.

Program title	Scope of the program
AIBS (American Institute of Biological Sciences)	Assess the basic data base, identify the need and type of additional testing for pesticide registration
Kodak (Eastman Kodak Company)	Human health and ecological exposures of new chemicals and consumer products
Monsanto	Fate and aquatic toxicity of new chemicals
Unilever (Port Sunlight Laboratory, UK)	Aquatic toxicity testing of new chemicals; utilizes dose-response relationship to predict evaluation of new chemical's impact.
Stufenplan (German Council for testing of dangerous substances)	Mammalian and aquatic testing of new chemicals
ASTM (American Society for Testing and Materials)	Aquatic testing for nontarget aquatic organisms
Lloyd	Aquatic risk assessment of new or existing chemicals and wastewater effluents
Hueck	Aquatic hazard assessment of new chemicals
FIFRA (Federal Insecticide, Fungicide and Rodenticide Act, US)	Specific data requirements and types of tests required for registering pesticides
Conservation Foundation (contracted by US Environmental Protection Agency)	Guidelines for testing priorities and requirements under US Toxic Substances Control Act (TSCA)
Dow Chemical Company	Mammalian and aquatic toxicity assessment of new chemicals
SDIA (Soap and Detergent Industries Association, UK)	Voluntary notification program primarily for domestic detergents and related products
TSCA (US Environmental Protection Agency)	Specific base data on physical/chemical properties and health and ecological tests required for premanufacture notification

Source: Maki (1979).

measures of acute toxicity on aquatic organisms are required. At this point, variation sets in among the schemes regarding the type of data, specific order, and conclusions. The programs vary in their objective decision for further testing (Table 11.7).

For all practical purposes, the last five programs do not seem to differ in their degree of objectivity.

Testing with model compounds has identified the essential points/stages in the hazard assessment program (Table 11.8).

Table 11.5. Results of the comparative hazard evaluation.

Plan	Hypothetical Material I	Hypothetical Material II	Hypothetical Material III
AIBS	Possible additional acute testing	Partial and full fish chronics	Full chronics, BCF, field tests needed
Kodak	Rating sum = 12, Tiers 0, I, II	Rating sum = 14, all tiers needed	Rating sum = 17, all tiers required
Monsanto	Phase I screening only	Testing through Phase III confirmation	Terminate testing after acute tests
Unilever	Acute screening tests only	Chronic testing required	Clear hazard imposed
Stufenplan	Complete testing based on use and volume	Complete testing based on use and volume	Complete testing based on use and volume
ASTM	Use following Phase II testing	Use material with monitoring	Discard use based on acute hazard
Lloyd	Phase I and II acute testing	Through Phase V field tests: omit BCF	Complete Phase I–V
Hueck	Initial phase only	Initial and main phase	Test through confirmatory phase
FIFRA	Acute testing and partial chronics	Simulated field testing	Actual field testing
Conservation Foundation	Through Tier III based on volume and use	Through Tier III based on volume and use	Through Tier III based on volume and use
Dow	Base data and range finding tests	Advanced studies to final key product	Final key product review
SDIA	Annex I–IV selected tests only	Annex I–IV	Annex I–IV
TSCA	Base data, health, and ecological effects	Base data, health, and ecological effects	Base data, health, and ecological effects

Source: Maki (1979).

Table 11.6. Variation in the implications of acute toxicity testing.

No further testing required
$LC_{50} > 1000$ mg L^{-1}
EEC > 0.002 (LC_{50})
EEC < 0.01 (LC_{50})
$LC_{50} \geq 100$ (EEC)
$LC_{50} > 5{,}000$ mg L^{-1}
Further testing required
LC_{50}/EEC = 1–1000
Use/volume/disposal patterns cause unusual concern
$LC_{50} < 1$ mg L^{-1}
LC_{50} highly variable among species
Material is a cumulative toxin
$LC_{50} \leq 10$ (EEC)
EEC > 0.01 (LC_{50})
Key effects in avian or mammalian efficacy tests
$LC_{50} < 5$ mg L^{-1}

Source: Maki (1979).

Toxicological testing mostly focuses on the more established single-species acute, embryo-larval, and full-life cycle testing to arrive at a no-effect concentration. Sophisticated tests are less attractive to both industry and the regulator because many sublethal effects are difficult to evaluate for their environmental significance. While such tests are sufficient during the catch-up phase of chemical testing, research on the development of new procedures to improve our projection from laboratory to field situations should be well supported. The current needs are:

Table 11.7. Ranking of decision criteria of various hazard assessment programs.

Plan	
(1) AIBS	Objective, clear decision criteria ↑
(2) Kodak	
(3) Monsanto	
(4) Unilever	
(5) Stufenplan	
(6) ASTM	
(7) Lloyd	
(8) Hueck	
(9) FIFRA	
(10) Conservation foundation	
(11) Dow	
(12) SDIA	
(13) TSCA	↓ Subjective, no clear decision criteria

Source: Maki (1979).

Table 11.8. Essential stages in a hazard assessment program.

Programs of aquatic hazard assessment
Design individuality
Consider usage and disposal patterns
Evaluate chemical and physical properties
Select toxicological tests
Relate effects to concentrations
Provide sequential assessment
Make decisions as early as possible

Source: Maki (1979).

(i) Establishment of safety factors to calculate an acceptable risk;
(ii) Improvement in the cross acceptability of several hazard assessment programs; and
(iii) Periodical review of data from these testing procedures to possibly identify the most cost-effective prediction of risk.

References

Branson, D.R. 1980. Prioritization of chemicals according to the degree of hazard in the aquatic environment. *Environmental Health Perspectives* **34**:133–138.

Crutzen, P.J., I.A.S. Isaken, and J.R. McAfee. 1978. The impact of the chlorocarbon industry on the ozone layer. *Journal of Geophysical Research* **83**:345–363.

Dawson, G.W., J.A. McNeese, and D.C. Christensen. 1978. An evaluation of alternatives for the removal/destruction of kepone residuals in the environment. *Proceedings of National Conference on Control of Hazardous Materials Spills*, 1978. pp. 244–249.

Dickson, K.L., A.W. Maki, and J. Cairns, Jr., 1979. *Analyzing the hazard evaluation process*. Water Quality Section, American Fisheries Society, Washington, D.C.

Dilling, W.L., C.J. Bredeweg, and N.B. Tefertiller. 1976. Organic photochemistry. Simulated atmospheric rates of methylene chloride, 1,1,1-trichloroethane, trichloroethylene, tetrachloroethylene, and other compounds. *Environmental Science and Technology* **10**:351–356.

Duthie, J.R., 1977. The importance of sequential assessment in test programs for estimating hazard to aquatic life. *In*: F.L. Mayer and J.L. Hamelink, (Eds.), *Aquatic toxicology and hazard evaluation*. ASTM STP 634. American Society for Testing Materials, pp. 17–35.

Hansch, C. 1980a. The role of partition coefficient in environmental toxicity. *In*: R. Haque (Ed.), *Dynamics, exposure and hazard assessment of toxic chemicals*. Ann Arbor Science Publishers, Ann Arbor, Michigan, pp. 273–286.

Hansch, C. 1980b. *Biological activity and chemical structure*. Elsevier Press, Amsterdam, 47 pp.

Haque, R., J. Falco, S. Cohen, and C. Riordan. 1980. Role of transport and fate studies in the exposure, assessment and screening of toxic chemicals. *In*: R. Haque (Ed.), *Dynamics, exposure and hazard assessment of toxic chemicals*. Ann Arbor Science Publishers, Ann Arbor, Michigan, pp. 47–67.

Hendry, D.G., T. Mill, L. Piskiewicz, J.A. Howard, and H.K. Eigenmann. 1974. A critical review of H-atom transfer in the liquid phase: chlorine atom, alkyl, trichloromethyl, alkoxy and alkylperoxy radicals. *Journal of Physical and Chemical Reference Data* **3**:937.

Kimerle, R.A. 1980. Aquatic hazard evaluation state of the art. *In*: R. Haque (Ed.), *Dynamics, exposure and hazard assessment of toxic chemicals*. Ann Arbor Science Publishers, Ann Arbor, Michigan, pp. 451–457.

Lu, P., and R.L. Metcalf. 1975. Environmental fate and biodegradability of benzene derivatives as studied in a model aquatic ecosystem. *Environmental Health Perspectives* **10**:269–284.

Mabey, W., and T. Mill. 1978. Critical review of hydrolysis of organic compounds in water under environmental conditions. *Journal of Physical and Chemical Reference Data* **7**:383–415.

Mabey, W.M., T. Mill, and D.G. Hendry. 1979. *Test protocols in environmental processes: Direct photolysis in water*. U.S. Environmental Protection Agency, Publication No. EPA-68-03-2227.

Maki, A.W. 1979. An analysis of decision criteria in environmental hazard evaluation programs. *In*: K.L. Dickson, A.W. Maki, and J. Cairns, Jr. (Eds.), *Analyzing the hazard evaluation program*. Proceedings of a Workshop, Waterville Valley, New Hampshire, August 14–18, 1978, American Fisheries Society, Washington, D.C. pp. 83–100.

Mill, T. 1980. Data needed to predict the environmental fate of organic chemicals. *In*: R. Haque (Ed.), *Dynamics, exposure and hazard assessment of toxic chemicals*. Ann Arbor Science Publishers, Ann Arbor, Michigan, pp. 297–322.

Mill, T., W.R. Mabey, and D.G. Hendry. 1978. *Test protocols for environmental processes: oxidation in water*. USEPA Draft Report. EPA contract 68-03-2227.

Miller, D.R., G. Butler, and L. Bramall. 1976. Validation of ecological system models. *Journal of Environmental Management* **4**:383–401.

Moghissi, A.A., R.E. Marland, F.J. Congel, and K.F. Eckerman. 1980. Methodology for environmental human exposure and health risk assessment. *In*: R. Haque (Ed.), *Dynamics, exposure and hazard assessment of toxic chemicals*. Ann Arbor Science Publishers, Ann Arbor, Michigan, pp. 471–489.

Muir, W.R. 1980. Chemical selection and evaluation: implementing the Toxic Substances Control Act. *In*: R. Haque (Ed.), *Dynamics, exposure and hazard assessment of toxic chemicals*. Ann Arbor Science Publishers, Ann Arbor, Michigan, pp. 15–20.

Murphy, S.D. 1980. Toxicological dynamics. *In*: R. Haque (Ed.), *Dynamics, exposure and hazard assessment of toxic chemicals*. Ann Arbor Science Publishers, Ann Arbor, Michigan, pp. 425–436.

National Academy of Sciences. 1976. *Halocarbons: effects on stratospheric ozone*. Washington, D.C., 352 pp.

Neely, W.B. 1977. A material balance study of polychlorinated biphenyls in Lake Michigan. *The Science of the Total Environment* **7**:117–129.

Neely, W.B. 1978. A method for selecting the most appropriate environmental experiments that need to be performed on a new chemical. Abstracts, American Chemical Society Meeting, Division of Environmental Chemistry, September, 1978.

Neely, W.B. 1979. An integrated approach to assessing the potential impact of organic chemicals in the environment. *In*: K.L. Dickson, A.W. Maki, and J. Cairns, Jr. (Eds.), *Analyzing the hazard evaluation process*. Proceedings of a Workshop, Waterville Valley, New Hampshire, August 14–18, 1978, American Fisheries Society, Washington, D.C. pp. 74–82.

Neely, W.B. 1980. A method for selecting the most appropriate environmental experiments on a new chemical. *In*: R. Haque (Ed.), *Dynamics, exposure and hazard assessment of toxic chemicals*. Ann Arbor Science Publishers, Ann Arbor, Michigan, pp. 287–296.

Neely, W.B., D.R. Branson, and G.E. Blau. 1974. Partition coefficient to measure bioconcentration potential of organic chemicals in fish. *Environmental Science and Technology* **8**:1113–1115.

O'Neill, R.V. 1971. Error analysis of ecological models. *In*: *Radionuclides in Ecosystems*. Proceedings of the Third National Symposium on Radioecology, May 10–12, Oak Ridge, Tennessee. Publication No. 547, Oak Ridge National Laboratory, pp. 898–908.

Roberts, J.R., M.F. Mitchell, M.J. Boddington, and J.M. Ridgeway—Part I; J.R. Roberts, M.T. McGarrity, and W.K. Marshall—Part II. 1981. *A screen for the relative persistence of lipophilic organic chemicals in aquatic ecosystems—an analysis of the role of a simple computer model in screening*. National Research Council of Canada, NRCC Publication No. 18570.

Smith, J.H., W.R. Mabey, N. Bohonos, B.R. Holt, S.S. Lee, T.W. Chou, D.C. Bomberger, and T. Mill. 1978. *Environmental pathways of selected chemicals in freshwater systems*. U.S. Environmental Protection Agency, Publication Nos. 600/7-77-113 and 600/7-78-074 (October 1977 and May 1978).

Stephenson, M.E. 1980. Setting toxicological and environmental testing priorities for commercial chemicals. *In*: R. Haque (Ed.), *Dynamics, exposure and hazard assessment of toxic chemicals*. Ann Arbor Science Publishers, Ann Arbor, Michigan, pp. 35–40.

Taft, R.W. 1956. *Steric effects in organic chemistry*. John Wiley, New York.

Thomann, R.V. 1981. Equilibrium model of fate of microcontaminants in diverse aquatic food chains. *Canadian Journal of Fisheries and Aquatic Sciences* **38**:280–296.

Versar. 1979. *Water-related environmental fate of 129 priority pollutants*. Vol. **I** and **II**. U.S. Environmental Protection Agency, Publication No. EPA-440/4-79-029b, Washington, D.C.

Well, P.R. 1963. Linear free energy relationships. *Chemical Reviews* **63**:171.

Zepp, R.G. 1980. Assessing the photochemistry of organic pollutants in aquatic environments. *In*: R. Haque (Ed.), *Dynamics, exposure and hazard assessment of toxic chemicals*. Ann Arbor Science Publishers, Ann Arbor, Michigan, pp. 69–110.

Appendix A
Chemical Formulae of Compounds Cited in this Book

NAME	MOLECULAR FORMULA	STRUCTURAL FORMULA
ACENAPHTHENE	$C_{12}H_{10}$	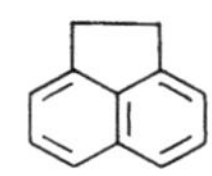
ACENAPHTHYLENE	$C_{12}H_8$	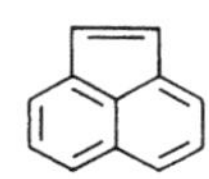
ACETONE	C_3H_6O	$CH_3-C(=O)-CH_3$
ACETYLENE	C_2H_2	$HC\equiv CH$
ACROLEIN	C_3H_4O	$CH_2=CH-CH=O$
ALDRIN	$C_{12}H_8Cl_6$	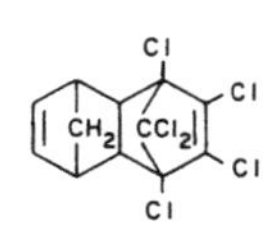
m-ALKYLPHENOLS	RC_6OH_5	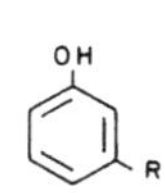
ALLYLALCOHOL	C_3H_6O	$CH_2=CH-CH_2-OH$
m-AMINOPHENOL	C_6H_7NO	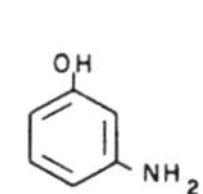

NAME	MOLECULAR FORMULA	STRUCTURAL FORMULA
m-AMYLPHENOL	$C_{11}H_{16}O$	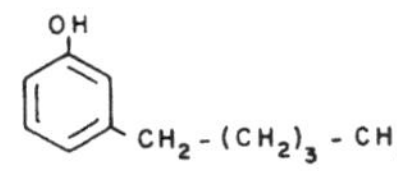
ANILINE	C_6H_7N	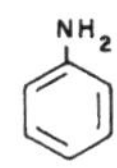
ANTHRACENE	$C_{14}H_{10}$	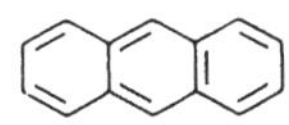
BENZ (A) ANTHRACENE	$C_{18}H_{12}$	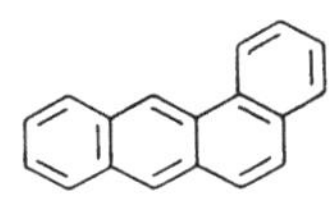
BENZENE	C_6H_{12}	
BENZYLCHLORIDE	C_7H_7Cl	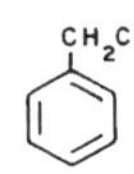
BENZO (B) FLUORANTHENE	$C_{20}H_{12}$	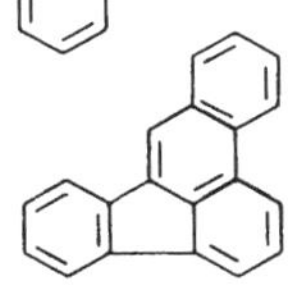
BENZO (J) FLUORANTHENE	$C_{20}H_{12}$	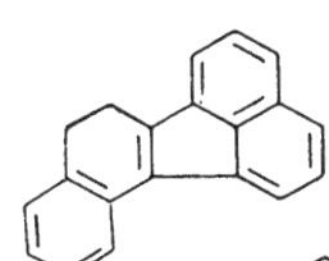
BENZO (K) FLUORANTHENE	$C_{20}H_{12}$	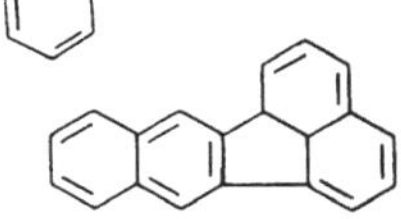
BENZOFURAN	C_8H_6O	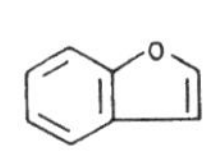
BENZOIC ACID	$C_7H_6O_2$	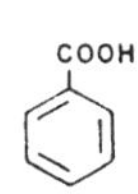

NAME	MOLECULAR FORMULA	STRUCTURAL FORMULA
iso-BUTANOL	$C_4H_{10}O$	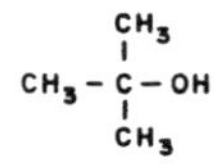
BUTENE	C_4H_8	$CH_3-CH=CH-CH_3$
BUTYLENE	C_4H_8	$CH_3-CH=CH-CH_3$
1,4-BUTYLPHENOL	$C_{10}H_{14}O$	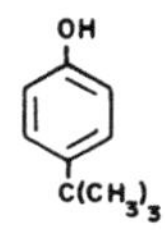
CAPROLACTAM	$C_6H_{11}NO$	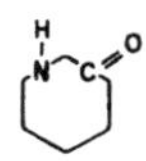
CARBON TETRACHLORIDE (Tetrachloromethane)	CCl_4	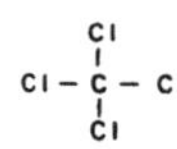
CARBONYL SULPHIDE	COS	S=C=O
CAROTENE	$C_{40}H_{56}$	
CHLORANIL	$C_6O_2Cl_4$	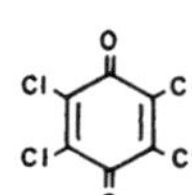
CHLORDANE	$C_{10}H_6Cl_8$	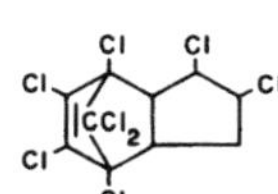

NAME	MOLECULAR FORMULA	STRUCTURAL FORMULA
BENZO (GHI) PERYLENE	$C_{22}H_{12}$	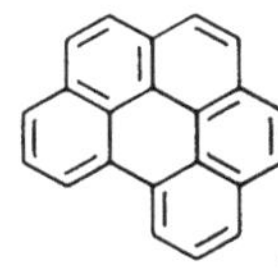
BENZO (A) PYRENE	$C_{20}H_{12}$	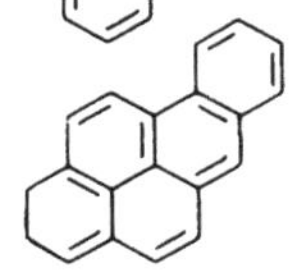
BENZO (E) PYRENE	$C_{20}H_{12}$	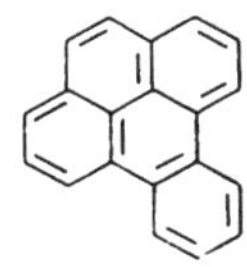
BENZOTHIOPHENE	C_8H_6S	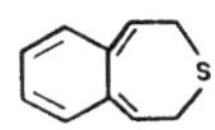
BICYCLOOCTANE	C_8H_{14}	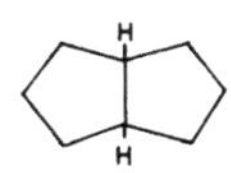
BISPHENOL A	$C_{15}H_{16}O_2$	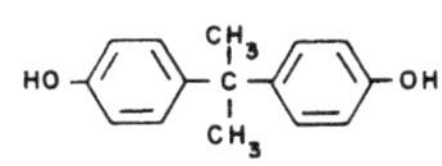
BROMODICHLOROMETHANE	$CHCl_2Br$	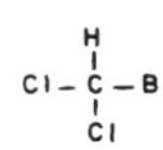
BROMOETHYLPROPANE	$C_5H_{11}Br$	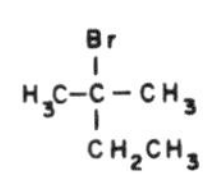
BROMOFORM	$CHBr_3$	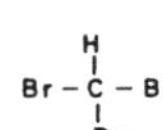
BROMOMETHANE	CH_3Br	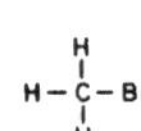

NAME	MOLECULAR FORMULA	STRUCTURAL FORMULA
CHLOROETHYLENE	C_2H_3Cl	$CH_2=CH-Cl$
CHLOROETHYLVINYLETHER	C_4H_7OCl	$Cl-CH_2-CH_2-O-CH=CH_2$
CHLOROFORM	$CHCl_3$	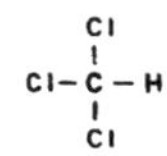
CHLOROMETHANE	CH_3Cl	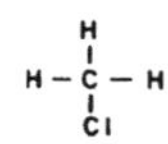
CHLOROMETHYLMETHYLETHER	C_2H_5OCl	$Cl-CH_2-O-CH_3$
CHLOROMETHYL GUAIACOL	$C_8H_9O_2Cl$	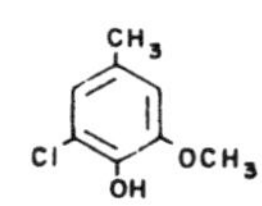
CHLOROMETHYL PHENOL	C_7H_7OCl	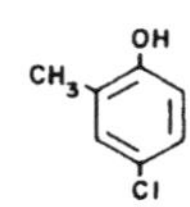
CHLOROMETHYLVERATROLE	$C_9H_{11}O_2Cl$	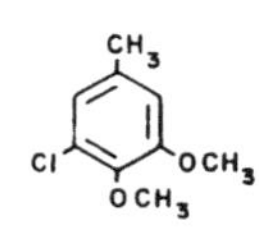
CHLORONAPHTHALENE	$C_{10}H_7Cl$	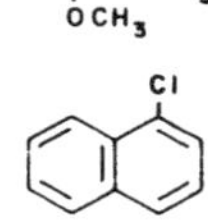
o-CHLORONITROBENZENE	$C_6H_4ClNO_2$	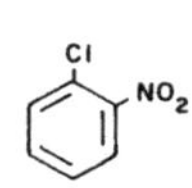

NAME	MOLECULAR FORMULA	STRUCTURAL FORMULA
CHLORDENE	$C_{10}H_4Cl_6$	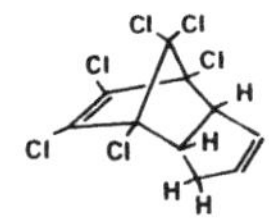
CHLORDENE EPOXIDE	$C_{10}H_6Cl_6O$	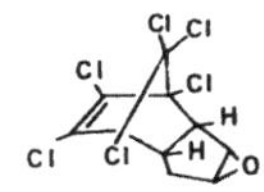
CHLOROBENZENE	C_6H_5Cl	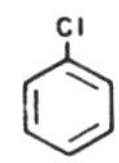
m-CHLOROBENZOIC ACID	$C_7H_5ClO_2$	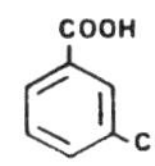
CHLOROBIPHENYL	$C_{12}H_9Cl$	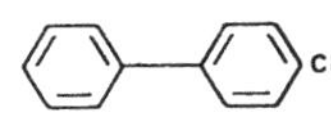
CHLORODECANE	$C_{10}Cl_{10}O$	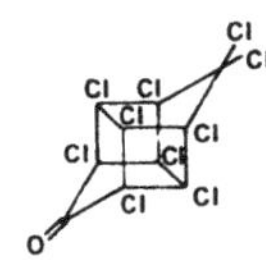
CHLORODIBENZOFURAN	$C_{12}H_8OCl$	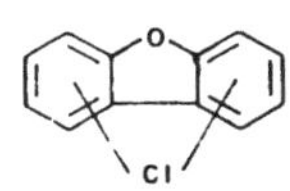
CHLORODIBROMOMETHANE	$CHBr_2Cl$	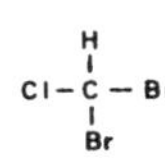
CHLORODIFLUOROMETHANE	$CHClF_2$	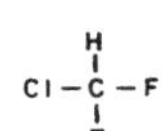
CHLOROETHANE	C_2H_5Cl	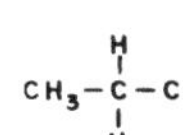

NAME	MOLECULAR FORMULA	STRUCTURAL FORMULA
o-CHLOROPHENOL	C_6H_5OCl	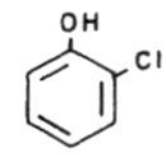
2-CHLOROPROPANE	C_3H_7Cl	$CH_3-CH(Cl)-CH_3$
CHLOROPROPENE	C_3H_5Cl	$Cl-CH=CH-CH_3$
CHRYSENE	$C_{18}H_{12}$	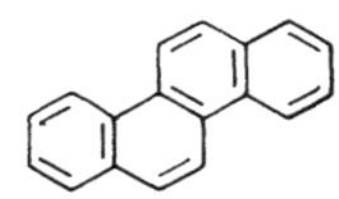
CORONENE	$C_{24}H_{12}$	
CREOSOTE	mixture of phenols	
m-CRESOL (methylphenol)	C_7H_8O	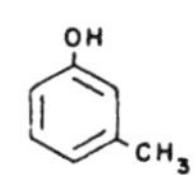
CYCLOHEPTANE	C_7H_{14}	
CYCLOHEXANE	C_6H_{12}	
CYCLOHEXANOL	$C_6H_{12}O$	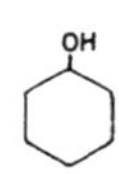

NAME	MOLECULAR FORMULA	STRUCTURAL FORMULA
CYCLOHEXANONE	$C_6H_{10}O$	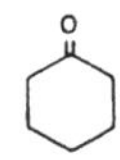
CYCLOPENTANE	C_5H_{10}	
CYMENE	$C_{10}H_{14}$	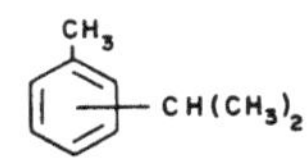
2,4-D (2,4-dichlorophenoxyacetic acid)	$C_8H_6Cl_2O_3$	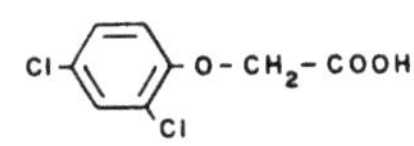
DBH (4,4'-dichlorobenzhydrol)	$C_{13}H_{10}Cl_2$	
DBP (4,4'-dichlorobenzophenone)	$C_{13}H_8OCl_2$	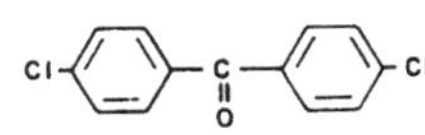
DDA (4,4'-dichlorodiphenyl acetic acid)	$C_{14}H_{10}O_2Cl_2$	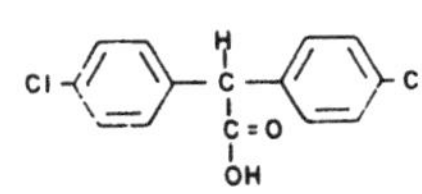
DDCN (4,4'-dichlorodiphenylacetonitrile)	$C_{14}H_9NCl_2$	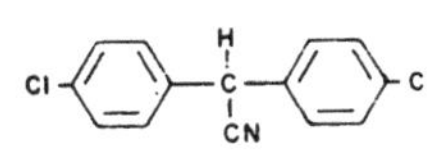
DDCO (dichloro diphenyl carbonyl)	$C_{13}H_8OCl_2$	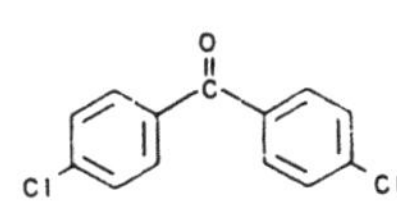
DDD (4,4'-dichlorodiphenyldichloroethane)	$C_{14}H_{10}Cl_4$	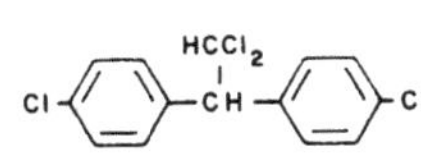

NAME	MOLECULAR FORMULA	STRUCTURAL FORMULA
DDE (4,4'-dichlorodiphenyldichloroethylene)	$C_{14}H_{8}Cl_{4}$	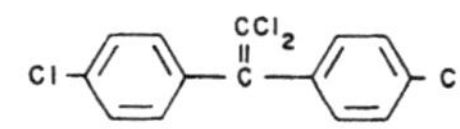
DDM (4,4'-dichlorodiphenylmethane)	$C_{13}H_{10}Cl_{2}$	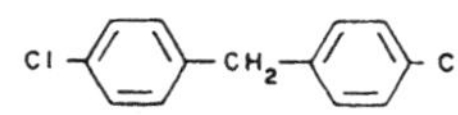
DDMS (4,4'-dichlorodiphenylchloroethane)	$C_{14}H_{11}Cl_{3}$	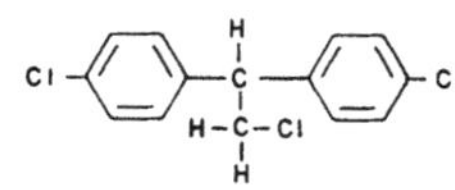
DDMU (4,4'-dichlorodiphenylchloroethylene)	$C_{14}H_{9}Cl_{3}$	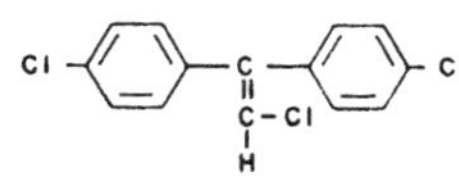
DDNS (4,4'-dichlorodiphenylethane)	$C_{14}H_{12}Cl_{2}$	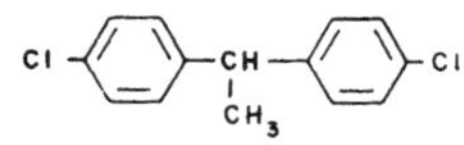
DDNU (4,4'-dichlorodiphenylethylene)	$C_{14}H_{10}Cl_{2}$	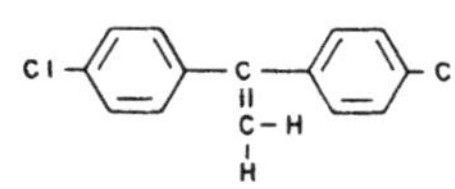
DDT (4,4'-dichlorodiphenyltrichloroethane)	$C_{14}H_{9}Cl_{5}$	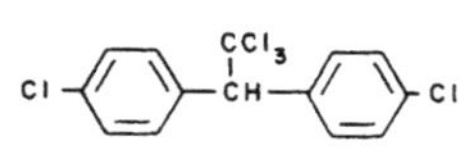
DECACHLOROBIPHENYL	$C_{6}Cl_{10}$	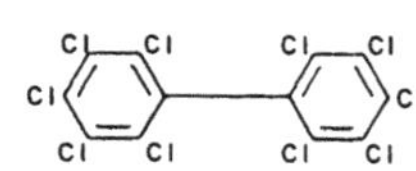
DECAHYDRONAPHTHALENE	$C_{10}H_{18}$	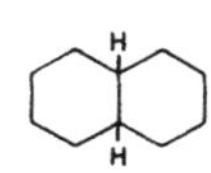
DIBENZ (A,H) ACRIDINE	$C_{21}H_{13}N$	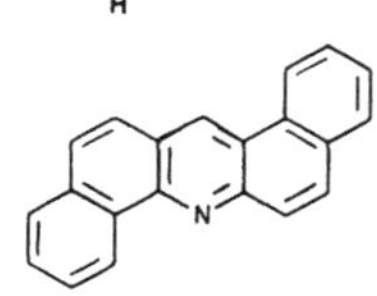

NAME	MOLECULAR FORMULA	STRUCTURAL FORMULA
DIBENZ (A, H) ANTHRACENE	$C_{22}H_{14}$	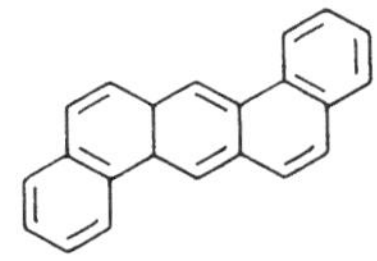
DIBENZOFURAN	$C_{12}H_8O$	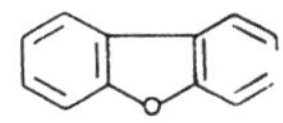
p-DIBROMOBENZENE	$C_6H_4Br_2$	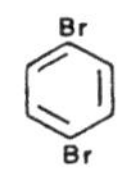
DIBROMOCHLOROMETHANE	$CHBr_2Cl$	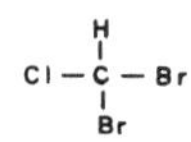
DIBROMOETHANE	$C_2H_4Br_2$	$Br-CH_2-CH_2-Br$
DIBROMOMETHANE	CH_2Br_2	$Br-CH_2-Br$
2,3-DIBROMOTOLUENE	$CH_3C_6H_3Br_2$	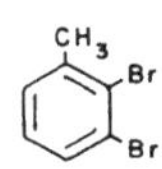
2,4-DIBUTYLPHENOL	$C_{14}H_{22}O$	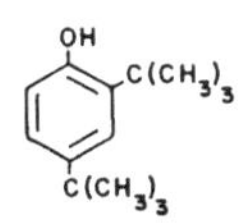
DICHLOBENIL	$C_7H_3Cl_2N$	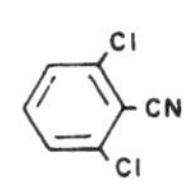
m-DICHLOROBENZENE	$C_6H_4Cl_2$	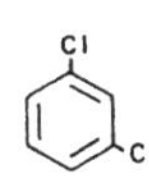

NAME	MOLECULAR FORMULA	STRUCTURAL FORMULA
DICHLOROBENZOFURAN	$C_8H_3Cl_2O$	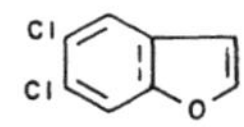
4,4'-DICHLOROBIPHENYL	$C_{12}H_8Cl_2$	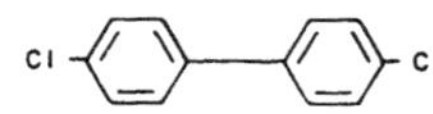
DICHLOROBROMOMETHANE	$CHCl_2Br$	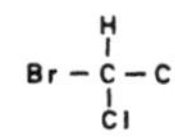
DICHLOROCATECHOL	$C_6H_4O_2Cl_2$	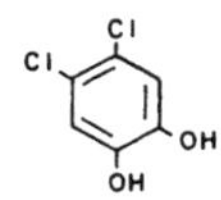
DICHLORODIFLUOROMETHANE	CCl_2F_2	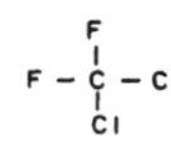
DICHLOROETHANE	$C_2H_4Cl_2$	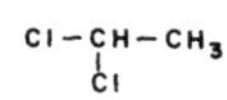
DICHLOROETHYLENE	$C_2H_2Cl_2$	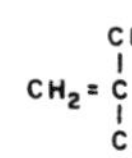
DICHLOROFLUOROMETHANE	$CHCl_2F$	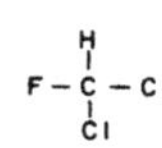
DICHLOROGUAIACOL	$C_7H_6O_2Cl_2$	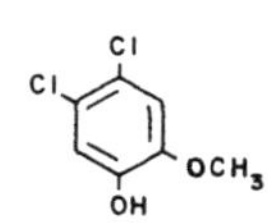
DICHLOROMETHANE (methylene chloride)	CH_2Cl_2	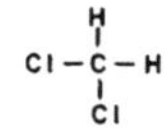

NAME	MOLECULAR FORMULA	STRUCTURAL FORMULA
DICHLOROMETHYL GUAIACOL	$C_8H_7O_2Cl_2$	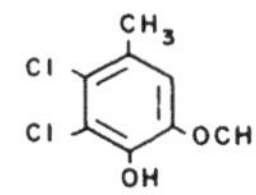
DICHLOROMETHYLPHENOL	$C_7H_6OCl_2$	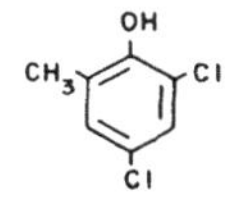
DICHLOROMETHYLVERATROLE	$C_9H_{10}O_2Cl_2$	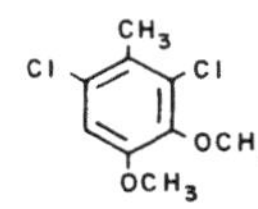
2,3-DICHLOROPHENOL	$C_6H_4OCl_2$	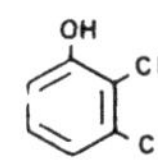
1,2-DICHLOROPROPANE	$C_3H_6Cl_2$	$\underset{Cl}{\underset{\mid}{CH_2}}-\underset{Cl}{\underset{\mid}{CH}}-CH_3$
1,3-DICHLOROPROPENE (dichloropropylene)	$C_3H_4Cl_2$	$Cl-CH=CH-CH_2-Cl$
2,4-DICHLOROTHIOPHENOL	$C_6H_3SHCl_2$	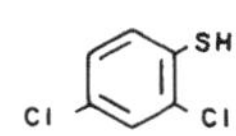
DIELDRIN	$C_{12}H_8Cl_6O$	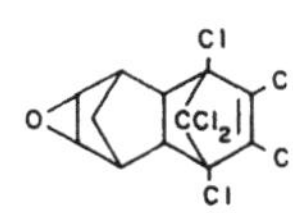
3,5-DIETHYLPHENOL	$C_{10}H_{14}O$	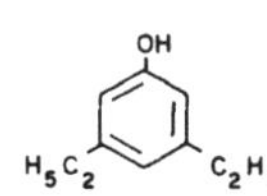
1,5-DIHYDRODIHYDROXY-BENZO (A) PYRENE	$C_{20}H_{14}O_2$	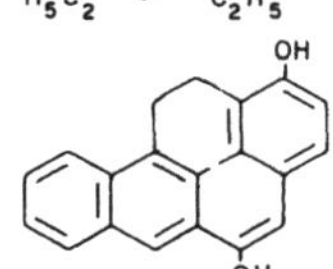

NAME	MOLECULAR FORMULA	STRUCTURAL FORMULA
ETHYLENE OXIDE	C_2H_4O	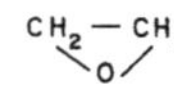
m-ETHYLPHENOL	$C_8H_{10}O$	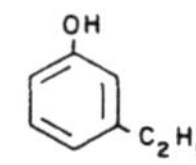
FENITROTHION	$C_9H_{12}NO_5PS$	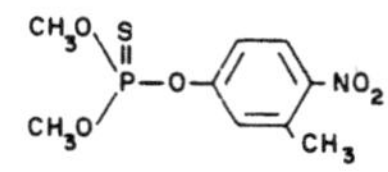
FLUORANTHENE	$C_{16}H_{10}$	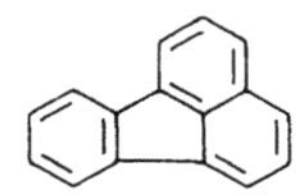
FLUORENE	$C_{13}H_{10}$	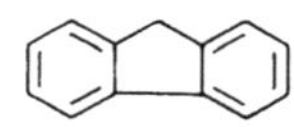
FLUOROTRICHLOROMETHANE	CCl_3F	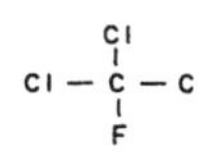
FORMALDEHYDE	H CHO	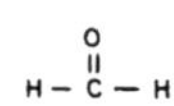
GLUCURONIC ACID	$C_6H_{10}O_7$	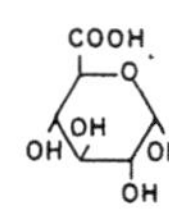
GLUTATHIONE	$C_{10}H_{17}N_3O_6S$	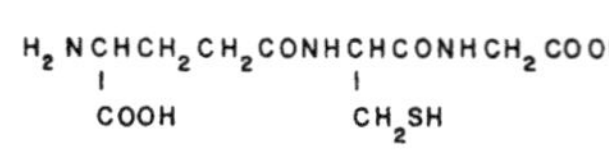
GLYCERINE (GLYCEROL)	$C_3H_8O_3$	$HOCH_2$ CHOH CH_2 OH

NAME	MOLECULAR FORMULA	STRUCTURAL FORMULA
ENDOSULFAN DIOL	$C_9H_8Cl_6O_2$	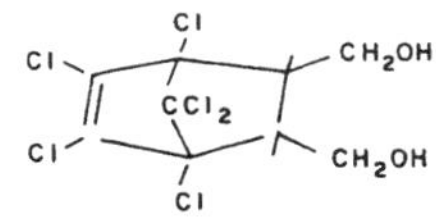
ENDOSULFAN SULPHATE	$C_9H_6Cl_6O_4S$	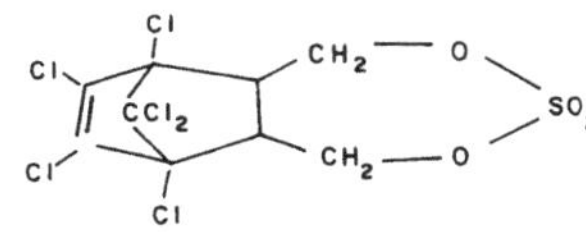
ENDRIN	$C_{12}H_8Cl_6O$	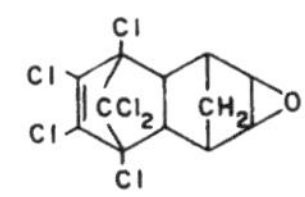
ENDRIN ALDEHYDE	$C_{12}H_6Cl_6O$	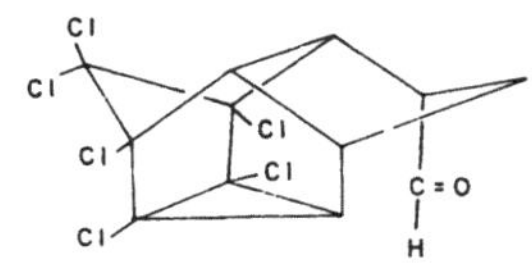
EPICHLOROHYDRIN	C_3H_5OCl	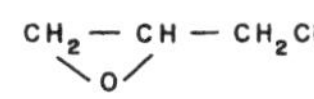
ETHANE	C_2H_6	CH_3-CH_3
ETHYLBENZENE	C_8H_{10}	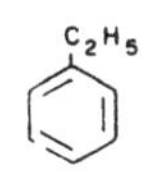
ETHYLENE	C_2H_4	$CH_2=CH_2$
ETHYLENE-1,1'-DIBROMIDE	$C_2H_4Br_2$	$Br-CH_2-CH_2-Br$
ETHYLENE DIAMINE	$C_2H_9N_2$	$NH_2-CH_2-CH_2-NH_2$

NAME	MOLECULAR FORMULA	STRUCTURAL FORMULA
2,4-DIMETHYL-3-ETHYLPHENOL (ethylxylenol)	$C_{10}H_{14}O$	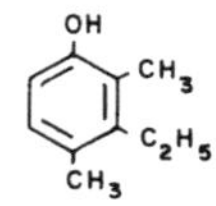
2,4-DIMETHYLPHENOL (XYLENOL)	$C_8H_{10}O$	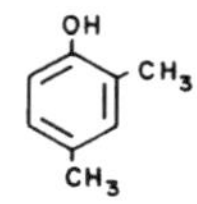
DINITROBENZENE	$C_6H_4N_2O_4$	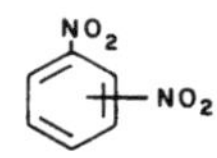
2,4-DINITRO-6-METHYLPHENOL	$C_7H_6O_5N_2$	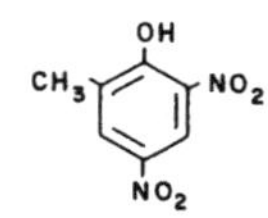
2,4-DINITROPHENOL	$C_6H_4N_2O_5$	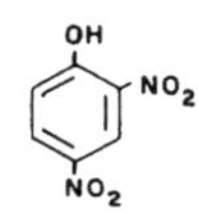
2,4-DINITROTOLUENE	$C_7H_6N_2O_4$	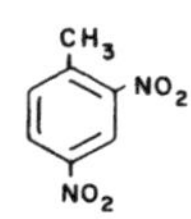
DIPHENYLISODECYL PHOSPHATE ESTER	$C_{22}H_{31}PO_4$	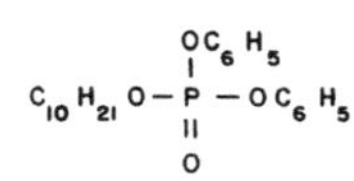
p-DODECYLPHENOL	$C_{18}H_{30}O$	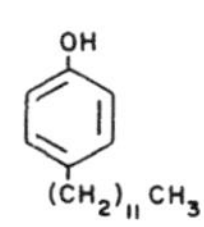
ENDOPEROXIDE	$C_{16}H_{14}O_2$	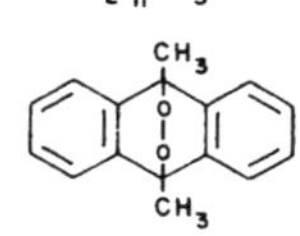
ENDOSULFAN	$C_9H_6Cl_9O_3S$	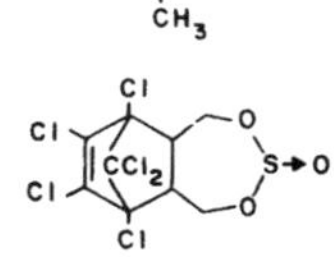

NAME	MOLECULAR FORMULA	STRUCTURAL FORMULA
1, 4 - DIHYDROXYANTHRAQUINONE	$C_{14}H_8O_4$	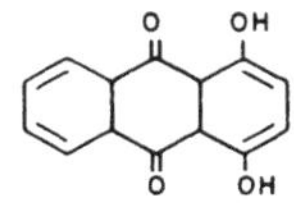
3,4 - DIHYDROXY-5-CHLOROTOLUENE (chloromethylcatechol)	$C_7H_7O_2Cl$	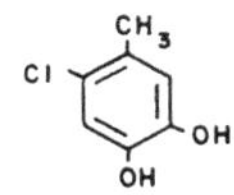
DIHYDROXYDIHYDROALDRIN	$C_{12}H_{10}O_2Cl_6$	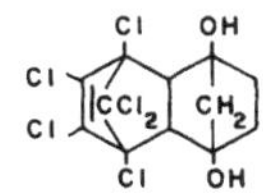
1,3 - DIHYDROXINAPHTHALENE	$C_{10}H_8O_2$	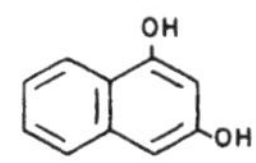
2,3 - DIHYDROXYTOLUENE (methylcatechol)	$C_7H_8O_2$	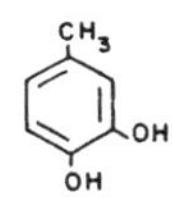
2,3 - DIHYDROXY-1,4,5-TRICHLOROTOLUENE	$C_7H_5O_2Cl_3$	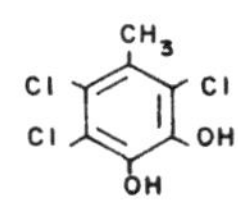
9,10 - DIMETHYLANTHRACENE	$C_{16}H_{11}$	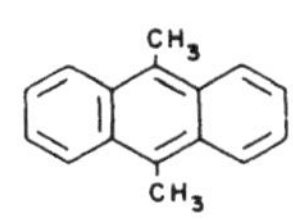
9,10 - DIMETHYLBENZ (A) ANTHRACENE	$C_{20}H_{16}$	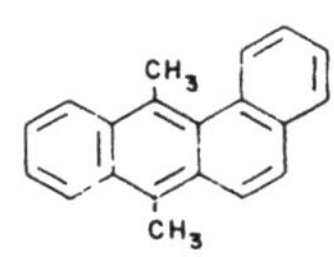
2,4 - DIMETHYL-6-tert-BUTYLPHENOL	$C_{12}H_{18}O$	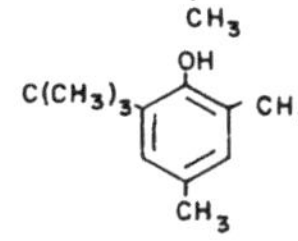
7,14 - DIMETHYLDIBENZ (A, H) ANTHRACENE	$C_{24}H_{18}$	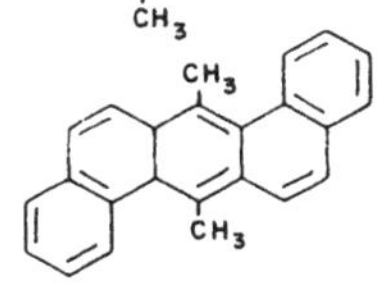

NAME	MOLECULAR FORMULA	STRUCTURAL FORMULA
METHYLETHYL KETONE	C_4H_8O	$CH_3-C(=O)-CH_2CH_3$
2-METHYL-4-ETHYLPHENOL	$C_9H_{12}O$	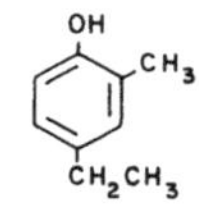
METHYL GUAIACOL	$C_8H_{10}O_2$	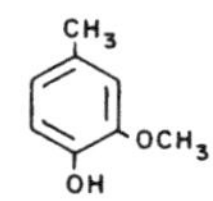
1-METHYLNAPHTHALENE	$C_{11}H_{10}$	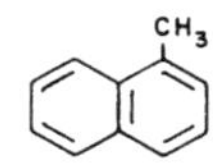
1-METHYLPHENANTHRENE	$C_{15}H_{12}$	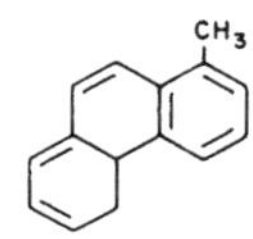
METHYLSTYRENE	C_9H_{10}	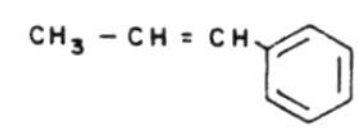
METHYLVERATROLE	$C_6H_{12}O_2$	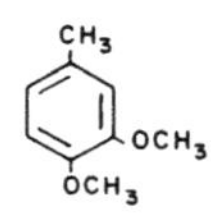
MIREX	$C_{10}Cl_{12}$	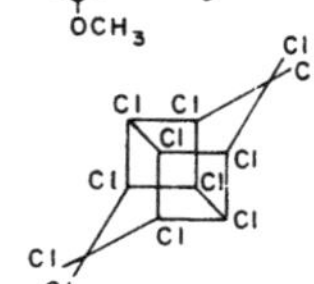
NAPHTHALENE	$C_{10}H_8$	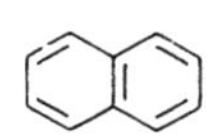
1-NAPHTHOIC ACID	$C_{11}H_8O_2$	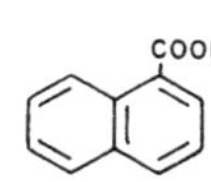

NAME	MOLECULAR FORMULA	STRUCTURAL FORMULA
KETOENDRIN	$C_{12}H_9OCl_6$	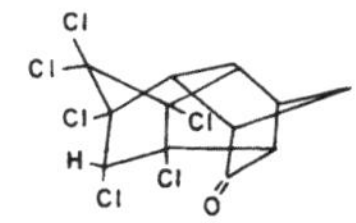
MALEIC ANHYDRIDE	$C_4H_2O_3$	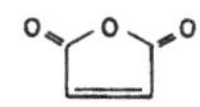
METHANE	CH_4	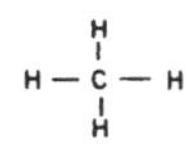
9-METHYLANTHRACENE	$C_{15}H_{12}$	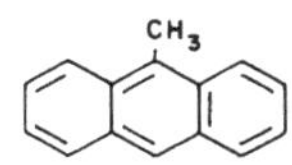
3-METHYL-4-tert-BUTYLPHENOL	$C_{11}H_{16}O$	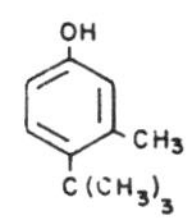
METHYLCHLORIDE (chloromethane)	CH_3Cl	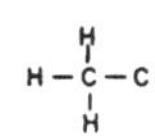
METHYL CHOLANTHRENE	$C_{21}H_{16}$	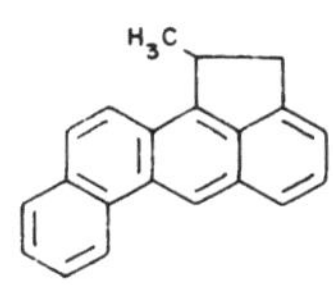
5-METHYL-2,6-ditert-BUTYLPHENOL	$C_{15}H_{24}O$	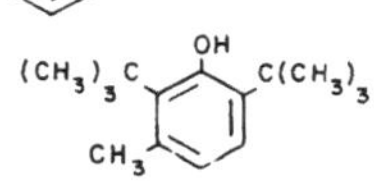
METHOXYCHLOR	$C_{16}H_{15}Cl_3O_2$	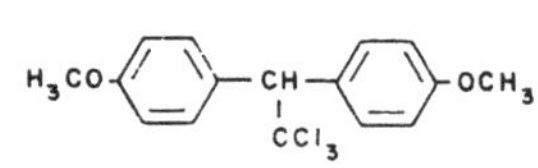
METHYLENE CHLORIDE (dichloromethane)	CH_2Cl_2	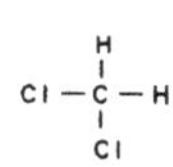

NAME	MOLECULAR FORMULA	STRUCTURAL FORMULA
HEXANE	C_6H_{14}	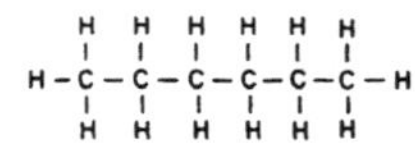
p-HEXYLPHENOL	$C_{12}H_{18}O$	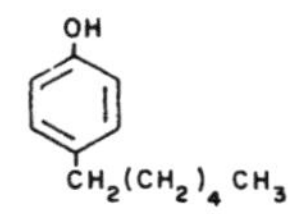
HYDROGEN SULPHIDE	H_2S	H – S – H
2-HYDROXYBIPHENYL	$C_{12}H_{10}O$	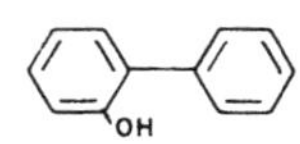
HYDROXYDIELDRIN	$C_{12}H_8O_2Cl_6$	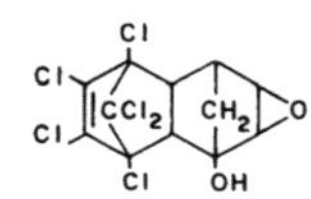
INDENO (1,2,3-cd) PYRENE	$C_{22}H_{12}$	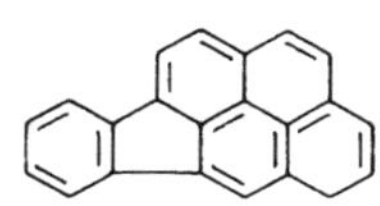
ISOPRENE	C_5H_8	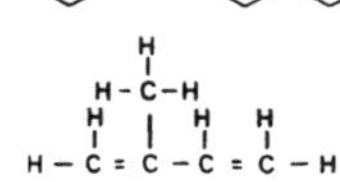
di-ISOPROPYLETHER	$C_6H_{14}O$	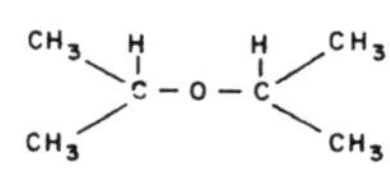
o-ISOPROPYLPHENOL	$C_9H_{12}O$	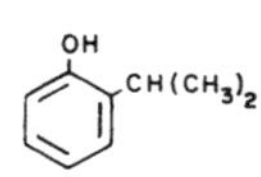
KETODIELDRIN	$C_{12}H_8OCl_6$	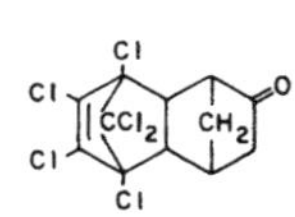

NAME	MOLECULAR FORMULA	STRUCTURAL FORMULA
HEPTACHLOR	$C_{10}H_5CL_7$	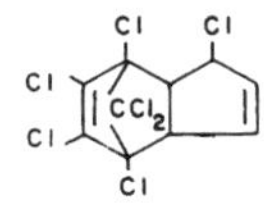
HEPTACHLOR EPOXIDE	$C_{10}H_4OCl_7$	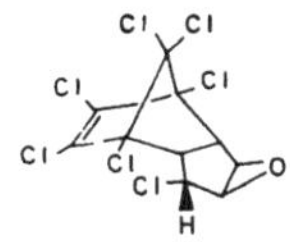
p-HEPTYLPHENOL	$C_{13}H_{20}O$	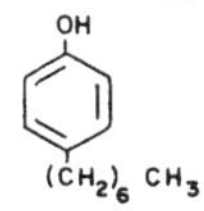
HEXACHLOROBENZENE	C_6Cl_6	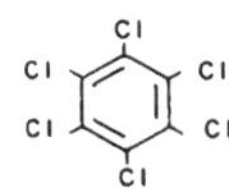
HEXACHLOROBIPHENYL	$C_6H_2Cl_6$	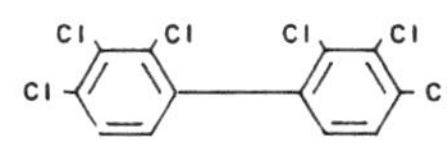
HEXACHLOROBUTADIENE	C_4Cl_6	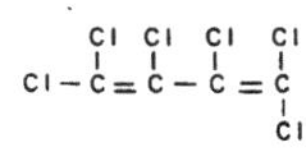
HEXACHLOROCYCLOHEXANE	$C_6H_6Cl_6$	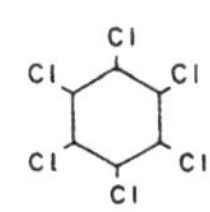
HEXACHLOROCYCLOPENTADIENE	C_5Cl_6	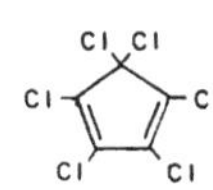
HEXACHLOROETHANE	C_2Cl_6	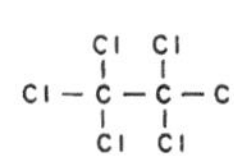
HEXADECANE	$C_{16}H_{34}$	$CH_3(CH_2)_{14}CH_3$

NAME	MOLECULAR FORMULA	STRUCTURAL FORMULA
NITROBENZENE	$C_6H_5NO_2$	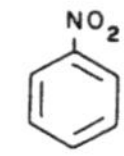
o-NITROPHENOL	$C_6H_5NO_3$	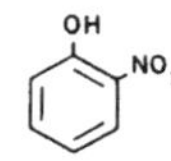
m-NITROTOLUENE	$C_7H_7NO_2$	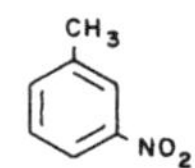
NONYLPHENOL	mixture of alkyl phenols	
OCTACHLOROBIPHENYL	$C_{12}H_2Cl_8$	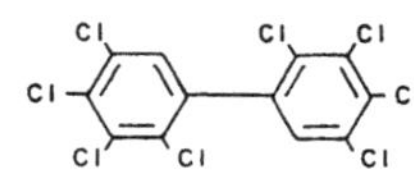
OCTACHLORONAPHTHALENE	$C_{10}Cl_8$	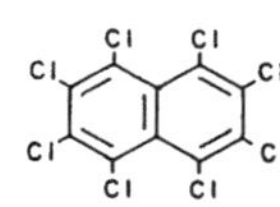
OCTADECANE	$C_{18}H_{38}$	$CH_3(CH_2)_{16}CH_3$
OCTANE	C_8H_{18}	$CH_3(CH_2)_6CH_3$
p-OCTYLPHENOL	$C_{14}H_{22}O$	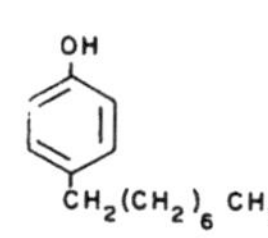
OXYCHLORDANE	$C_5H_6Cl_2O$	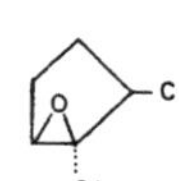

NAME	MOLECULAR FORMULA	STRUCTURAL FORMULA
PARAFFIN	$C_n H_{2n+2}$	—
PARATHION	$C_{10}H_{14}NO_5P_5$	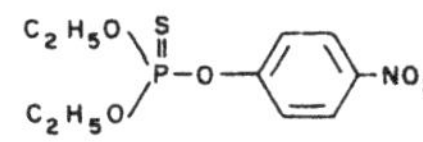
PENTACHLOROANISOLE	C_7H_3O	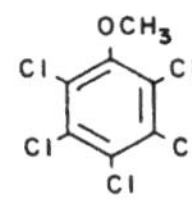
PENTACHLOROBENZENE	C_6HCl_5	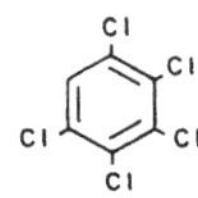
PENTACHLOROBIPHENYL	$C_{12}H_5Cl_5$	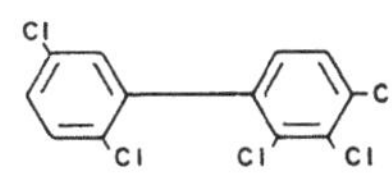
PENTACHLOROETHANE	C_2HCl_5	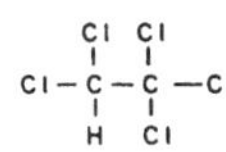
PENTACHLOROPHENOL	C_6Cl_5OH	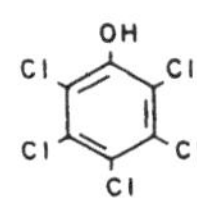
PENTANE	C_5H_{12}	$CH_3-CH_2-CH_2-CH_2-CH_3$
p-PENTYLPHENOL	$C_{11}H_{16}O$	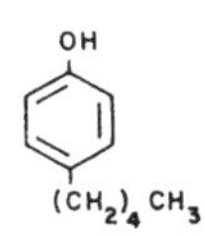
PERYLENE	$C_{20}H_{12}$	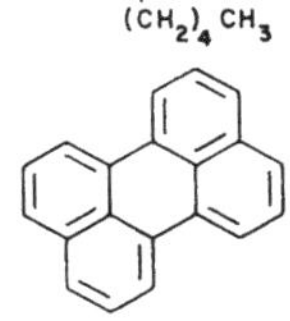

NAME	MOLECULAR FORMULA	STRUCTURAL FORMULA
PHENANTHRENE	$C_{14}H_{10}$	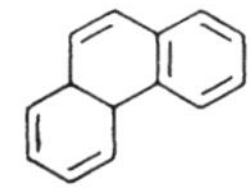
PHENOL	C_6H_5OH	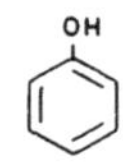
o-PHENYLPHENOL	$C_{12}H_{10}O$	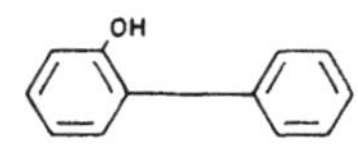
PHOSGENE	$COCl_2$	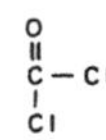
PHOTOALDRIN	$C_{12}H_8Cl_6$	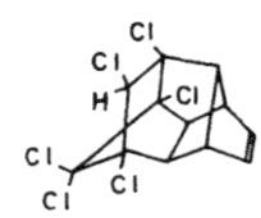
PHOTODIELDRIN	$C_{12}H_8Cl_6O$	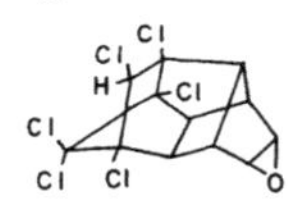
PHTHALIC ANHYDRIDE	$C_8H_4O_3$	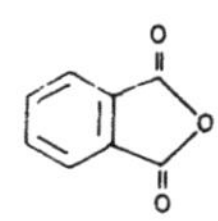
PHYTOL	$C_{20}H_{40}O$	—
PICLORAM	$C_6H_3Cl_3N_2O_2$	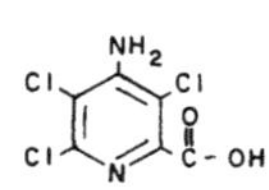
PORHYRIN	$C_{40}H_{48}N_6O_8S_2$	—

NAME	MOLECULAR FORMULA	STRUCTURAL FORMULA
PROPACHLOR	$C_{11}H_{14}ClNO$	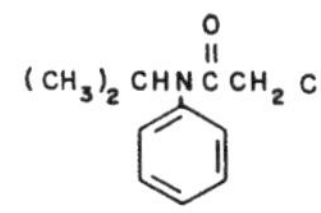
PROPYLENE	C_3H_6	$H_2C = CH - CH_3$
PYRENE	$C_{16}H_{10}$	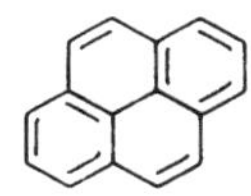
QUINONE	$C_6H_4O_2$	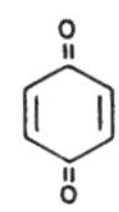
SALICYLIC ACID	$C_7H_6O_3$	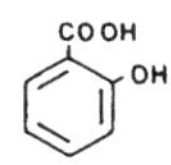
SILVEX	$C_9H_7Cl_3O_3$	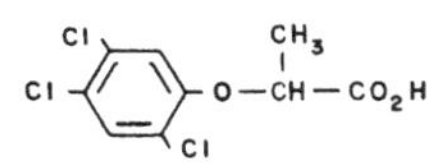
SODIUM DIOCTYL SULFOSUCCINATE	$C_{19}H_{35}O_7SNa$	—
STYRENE	C_8H_8	$C_6H_5CH = CH_2$
2,4,5-T [(2,4,5-trichlorophenoxy) acetic acid]	$C_8H_5Cl_3O_3$	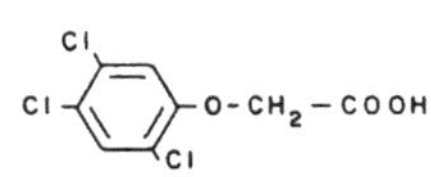
TERPENOIDS	approx. $C_{10}H_{16}$ & up	—

NAME	MOLECULAR FORMULA	STRUCTURAL FORMULA
TETRABROMOBENZENE	$C_6H_2Br_4$	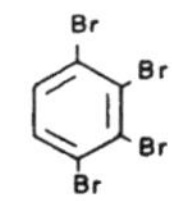
TETRACHLOROBENZENE	$C_6H_2Cl_4$	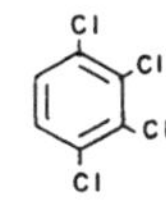
TETRACHLOROBIPHENYL	$C_{12}H_6Cl_4$	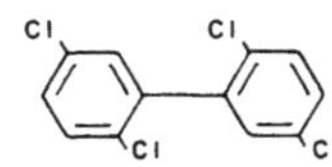
TETRACHLOROCATECHOL	$C_6H_2O_2Cl_4$	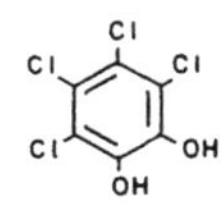
TETRACHLORODIBENZO-P-DIOXIN	$C_{12}H_4Cl_4O_2$	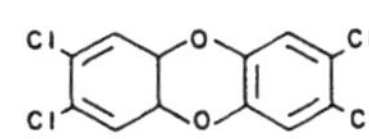
TETRACHLORODIBENZOFURAN	$C_{12}H_4Cl_4O$	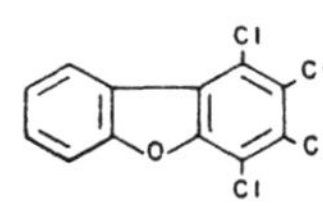
TETRACHLOROETHANE	$C_2H_2Cl_4$	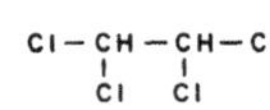
TETRACHLOROETHYLENE	C_2Cl_4	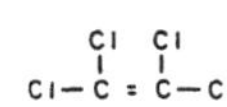
TETRACHLOROGUAIACOL	$C_7H_4O_2Cl_4$	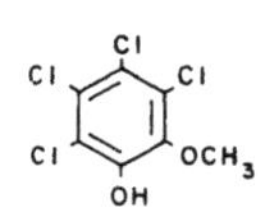
TETRACHLOROMETHANE (carbon tetrachloride)	CCl_4	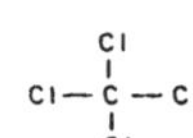

NAME	MOLECULAR FORMULA	STRUCTURAL FORMULA
TETRACHLOROPHENOL	$C_6H_2OCl_4$	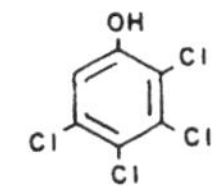
T F M (trifluoromethylnitrophenol)	$C_7H_4F_3NO_3$	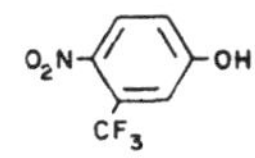
THIOPHENOL	C_6H_6S	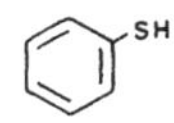
TOLUENE	C_7H_8	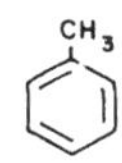
TOXAPHENE	approx. $C_{10}H_{10}Cl_8$	
TRIALLATE	$C_{10}H_{16}Cl_3NOS$	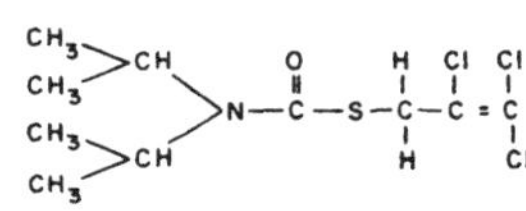
TRIBROMOBENZENE	$C_6H_3Br_3$	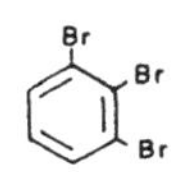
TRIBROMOMETHANE (bromoform)	$CHBr_3$	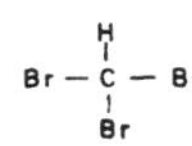
TRICHLOROBENZENE	$C_6H_2Cl_3$	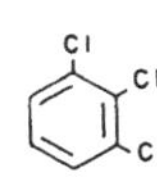
TRICHLOROBIPHENYL	$C_{12}H_7Cl_3$	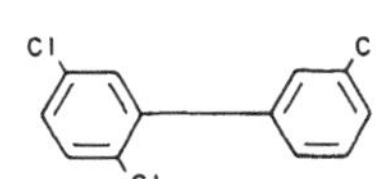

NAME	MOLECULAR FORMULA	STRUCTURAL FORMULA
TRICHLOROCATECHOL	$C_6H_3O_2Cl_3$	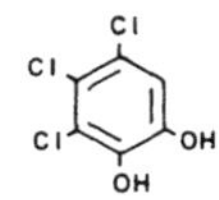
TRICHLOROETHANE	$C_2H_3Cl_3$	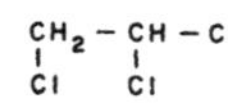
TRICHLOROETHYLENE	C_2HCl_3	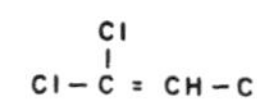
TRICHLOROFLUOROMETHANE	CCl_3F	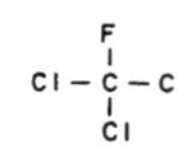
TRICHLOROMETHANE (chloroform)	$CHCl_3$	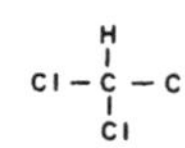
TRICHLOROMETHYL GUAIACOL	$C_8H_7O_2Cl_3$	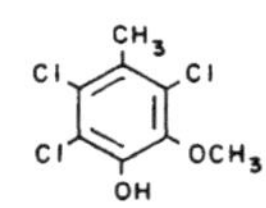
TRICHLOROMETHYLVERATROLE	$C_9H_9O_2Cl_3$	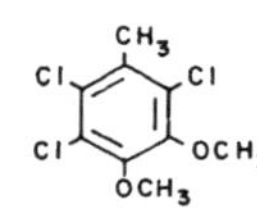
TRICHLOROGUAIACOL	$C_7H_5O_2Cl_3$	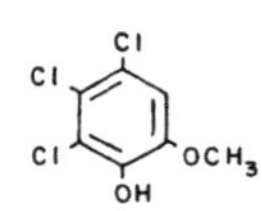
TRICHLOROPHENOL	$C_6H_3Cl_3O$	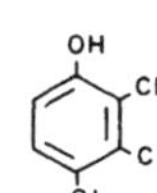
TRIMETHYL BENZENE	C_9H_{12}	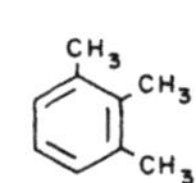

NAME	MOLECULAR FORMULA	STRUCTURAL FORMULA
TRIMETHYLNAPHTHALENE	$C_{13}H_{14}$	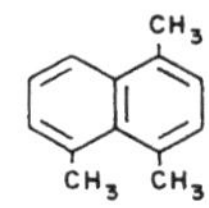
TRIMETHYLPHENOL	$C_9H_{12}O$	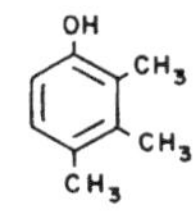
TRINITROPHENOL	$C_6H_3N_3O_7$	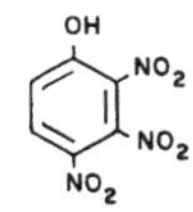
TRINITROTOLUENE	$C_7H_5N_3O_6$	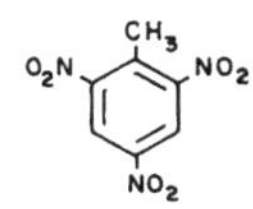
VINYL CHLORIDE (chloroethylene)	C_2H_3Cl	$H_2C = CH - Cl$
VINYLIDENE CHLORIDE (dichloroethylene)	$C_2H_2Cl_2$	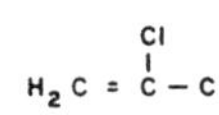
XYLENE	C_8H_{10}	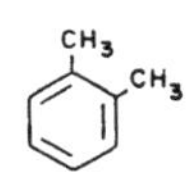
XYLENOL (dimethylphenol)	$C_8H_{10}O$	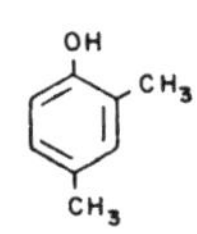

Appendix B
Physical and Chemical Terms Cited in This Book

Alkylation Transfer of an alkyl group (C_nH_{2n+1}) to a metal atom
Anthropogenic Any material originating from human activity
Arylation Transfer of an aryl group (C_6H_5) to a metal atom
Chelator A ligand with more than one binding site
Chromophore A chemical group that gives rise to color in a molecule
Colloids Particles of size range 0.1–0.45 μm
Complex A simple ligand with one binding site
Concentration Factor (CF) Quotient relating concentrations of pollutant in two different phases. For example, biota/water
Conditional Stability Constant Stability constant valid for a given set of conditions, such as pH, and ionic strength. cf Stability Constant
Contamination Natural and/or man-induced adulteration of abiotic and/or biotic substrates
Cracking Decomposition of petroleum with heat, steam, or other agents
ΣDDT All DDT analogs combined
Dehalogenation Loss of one or more halide atoms from a compound
Desorption Release of surface bound pollutants from solid matrices
Disproportionation Nonstoichiometric breakdown of a compound
Eh Redox potential
Evaporation Transport of a compound from the liquid phase to the gas phase
Fractionation Separation of a compound(s) into different physical and chemical components
Fulvic Acid An acid-alkali–soluble humic material, originating from the breakdown of lignin and tannins

Glycoside Acetals derived from hydroxy compounds and sugars
Halogenation Combination of Cl, Br, I, or F with a compound
Humic Acid An acid-insoluble component of humic material, with molecular weight greater than fulvic acid
Hydrolysis Interaction of a compound with hydrogen, hydroxyl radicals, or water molecules
IUPAC International Union of Pure and Applied Chemistry
K_{oc} Partition coefficient normalized to organic carbon
Lactone Internal cyclic monoester of a hydroxycarboxylic acid
Ligand A molecule containing a donor atom capable of forming a bond with a metal
Mercaptan Compounds resembling alcohols but having the oxygen of the hydroxyl group replaced by sulfur
Nucleophile Atoms or groups with an excess of nonbonding electrons having an affinity for positively charged sites
Parachor Molecular volume
Partition Coefficient Distribution of a compound in different phases or matrices (tissue, water, sediments, etc.)
Photolysis Chemical decomposition by the action of radiant energy
Photooxidation Oxidation induced by radiant energy
Refracture Index Measure of the biodegradability of a compound
Sorption Reversible binding of a pollutant to a solid matrix
Substituted Compound Replacement of one or more hydrogen atoms with other atoms or groups
Unsubstituted Compound No replacement of hydrogen atoms with other atoms
Uptake Nonreversible accumulation of a pollutant
Volatilization Transport of a compound from the surface of a liquid to the gas phase
Vapor Pressure Solubility of a compound in air from the liquid phase
Xenobiotic Foreign substance in a living system

Appendix C
Common and Scientific Names of Fish Cited in This Book

Alewife *Alosa pseudoharengus*
American shad *Alosa sapidissima*
Atlantic cod *Gadus morhua*
Atlantic salmon *Salmo salar*

Bass *Micropterus* sp.
Black bullhead *Ictalurus melas*
Blenny *Blennius pavo*
Bluefin tuna *Thunnus thynnus*
Bluegill *Lepomis machrochirus*
Brook trout *Salvelinus fontinalis*
Brown bullhead *Ictalurus nebulosus*
Brown trout *Salmo trutta*
Burbot *Lota lota*

Carp *Cyprinus carpio*
Channel catfish *Ictalurus punctatus*
Chinook salmon *Oncorhynchus tshawytscha*
Coho salmon *Oncorhynchus kisutch*
Cunner *Tautogolabrus adspersus*

Dolly varden *Salvelinus malma*

Eel *Anguilla anguilla*

Fathead minnow *Pimephales promelas*
Flounder *Pleuronectes flesus*

Goldfish *Carassius auratus*
Greenland halibut *Reinhardtius hippoglossoides*
Gulf killifish *Fundulus grandis*
Guppy *Lebistes reticulata*

Herring *Clupea harengus*

Lake trout *Salvelinus namaycush*
Lake whitefish *Coregonus clupeaformis*
Largemouth bass *Micropterus salmoides*
Largescale sucker *Catostomus macrocheilus*
Little skate *Raja erinacea*

Mosquito fish *Gambusia affinis*
Mummichog *Fundulus heteroclitus*

Nase *Chondrostoma nasus*
Northern squawfish *Ptychocheilus oregonensis*

Ocean perch *Sebastes marinus*

Pacific herring *Clupea harengus pallasi*
Perch *Perca fluviatilis*
Pike *Esox lucius*
Pilchard *Sardina pilchardus*
Plaice *Pleuronectes platessa*

Rainbow trout *Salmo gairdneri*
Roach *Rutilus rutilus*

Sheepshead minnow *Cyprinodon variegatus*
Smallmouth bass *Micropterus dolomieui*
Smelt *Osmerus eperlanus*
Spottail shiner *Notropis hudsonius*
Staghorn sculpin *Leptocottus armatus*
Striped mullet *Mugil cephalus*
Sucker *Castostomus* sp.
Surf smelt *Hypomesus pretiosus*

Tomcod *Microgadus proximus*

Walleye *Stizostedion vitreum*
White perch *Morone americana*
White sucker *Castostomus commersoni*
Whiting *Merlangius merlangus*
Winter flounder *Pseudopleuronectes americanus*

Yellow perch *Perca flavescens*

Appendix D
Equations for the Evalution of Physico-Chemical Fate Processes

D-1. Calculation of the aqueous solubility of an organic compound using Quayle's parachor value.

$$\log_{10}(1/S) = (1.50)\cdot(p_r)\cdot(10^2)-(1.51)\cdot(E_w)-1.01$$

where S = aqueous solubility in molal concentration, P_r = Quayle's parachor, and E_w = hydrophilic group factor. The validity of this equation was tested with 156 compounds of known solubility, and the correlation coefficient was found to be 0.962 (Moriguchi, 1975).

D-2. Calculation of vapor pressure by the equation of Weast (1974)

$$\log_{10} P = (-0.2185\ A/K) + B$$

where P = vapor pressure in torr, A = molar heat of vaporization, K = temperature in degrees Kelvin, and B = constant. For a given compound, values of A and B are constant over a moderate range of temperature. Values of A and B for several pollutants have been listed by Weast (1974) and can be used to calculate vapor pressures directly. Where A and B are not available and two or more vapor pressure values are given for temperatures bracketing 25°C, this equation can be used to calculate A and B from the known two sets of ordered pairs (K_1, P_1) and (K_2, P_2). Then the constants A and B are substituted in the equation to calculate P at T = 298°K (=25°C).

D-3. Calculation of vapor pressure by the Clausius-Clapeyron equation

$$\ln \frac{p_2}{P_1} = \frac{-\Delta H_v}{R} \frac{T_2 - T_1}{(T_1 T_2)}$$

where P = vapor pressure in torr, ΔH_v = molal heat of vaporization, T = temperature in degrees Kelvin, R = gas law constant (1.99 cal/mole°K), and subscripts 1 and 2 refer to two different temperatures. The solution of this equation requires the knowledge of boiling point and heat of vaporization. Calculations from this equation provide only a rough estimate of vapor pressure.

D-4. Calculation of volatilization rate constant

$$R_v = -\frac{d[c_w]}{dt} = k_v[c_w]$$

where

$$k_v = \frac{1}{L}\left[\frac{1}{k_l} + \frac{RT}{H_c k_g}\right]^{-1}$$

and

R_v = volatilization rate of a chemical, C (moles L^{-1} hr^{-1})
C_w = aqueous concentration of C (moles L^{-1} (=M))
k_v = volatilization rate constant (hr^{-1})
L = depth (cm)
k_ℓ = mass transfer coefficient in the liquid phase (cm hr^{-1})
H_c = Henry's law constant (torr M^{-1})
k_g = transfer coefficient in the gas phase (cm hr^{-1})
R = gas constant (liter-atm-mole^{-1} degree^{-1}) and
T = absolute temperature (degrees Kelvin).

In both phases,

$$k_\ell = D_{\ell/\partial\ell} \text{ and } k_g = D_{g/\partial g}$$

where D = diffusion coefficient and
∂ = boundary layer thickness.

D-5. Calculation of the volatilization loss of an organic compound

$$(k_v^c)_{env} = (k_v^c)_{\ell ab}\,(k_v^o)\text{env}$$

where k_v^c = volatilization rate constant for the chemical (hr^{-1})

and k_v^o = oxygen reaeration constant (hr^{-1}) in the laboratory or environment.

For example, the quotient k_v^c/k_v^o for benzene was independent of turbulence, salt concentration, temperature (4–50°C), and presence of surface active compounds (Smith *et al.*, 1980).

D-6. Equations for electron and proton changes

Equations for Electron and Proton Changes

Concentration of proton, H	Concentration of electron, ε
$pH = -\log [H]$	$P\varepsilon = \log [e^-]$
High pH = low H^+ activity and conversely	High $P\varepsilon$ = low ε^- activity and conversely
$pH = pK_a + \log [A^-]/[HA]$	$P\varepsilon = P\varepsilon^\circ + \log \dfrac{[\text{oxidized}]}{[\text{reduced}]}$
$pH = pK_a$ when $[A^-] = [HA]$	$P\varepsilon = P\varepsilon^\circ$ when [oxidized] = [reduced]
K_a = acid dissociation constant, $HA \rightleftharpoons H^+ + A^-$	$P\varepsilon^\circ$ = equilibrium potential

D-7. Calculation of the rate of hydrolysis of a chemical compound

$$-\frac{dc}{dt} = k_A\,[H^+]\,[C] + k_B\,[OH]\,[C] + k_N\,[C]$$

where k_h = first-order hydrolysis rate constant at a specific pH; k_A and k_B = second-order acid and base hydrolysis constants, respectively; and k_N = first-order hydrolysis rate constant for pH independent reaction.

D-8. Rate of disappearance of an organic compound by direct photolysis

$$-\frac{dc}{dt} = K_p\,[C] = k_a\,\phi\,[C]$$

where k_p = first-order rate constant, ϕ = reaction quantum yield, and k_a = rate constant for light absorption by the chemical that depends on the light intensity, chromaticity of light, and extinction coefficient of the chemical.

D-9. Rate of disappearance of an organic compound by indirect photolysis

$$-\frac{dc}{dt} = k_2 [C] [X] = k_p' [C]$$

where k_2 = second-order constant for the interaction between the chemical and the intermediate, X; for a photosensitized reaction the k_p' would be a combined term including the concentration of the excited state species and the quantum yields for the energy transfer to and subsequent reaction of the chemical. In any estimate of k_p or k_p', values of k_a or $[X]$ should be specific, taking into account the variation of the intensity of sunlight with time of the day, season, and latitude.

D-10. Rate of substrate utilization

$$-\frac{dc}{dt} = \frac{\mu X}{Y} = \frac{(\mu_m)}{(Y)} \cdot \frac{(CX)}{(K_s + C)} = (k_b) \cdot \frac{(CX)}{(K_s + C)}$$

where μ = specific growth rate, X = biomass per unit volume, μ_m = maximum specific growth rate, K_s = concentration of the substrate to support half-maximum specific growth rate ($0.5\mu_m$), k_b = substrate utilization constant or biodegradation constant, ($=\mu_m/Y$), and Y = biomass produced from a unit amount of substrate consumed. These constants μ_m, K_s, and Y are dependent on the characteristics of the microbes, pH, temperature, and media.

D-11. Reduced equation for the rate of substrate utilization

When the substrate concentration $C \gg K_s$, the equation D-10 reduces to:

$$-\frac{dc}{dt} = k_b X$$

This means that the biodegradation rate is first order with respect to all biomass concentration and zero order with respect to chemical concentration.

D-12. Reduced equation for the rate of substrate utilization

In actual environmental situations for many pollutants, $C << K_s$, hence equation D-10 becomes:

$$-\frac{dc}{dt} = (k_b) \cdot \frac{(CX)}{(K_s)} = k_{b2} [C] [X]$$

where k_{b2} is a second-order rate constant.

D-13. Reduced equation for the degradation rate of a chemical

When the biomass concentration is relatively large compared with the pollutant concentration, the degradation rate is pseudo-first order and given by:

$$-\frac{dc}{dt} = k'_b C$$

where k'_b is the pseudo-first-order rate constant and dependent on the cell concentration (X_o).

D-14. Calculation of the half-life of a chemical under degradation

The half-life of the chemical under degradation ($t_{½}$ at a given X_o) will be

$$t_{1/2} = \frac{\ln 2}{k_{b2} X_o} = \frac{0.693}{k_{b2} X_o}$$

where

$$k_{b2} = \frac{k'_b}{X_o}$$

(k'_b = pseudo-first-order rate constant and X_o = cell concentration.)

D-15. Kinetic half-lives of chemicals

Half-lives of organic compounds are calculated from the respective rate constants and their dependence on physical parameters such as temperature.

For a first-order kinetic reaction,
$A \xrightarrow{j}$ products at a constant volume.

The rate of disappearance of A is given by:

$$-\frac{dC_A}{dt} = k_j\, C_A$$

where C_A = concentration of A in moles L^{-1}
t = time in appropriate units
k_j = reaction rate for the process j in units of inverse time, and

$$\frac{dC_A}{dt} = \text{rate of change of } C_A \text{ with time.}$$

Integrating the equation between the limits of t_0 (initial time) and t, yields:

$$k_j = \frac{1}{(t - t_0)} \ln \frac{(C_{A_0})}{(C_A)}$$

where C_{A_0} = initial concentration of C_A at t_0.
For $C_A = 0.5\ C_{A_0}$, the half-life is given by:

$$t_{1/2} = \frac{1}{k_j} \ln \frac{(2C_{A_0})}{(C_{A_0})}$$

or

$$t_{1/2} = 0.693 \left(\frac{1}{k_j}\right)$$

If all the transformation processes are expressed as a first-order or pseudo first-order kinetic process, the net half-life for the chemical is given by:

$$t_{1/2} = \frac{\ln 2}{\Sigma_j k_j}$$

References

Moriguchi, I. 1975. Quantitative structure-activity studies. I. Parameters relating to hydrophobicity. *Chemical and Pharmaceutical Bulletin (Tokyo)* **23**:247–257.
Smith, J.H., D.C. Bomberger, Jr., and D.L. Haynes. 1980. Prediction of the volatilization rates of high-volatility chemicals from natural water bodies. *Environmental Science and Technology* **14**:1332–1337.
Weast, R.C. (Ed.). 1974. *CRC handbook of chemistry and physics*. 54th edition. CRC Press, Cleveland, Ohio, pp. D-162–D-188.

Index

Springer Series on Environmental Management
Robert S. DeSanto, Series Editor

Natural Hazard Risk Assessment and Public Policy
Anticipating the Unexpected
by **William J. Petak** and **Arthur A. Atkisson**

This volume details the practical actions that public policy makers can take to lessen the adverse effects natural hazards have on people and property, guiding the reader step-by-step through all phases of natural disaster.

1982/489 pp./89 illus./cloth
ISBN-13: 978-1-4613-9540-9

Gradient Modeling
Resource and Fire Management
by **Stephen R. Kessell**

"[The] approach is both muscular enough to satisfy the applied scientist and yet elegant and deep enough to satisfy the aesthetics of the basic scientist ... Kessell's approach ... seems to overcome many of the frustrations inherent in land-systems classifications."
—*Ecology*

1979/432 pp./175 illus./27 tables/cloth
ISBN-13: 978-1-4613-9540-9

Disaster Planning
The Preservation of Life and Property
by **Harold D. Foster**

"This book draws on an impressively wide range of examples both of man-made and of natural disasters, organized around a framework designed to stimulate the awareness of planners to sources of potential catastrophe in their areas, and indicate what can be done in the preparation of detailed and reliable measures that will hopefully never need to be used."
—*Environment and Planning A*

1980/275 pp./48 illus./cloth
ISBN-13: 978-1-4613-9540-9

Air Pollution and Forests
Interactions between Air Contaminants and Forest Ecosystems
by **William H. Smith**

"A definitive book on the complex relationship between forest ecosystems and atomospheric deposition ... long-needed ... a thorough and objective review and analysis."
—*Journal of Forestry*

1981/379 pp./60 illus./cloth
ISBN-13: 978-1-4613-9540-9